SCHAUM'S A–Z

BIOLOGY

D0057712

SCHAUM'S A-Z

BIOLOGY

BILL INDGE

SCHAUM'S A-Z SERIES

McGraw·Hill

New York   Chicago   San Francisco   Lisbon   London   Madrid   Mexico City
Milan   New Delhi   San Juan   Seoul   Singapore   Sydney   Toronto

Originally published by Hodder & Stoughton Educational, a division of Hodder Headline Plc, 338 Euston Road, London NW1 3BH.

1 2 3 4 5 6 7 8 9 0   2 1 0 9 8 7 6 5 4 3

ISBN 0-07-141934-9

HOW TO USE THIS BOOK

This book provides an explanation of the main ideas and important concepts central to all intermediate/advanced level biology courses. It does not set out to give an exhaustive list of obscure technical terms. A decision has been made to select a range of basic terms which are likely to meet the needs of most students following an intermediate/advanced level course.

Each item starts with a simple, one-line definition and progresses to explain the term in a little more detail, showing how it relates to other, associated, areas of biology. A conscious effort has been made to strike a balance between providing a simple explanation that should be understood by someone encountering the term for the first time, and producing an entry which is sufficiently detailed to meet the requirements of an examiner who asks the question "What is meant by the term ...?" Entries are extensively cross-referenced. Italics have been used to identify terms which have separate entries.

Some entries contain Hint boxes. These boxes are meant to highlight some of the more common errors and shortcomings shown by candidates in examinations. Read these boxes carefully. They should help you to avoid losing examination points needlessly.

During your course, the Schaum's A–Z is essentially a book to pick up and put down, a book to browse through and use to add to your understanding of basic ideas. It is also a book which will enable you to clarify new ideas. In addition, it should prove useful at the end of the course and for revision.

A-band: one of the dark bands which can be seen running across a *myofibril* in a *skeletal muscle*. It corresponds to the location of filaments of the protein *myosin*.

abiotic: an ecological factor which is concerned with the non-living part of the environment. Rainfall, soil pH, temperature and humidity are all examples of abiotic factors which will affect the distribution of various organisms. *Competition* and predation, on the other hand, are concerned with the living part of the environment and are *biotic* factors. Abiotic factors are usually *density-independent*. That is, it doesn't matter how many organisms you have in a particular area, abiotic factors will affect them all. If conditions become very hot and dry, for example, it is likely that all the plants of a particular species growing in the area concerned with be affected. It will not matter if the species is extremely abundant or very rare.

abscisic acid (ABA): a *plant growth substance* which acts mainly as a growth inhibitor. It is present in large quantities, for example, in dormant buds. It is also involved in controlling the opening of *stomata*. A rise in the level of abscisic acid has been shown to be associated with the closing of stomata.

absolute growth rate: see *growth rate*

absorption: the process by which the small, soluble molecules produced by digestion are taken up from the gut. There are three basic stages involved. These are:

- the movement of molecules to the wall of the intestine. This is helped by the continual mixing of the gut contents. Muscles in the gut wall bring about movements of the gut which achieve this
- transport across the plasma membrane into the cytoplasm of the cells which line the gut. A number of mechanisms are involved. When the concentration of digested molecules is higher in the gut than in the cytoplasm of the lining cells, then movement by either simple *diffusion* or by *facilitated diffusion* is possible. *Active transport* is necessary when movement is against a concentration gradient and this is the most important of the mechanisms involved. Larger molecules and fluids may be taken up by the membrane by the process of *pinocytosis*
- transport away from the gut is by the blood and lymphatic systems. Molecules such as simple sugars and amino acids are absorbed into the blood system. They are transported in the hepatic portal vein to the *liver* where they are processed. The products of fat digestion enter the lacteals, branches of the lymphatic system in the *small intestine*.

Most absorption in the gut of a mammal takes place in the small intestine which shows a number of adaptations. The surface area is increased enormously by the possession of small finger-like processes known as *villi*. In addition, the epithelial cells which line these are covered in *microvilli*. The cytoplasm of the epithelial cells contains large numbers of

1

mitochondria which provide the *ATP* necessary for active transport of substances from the epithelium of the gut.

> **hint** Food must be digested before it can be absorbed so we should not write about absorption of "food." Absorption of "digested food" is more accurate.

absorption spectrum: a graph which shows the relative amounts of light of different wavelengths that is absorbed by a particular pigment. The diagram shows the absorption spectrum for one particular type of *chlorophyll*, chlorophyll a.

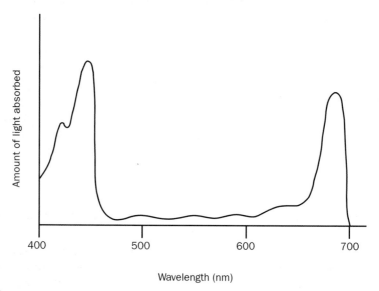

Absorption spectrum for chlorophyll a

Note that the curve shows two peaks, one between 400 and 500 nm, the other between 600 and 700 nm. This corresponds to the fact that chlorophyll a strongly absorbs light at the blue and red ends of the spectrum. This is the reason why most plant leaves appear green in color. The chlorophyll that they contain does not absorb green light. It reflects it.

accommodation: the ability of the eye to change focus so that near and distant objects can be seen clearly. When a person looks at a distant object, the rays of light are bent or refracted by the cornea and the lens and focused on the *retina*. If the object is brought closer, the light rays have to be refracted more if they are still to be focused. This is done by changing the shape of the lens. The ciliary muscles contract and the lens becomes more convex as a result.

acetylcholine (ACh): one of the main substances responsible for the transmission of an impulse across a *synapse*. Acetylcholine is an important *neurotransmitter* in the central nervous system and in synapses in the *parasympathetic* branch of the autonomic nervous system. Neurones which release acetylcholine as transmitters are known as cholinergic neurones. As a result of a nerve impulse arriving at the presynaptic membrane, acetylcholine is released into the synaptic cleft. It diffuses across the cleft where it acts on receptors on the postsynaptic membrane. These trigger another nerve impulse.

acetylcholinesterase: an enzyme found in large quantities in *synapses* which causes the rapid breakdown of *acetylcholine*. Acetylcholine must be removed rapidly from the synapse once it has triggered a nerve impulse or there would be continuous transmission along the postsynaptic neurone.

acetylcoenzyme A: the substance produced in the *link reaction* of respiration when coenzyme A accepts the two-carbon fragment formed from pyruvate. As well as being the molecule which feeds this two-carbon fragment into the *Krebs cycle*, acetylcoenzyme A is very important in other biochemical pathways involving fatty acids and amino acids.

acid rain: rain which has been made more acidic than normal by the solution of various gases in the atmosphere. Because of dissolved carbon dioxide, normal unpolluted rain water is slightly acid with a pH of about 5.6. In industrial areas, however, the combustion of fossil fuels releases oxides of nitrogen and sulphur into the air. These produce nitric and sulphuric acids which may lower the pH of the rain to around 4. Although there is some uncertainty over the way in which acid rain affects ecosystems, it is definitely known to be associated with the following:

* a fall in pH of freshwater lakes leading to the death of fish and many invertebrates

* death of trees, particularly of conifers, where the first signs involve die-back of the crowns

* release of more aluminium into solution. Aluminium is known to be toxic to many organisms.

acoelomate: an animal without a coelom. Flatworms and tapeworms which belong to the phylum *Platyhelminthes* are examples of acoelomates.

acrosome: a cap-like *vesicle* at the front of a sperm that contains enzymes. Just before fertilization the membrane that surrounds this vesicle bursts and the enzymes are released. Their digestive action helps to separate the cells which still surround the egg. They also make it possible for the sperm to penetrate the membranes surrounding the egg and fertilization to take place.

actin: a protein found in many cells but of particular importance in muscle. In the *myofibrils* of skeletal muscle, actin molecules form the thin filaments in the *I-band*. The *sliding-filament hypothesis* suggests that these thin actin filaments slide between the thicker *myosin* filaments when a muscle contracts.

action potential: the change which occurs in the electrical charge across the plasma membrane of a nerve cell during the passage of a *nerve impulse*. If a nerve cell is stimulated electrically, special sodium ion-channel proteins in the membrane open and allow Na^+ ions to enter. Since the inside of the cell is normally negatively charged, the entry of positive ions will make it less negative. In other words, there is a decrease in the potential difference across the membrane. This change in potential difference allows even more sodium ions to enter, changing the potential difference still more. The end result is that the potential difference changes from its resting potential of around -60 mV to a value of approximately $+40$ mV. This is summarized in the flow chart on page 4.

Once it has reached this value, the sodium ion-channel proteins close leaving the potassium ion-channel proteins fully open. Sodium ions can no longer enter but potassium ions can leave. The result of positive potassium ions moving out is that the potential difference falls back to its resting level. The overall change in the electrical charge across the membrane, lasting only a few milliseconds, is called an action potential. The spread of an action potential along the nerve cell is the basis of a nerve impulse. After the impulse passes, the sodium ions which entered and the potassium ions which left the nerve cell during the action potential are pumped back to where they began by active transport.

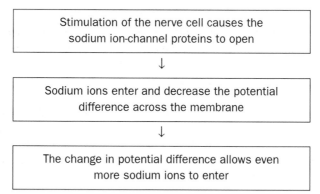

Stimulation of the nerve cell causes the
sodium ion-channel proteins to open

↓

Sodium ions enter and decrease the potential
difference across the membrane

↓

The change in potential difference allows even
more sodium ions to enter

action spectrum: a graph which shows the proportion of light of a particular wavelength that is actually used in a particular process. The action spectrum for photosynthesis is very similar to the *absorption spectrum* for *chlorophyll*. This shows the importance of chlorophyll in absorbing the light used in photosynthesis.

activation energy: the amount of energy necessary to bring molecules together so that they will react with each other. The graph shows the changes in energy level during the course of a chemical reaction.

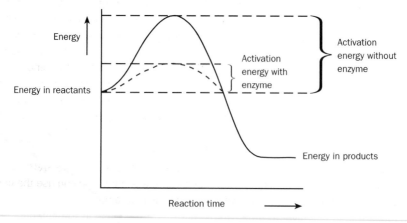

Think about burning some ethanol. This is a chemical reaction in which ethanol and oxygen combine to produce carbon dioxide and water. There is a lot of energy in ethanol and that is why it is a good fuel. There is a lot less energy in the products, carbon dioxide and water. However ethanol does not combine with oxygen to produce carbon dioxide and water on its own. To make it do this we need to supply some extra energy, the activation energy, in the form of a lighted match. *Enzymes* work by lowering the activation energy by splitting the reaction into stages. In this way the reaction can take place at the temperatures and pressures which exist inside living cells.

hint You should be able to draw the graph showing the energy changes during a chemical reaction. Look carefully at the position of the curve representing the addition of the enzyme. Adding an enzyme does not affect the amount of energy in either the reactants or the products, so the curve will start and finish at the same energy levels as with no enzyme present.

active site: the part of an *enzyme* molecule into which a substrate molecule fits during the course of a biochemical reaction. Enzymes are *proteins*. Each individual enzyme has its own sequence of amino acids which results in the molecule folding into a particular shape. In other words, each enzyme has a specific *tertiary structure*. A small group of the amino acids which make up the molecule is important in forming the place where the substrate molecule will fit to form an *enzyme–substrate complex*. The shape of this active site is very important in determining enzyme specificity. Only a substrate with the corresponding shape will fit the active site. In addition, anything which alters the shape of the active site will mean that the enzyme will not function any more. Changes in pH, high temperatures and the presence of certain types of *inhibitor* all affect enzymes in this way.

hint The active site is the part of the enzyme molecule to which the substrate binds. The enzyme has the active site, not the substrate. When you answer questions about active sites and enzyme action, you need to make sure that you bring in two important ideas – shape and fit.

active-site directed inhibition: see *competitive inhibition*

active transport: the movement of substances from where they are less concentrated to where they are more concentrated; in other words, it is the movement of substances against a concentration gradient. Like *facilitated diffusion*, active transport involves the presence of specific carrier proteins in cell membranes to transport the molecule or ion concerned. The processes are different, however, as moving a substance against a concentration gradient requires an input of energy. This is supplied by ATP. Active transport is extremely important in living organisms. Most cells need at some stage to take up substances which are only present in their environment at very low concentrations. For example, it is by active transport mechanisms that ions are taken up from very dilute solution in the soil into the more concentrated solution inside the cells of a plant root, that sugars are loaded into *phloem* sieve tubes and that much reabsorption takes place in the *kidney*.

Active transport is one of the important synoptic themes that links together different aspects of biology. The spider diagram on page 6 shows some of these links. You can use the cross-referencing system in this book to add greater detail to the diagram.

hint Note that in this entry active transport is described as involving the movement of substances "against" a concentration gradient. Against a concentration gradient obviously involves movement from a low concentration to a higher concentration. When you are describing active transport do not use words like "across" or "along" a concentration gradient. Their meanings are not clear enough.

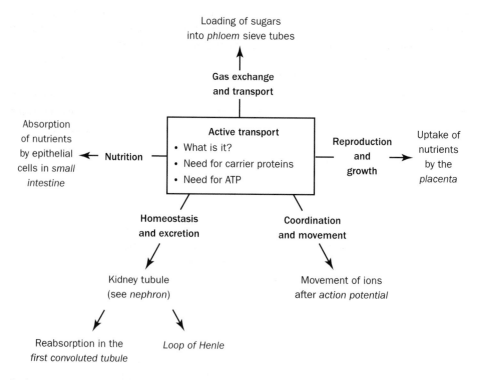

Active transport

acuity: the ability of the eye to distinguish between objects close together. It depends in part on the connections which the *cone cells* in the fovea of the eye have with sensory neurones. In simple terms, each cone cell has its own sensory neurone. Stimulation of cone cells, even if they are very close together, results in separate impulses in separate neurones in the optic nerve. Because of this, objects very close together can be seen as separate. The idea of acuity can be extended to other senses such as that of touch.

adaptive radiation: the way in which a single ancestral type of organism can give rise to different forms, each occupying a different *niche*. Adaptive radiation is often seen among species that live on islands. A familiar example is provided by Darwin's finches, a group of birds found on the Galapagos islands. These are volcanic islands about six hundred miles off the coast of Ecuador. It is thought that, at some stage in the past, they were colonized by finch-like birds from the South American mainland. There were no other small land birds on these islands so there was no competition from other species. Over a period of time, different groups of finches adapted to different ecological *niches*. As a result, there are now thirteen different species of finch found on the Galapagos islands.

adenine: one of the *nucleotide bases* found in nucleic acid molecules. Adenine is a *purine* which means that it has two rings of atoms in each of its molecules. When two polynucleotide chains come together in a molecule of *DNA*, adenine always bonds with the base *thymine*. The atoms of these two bases are arranged in such a way that two hydrogen bonds are able to form between them.

adenosine triphosphate: see *ATP*

ADH: a *hormone* involved in the control of water balance in the body. ADH or antidiuretic hormone is produced by a group of cells in the *hypothalamus* of the brain. It is stored in the posterior lobe of the *pituitary gland*. The flowchart below summarizes the way in which this hormone helps to control water balance when the *water potential* of the body fluids falls. This is particularly important in hot conditions.

If the water potential becomes less negative, the release of ADH is inhibited. The cells of the second convoluted tubule and the collecting duct become less permeable to water. Therefore less water is reabsorbed and large quantities of dilute urine are produced.

A fall in the water potential of the blood is detected by osmoreceptors in the hypothalamus

↓

Nerve impulses from the hypothalamus bring about the release of ADH from where it is stored in the posterior lobe of the pituitary gland

↓

ADH is released into the blood. It travels to the kidney where it makes the cells of the second convoluted tubule and the collecting duct more permeable to water

↓

More water is reabsorbed and less is lost in the urine

adrenal gland: one of a pair of endocrine glands situated near the kidneys. Each adrenal gland consists of two parts, an outer cortex and an inner medulla. The outer cortex produces steroid hormones, which have a number of functions in the body but are mainly involved in the control of carbohydrates, proteins and mineral ions. The medulla is responsible for the secretion of adrenaline, a hormone which is released at times of stress, and which affects many organs and helps the body respond to emergencies.

adrenaline: a hormone produced by the adrenal glands and released at times of stress. It affects many organs and helps the body to respond to an emergency. Some of these effects are shown below.

Effect on the body	Advantages in preparing the body for action
• Increases stroke volume and heart rate	Increases the supply of essential substances such as glucose and oxygen to muscles
• Constriction of blood vessels to many of the internal organs	Increases blood pressure

- Dilates blood vessels supplying the muscles — Increases the supply of essential substances to the muscles and the removal of waste products from them

- Stimulates the conversion of glycogen to glucose — Makes more glucose available for increased respiration

- Causes contraction of the muscles at the base of hairs — Makes the hair stand on end which, in many animals, may serve to make them look bigger and fiercer than they actually are

adrenergic: a nerve cell or *synapse* in which the *neurotransmitter* is *noradrenaline*. In mammals and other vertebrate animals, adrenergic synapses are found in the *sympathetic nervous system*. Here, the neurotransmitter in the synapse between the motor neurone and the effector to which it goes is noradrenaline. The sympathetic nerves which supply the heart, for example, release noradrenaline which brings about an increase in heart rate.

aerenchyma: a tissue found in the stems of plants that grow in water, such as rice and water lilies. It is a type of *parenchyma* and contains a mixture of cells and large, air-filled spaces. These spaces have two important functions:

- The air they contain provides buoyancy. This helps to ensure that the leaves and stems reach the surface where they will receive sufficient light for *photosynthesis*.

- They form a pathway allowing the diffusion of oxygen and carbon dioxide to parts of the plant that are under water.

aerobic respiration: *respiration* which requires the presence of oxygen. The equation which represents this process is shown below:

$$C_6H_{12}O_6 + 6O_2 \rightarrow 6H_2O + 6CO_2$$

This equation can be rather misleading as it only summarizes the process. Respiration is much more complex than this and consists of a series of biochemical reactions which may be summarized as a diagram.

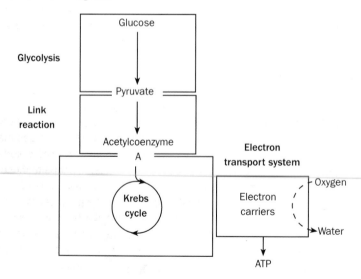

The biochemical pathway of aerobic respiration

The six-carbon glucose molecule is first split into two molecules of pyruvate. Each of these contains three carbon atoms. This process, known as *glycolysis*, generates a small amount of *ATP*. The pyruvate produced in glycolysis is then converted into a substance known as acetylcoenzyme A in what is conveniently called the *link reaction*. Acetylcoenzyme A then enters a cycle of biochemical reactions which is known as *Krebs cycle*. In this cycle, hydrogen is lost and passed along the *electron transport system* or respiratory chain until it reaches the end where it combines with oxygen to produce water. During the passage along this chain a considerable amount of energy is released and this is used to produce ATP.

afferent: means going to. In a *reflex arc*, for example, the afferent neurone transmits a nerve impulse to the spinal cord. In the *kidney*, the afferent arteriole takes blood to the *glomerulus*.

agar: a *polysaccharide* obtained from seaweed. It is important in making solid media for the culture of microorganisms. After heating and mixing with various nutrients, it cools to form a stiff jelly on which microorganisms such as bacteria and fungi may be grown. Agar is an inert substance and therefore provides no nutrients. Different nutrients may be added to make media suitable for different microorganisms. Blood agar, for example, is made by mixing horse blood with agar and is used for culturing the hemolytic bacteria that destroy blood cells. As very few microorganisms have the ability to break down agar jelly it remains firm in consistency but, at the same time, it allows substances to diffuse freely.

agglutination: the sticking together or clumping of red blood cells, caused by antibodies found in the plasma. Red blood cells have protein molecules on their plasma membranes which act as *antigens*. Group B individuals, for example, have red blood cells with antigen B. If, during a blood transfusion, group B blood were given to someone with antibody b in their blood plasma, the antigens and antibodies would combine, causing the red blood cells to stick to each other. This is agglutination and it can lead to blocked blood vessels and death. Agglutination must be distinguished from the completely different process of *blood clotting*.

agonistic: a term which may be used to describe the action of substances which mimic *neurotransmitters* at *synapses*. A synapse is a junction between two nerve cells. A small quantity of a chemical transmitter substance is released by the presynaptic nerve cell. This diffuses across the gap between the cells and fits into a receptor molecule in the plasma membrane of the postsynaptic nerve cell. This leads to the production of a nerve impulse. The molecules of the neurotransmitter have a specific shape which enables them to fit into the receptor molecule. There are other substances which have similar shapes and they can also fit into these receptors. Neurotransmitters only exert their effects for a very short amount of time as they are usually rapidly broken down by enzymes. Agonistic substances, on the other hand, are not broken down and can continue to stimulate the postsynaptic nerve cell. If freshwater is contaminated with large amounts of nitrates or phosphates, algae may grow in enormous numbers. Some of these algae produce a poisonous substance called anatoxin. Anatoxin is an agonistic substance. Its molecules are very similar in shape to those of *acetylcholine* but are not broken down by the enzyme *acetylcholinesterase*. One of the signs of anatoxin poisoning is that the affected animal produces very large amounts of saliva. The anatoxin fits into the receptor sites which would normally have been occupied by acetylcholine molecules. They are not broken down and continue to cause the production of nerve impulses. This continuous stream of impulses leads to the production of excess saliva.

agricultural ecosystem: an ecosystem based on domestic animals or crop plants. The principles that apply to natural ecosystems also apply to those based on agriculture. Energy is transferred from producers to consumers and nutrients are recycled in the same way.

9

Agricultural ecosystems, however, have a number of characteristics. These include:

* Agricultural ecosystems have a low species diversity. Not only is modern agriculture often based on monoculture, where large areas of land are devoted to single crops, but the use of *pesticides* reduces the numbers of other organisms present.

* The search for more productive animals and plants results in little genetic variation. There is a lack of genetic diversity.

* Productivity is increased by the management of *abiotic* conditions such as applying fertilizer and irrigating field crops, or enhancing light, temperature and carbon dioxide concentration in glasshouse crops. This increase usually involves single species, however. The productivity of the ecosystem as a whole may be lower in agricultural ecosystems because fewer ecological *niches* are filled.

* Agriculture demands a high energy input. The manufacture of agricultural machinery, its use in cultivating crops, the production of fertilizers and pesticides and transport all add in energy terms to the real cost of growing a crop.

Agrobacterium: a group, or genus, of bacteria found naturally in the soil. These bacteria enter plant cells and cause cancer-like growths known as a crown galls to form. *Agrobacterium* possesses tumor-forming genes which are contained in small rings of DNA known as *plasmids*. The plasmids are incorporated into the DNA of the plant and the tumor-forming genes bring about abnormal growth of the infected cells.

Genetic engineers can use enzymes to cut open these plasmids and insert particular genes such as that, for example, which gives resistance to the herbicide, glyphosate. When the bacterium invades the plant cells it will introduce this gene. The plant will then be resistant to the herbicide. This means that weeds can be controlled by spraying fields with glyphosate. The weeds will be killed but the crop will be resistant and, therefore, will not be affected.

aldosterone: a *hormone* involved in controlling the sodium chloride concentration of the blood. If there is a fall in the salt concentration of the blood, its *water potential* will rise. As a result of osmosis, more water will now move into the body tissues with the result that the blood volume falls. A group of cells in the *kidney* detect this fall in blood volume and as a result stimulate the secretion of aldosterone from the adrenal gland. Aldosterone increases the *active transport* of sodium ions from the kidney tubule. The uptake of sodium ions is accompanied by the reabsorption of water.

algal bloom: the rapid growth of a population of algae resulting from an increased nutrient supply. Pollution of freshwater lakes and rivers may result in an increase in the concentration of phosphate and nitrate ions. A low concentration of these ions normally limits the growth of the microscopic algae which float at the surface and form the plankton. The addition of more phosphates and nitrates results in very rapid growth of the algal population. Some species of algae produce toxic substances so algal blooms may prove dangerous to the health of both humans and domestic animals. More frequently, however, they lead to an increase in the activity of decomposers and the depletion of dissolved oxygen (see *eutrophication*).

> **hint** Algae are members of the *Protoctista*. They are not plants so they do not have flowers. "Bloom" in this context does not mean flower, it simply means a huge increase in numbers.

alimentary canal: an alternative name for the gut of an animal. It runs from the mouth at one end to the anus at the other. The alimentary canal has a number of different functions. These are:

* ingestion – taking food into the mouth

* digestion – breaking large insoluble molecules like starch and protein into smaller soluble ones like glucose and amino acids

* absorption – the process by which the soluble molecules which result from digestion pass through the wall of the alimentary canal into the body

* egestion – the removal of waste material from the gut in the form of feces.

In a human, the main regions of the alimentary canal are the *esophagus*, *stomach*, *ileum* and *colon*. Although all these regions have a similar basic structure, there are obvious variations. Each has a structure which reflects its function. In addition, the detailed structure of the gut differs in different species. Even among mammals, these differences can be quite marked. The gut of a *ruminant* like a cow, for example, is very distinctive and quite unlike that of a carnivore such as a dog.

all or nothing: an expression used to describe a *nerve impulse*. When an *action potential* is produced in a nerve cell, it is always the same size. It does not matter how big the initial stimulus, the action potential will always involve the same change in potential difference across the plasma membrane. Because of this, the only way that information about the strength of a stimulus can be carried is to change the number of nerve impulses in a given time. It is not possible to have a small action potential for a small stimulus and a larger one for a large stimulus.

allele: one of the different forms of a particular *gene*. For example, snails which belong to the species *Cepea nemoralis* have shells which may be either plain or banded. This shell pattern is controlled by a gene. There are two different forms of this gene. One form or allele produces a plain shell while the other produces a banded shell. If the letter **B** is used to represent the gene, then **B** represents the dominant allele which results in a plain shell and **b** represents the recessive allele which produces a banded shell. Although many genes have two different alleles, it is possible for a single gene to have more than two forms in which case they are known as *multiple alleles*.

 It is important to distinguish carefully between the terms *gene* and *allele* and to use them correctly.

allergy: a condition in which the body becomes extremely sensitive to a particular substance that produces characteristic symptoms every time it is encountered. People encounter many foreign substances, bearing *antigens*, during the course of their lives. Under normal conditions, *antibodies* in the blood and tissues get rid of these antigens from the body and there are no side effects. In a person suffering from an allergy, however, the reaction between a particular antigen (known as an allergen) and antibodies in the tissues leads to the release of substances such as histamine. These substances produce the characteristic symptoms of the particular allergy. Some allergens produce localized effects such as occur with hay fever and asthma. The effect with others is more generalized and can be extremely serious. Anaphylactic shock, which may occur, for example, in people allergic to nuts, affects the lungs and circulatory system. Unless medical help is given immediately death may result.

11

allopatric speciation: the formation of new *species* which occurs when *populations* of the parent species become separated from each other by some form of geographical barrier. If a species is spread over a wide area, then features such as mountain ranges, seas, rivers or deserts may divide it into a number of populations which will be unable to breed with each other. These populations can be described as being reproductively isolated from each other. *Natural selection* will result in the evolution of slight differences in the characteristics of the organisms in the different populations. Gradually, over a period of time, these differences will become more and more pronounced. Finally, the separated populations will be sufficiently different to be recognized as separate species. Even if they mix with each other, individuals will be unable to interbreed. Allopatric speciation is shown in the diagram on page 13.

allosteric: refers to a molecule whose shape can be altered by something in its environment. This usually leads to a change in its function. Allosteric changes are important in enzyme-controlled reactions as they allow the rate of the reaction to be controlled. *Enzymes* have *active sites* which bind to substrate molecules. There are also other places on the surface of the enzyme to which different molecules can bind. Binding with one of these other molecules alters the shape of the active site so that the enzyme is stopped from working. This is really an example of *noncompetitive inhibition*.

Another molecule which undergoes an allosteric change is *hemoglobin*. A change in pH of the blood alters the shape of the molecule and affects the way in which it binds with oxygen. This gives rise to the *Bohr effect*.

alternation of generations: the alternation in the life cycle of a plant of two distinct stages, one with the *haploid* number of chromosomes, the other with the *diploid* number. The diagram shows this basic cycle with the haploid gamete-producing stage known as the *gametophyte* alternating with the diploid spore-producing stage called the *sporophyte*.

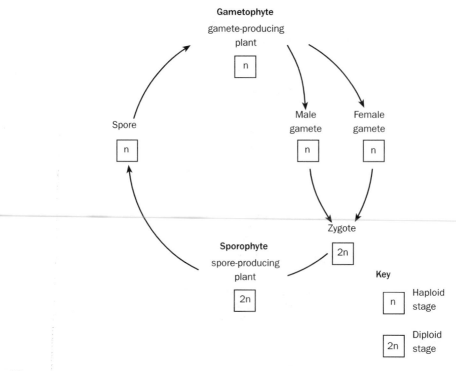

In the simplest plants, such as mosses, the gametophyte is the dominant stage. The adult green plant is a gametophyte which will produce male and female gametes. In flowering plants, on the other hand, the adult plant, be it a daisy or an oak tree, is a sporophyte. The gametophyte stages in flowering plants are reduced to tiny structures found within the flowers. The change from the gametophyte being the most important stage to the sporophyte being dominant is an important evolutionary trend in plants.

Alzheimer's disease: a disease which occurs in older people and involves the progressive deterioration of mental processes. It is a form of *senile dementia*. In its early stages, it is characterized by loss of short-term memory and mental confusion, but as the disease runs its course, these effects become more and more pronounced. Postmortem examinations of affected patients show that the condition is associated with structural and chemical changes in the brain tissue. These include the presence of plaques of an abnormal protein which form outside the brain cells and of tangled fibers of another sort of protein found inside them. Although the precise cause of Alzheimer's disease is unknown, it is thought that certain people inherit genes which make them more susceptible to the range of environmental factors which lead to development of the condition.

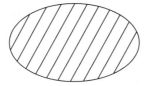

Single species spread over wide area

Geographical features divide species into different populations. Each population adapts to its environment

River

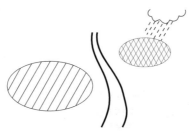

The separate populations are so different that they may be considered as different species

Allopatric speciation

alveolus: one of the air sacs in the lungs where gas exchange takes place. Oxygen enters the blood, and carbon dioxide is removed from it by *diffusion*. The alveoli show a number of adaptations which make them very efficient gas exchange surfaces:

* there is a large surface area over which diffusion can take place. This is provided by the huge number of alveoli. There are an estimated 300 million alveoli in each human lung giving a total surface area of between 50 and 80 square meters

* the difference in the concentration of respiratory gases is maintained as high as possible. Oxygen, for example, is continually being removed by combining with hemoglobin. *Ventilation* mechanisms replace the air in the alveoli

* the diffusion pathway is very short. The alveoli are lined with very thin, flat epithelial cells known as *squamous epithelium*. The total distance from the air in the alveoli, through these epithelial cells and those in the capillary wall, into the blood is considerably less than one micrometer.

amino acid: the basic unit or monomer from which *proteins* are formed. The diagram shows the general structure of an amino acid.

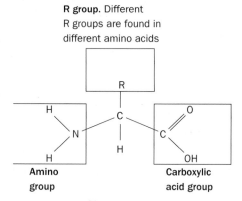

The basic structure of an amino acid

There are twenty different amino acids which may be linked together by *condensation* reactions to form proteins. All of these amino acids have the same basic chemical structure. It is only the R-group that differs. In the amino acid glycine, for example, it is H while in alanine it is CH_3. The amino acids cysteine and methionine are a little different as their R-groups contain sulphur. This is important in forming bonds which help to give proteins their *tertiary structure*.

Plants are able to make amino acids by combining some of the molecules produced by photosynthesis with nitrogen obtained from nitrates taken up from the soil. Animals are unable to do this and need to obtain many of the amino acids they require from their diet. From these, they can produce others by the process of *transamination*. Animals cannot store excess amino acids in their bodies until they are required. The surplus molecules are broken down by *deamination* and converted to other products which are excreted.

hint Proteins and nucleic acids are totally different substances. *Amino acids* are the monomers which make up proteins. *Nucleotides* are the monomers which make up nucleic acids.

ammonification: the stage in the *nitrogen cycle* in which saprophytic microorganisms break down organic, nitrogen-containing substances such as proteins and produce ammonium compounds. These are then oxidized to nitrites and nitrates in the process of *nitrification*.

amniocentesis: a technique used for the *genetic screening* of a *fetus* while still inside the uterus of its mother. *Ultrasound* is used to determine the precise position of the fetus and the *placenta* within the uterus. A fine needle is then inserted through the abdominal wall of the mother into the amniotic cavity. A sample of amniotic fluid is removed. This will contain some fetal cells. These cells can be examined and their *chromosomes* observed or the DNA that they contain investigated.

amnion: one of the membranes that surrounds a developing mammalian *fetus* in the *uterus* of its mother. The cells which make up the amnion secrete amniotic fluid which fills the space between the membrane and the fetus. This fluid provides protection and support for the delicate fetal tissues.

amylase: an enzyme which digests starch. Like all digestive enzymes, amylases are *hydrolases* and break down starch into soluble sugars by means of *hydrolysis*. Although amylases are familiar as important mammalian digestive enzymes found in saliva and pancreatic juice, they are also found in many microorganisms and in plant tissues.

hint Amylase and amylose are different substances. Amylase is an enzyme; amylose is a polysaccharide. Make sure that you write clearly enough to distinguish between them.

amylopectin: a *polysaccharide* that is an important constituent of starch. It consists of branched chains of α-glucose molecules joined together by *condensation*.

amylose: a *polysaccharide* that is an important constituent of starch. It consists of long straight chains of α-glucose molecules joined together by *condensation*.

anabolic reaction: a chemical reaction which is concerned with combining smaller molecules to form larger ones. Examples are the reactions which are involved in *protein synthesis* and in *photosynthesis*. Anabolic reactions require energy and, in living organisms, this is usually provided by *ATP*.

Hormones such as *testosterone* work by switching on genes which are responsible for particular anabolic reactions. The increased muscle and bone development at puberty, for example, is controlled by testosterone in this way. Testosterone is an anabolic steroid because it is a *steroid* which influences the building up of substances in the body. Anabolic steroids have unfortunately been the subject of abuse by some athletes.

anaerobic respiration: respiration which takes place in the absence of oxygen. Most organisms rely on the presence of oxygen in order to respire. Occasionally, however, there is insufficient oxygen available to meet their respiratory requirements and the whole organism, or part of it, is able to respire anaerobically for a period of time. Basically, anaerobic respiration relies on *glycolysis* to produce *ATP*. Since the net amount of ATP produced in glycolysis is only two molecules for each molecule of glucose as opposed to the 38 molecules of ATP produced in aerobic respiration, it can be seen that the anaerobic process is nowhere near as efficient.

Glycolysis involves oxidation and in this process the coenzyme, *NAD*, is reduced. Reduced NAD is normally reconverted to NAD in the reactions of aerobic respiration. Clearly there is

15

a problem. Quite simply, the NAD would run out. The biochemical pathways involved in anaerobic respiration in animals and in plants and microorganisms such as yeast are shown below.

In the conversion of pyruvate either to lactate or to ethanol and carbon dioxide, NAD is reformed from reduced NAD. This allows the first part of the reaction to continue in the absence of oxygen.

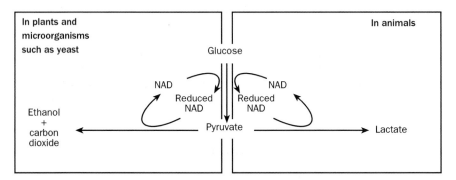

> **hint** Get the pathways the right way round. Plants and microorganisms such as yeast produce ethanol and carbon dioxide. Animals produce lactate.

anemia: a condition in which there is a reduced amount of hemoglobin in the blood. People suffering from anemia tire easily. They have pale skin and get out of breath if they exert themselves. There are a number of causes of anemia which include:

* a shortage of iron in the diet; iron is an important part of *hemoglobin* molecules

* conditions which lead to loss of blood. This might either be the result of an accident or come from something like an ulcer which results in chronic bleeding over a period of time

* conditions which result in the destruction of red blood cells. *Sickle-cell anemia* is an inherited condition in which the affected person has a different type of hemoglobin present in his or her body which is less efficient at transporting oxygen. The red blood cells of someone with this condition have a shorter lifespan than normal red blood cells.

aneurysm: a weakening of the wall of an *artery* which produces a balloon-like swelling. The wall of the blood vessel is weakened and, clearly, the consequences for the patient are very serious if it bursts.

Angiospermophyta: the plant phylum which contains the flowering plants. It includes a great variety of plants ranging from daisies to oak trees and includes most plants which are of economic importance. Members of the Angiospermophyta share the following features:

* the reproductive organs are found within *flowers*

* *seeds* are formed as a result of fertilization and these are enclosed within fruits.

animal: a member of the kingdom *Animalia*.

> **hint** Care should be taken to use the word animal only when referring to a member of the kingdom *Animalia*. Single-celled organisms such as amoeba are members of the kingdom Protoctista. They are not animals.

16

Animalia: the kingdom containing animals (see *classification*). It is important to distinguish this kingdom from the *Protoctista*. Unicellular organisms such amoeba and paramecium are not members of Animalia. Animals share the following features:

* They are multicellular organisms whose cells do not possess cell walls.

* They obtain their food by taking it into a digestive cavity in a process known as *ingestion*. Once inside the digestive cavity, food is digested and absorbed.

* After *fertilization* a *zygote* is formed. This zygote develops into a hollow ball of cells called a *blastula*.

Animals differ from plants in a number of important ways, and since these concern various aspects of biology, the differences between animals and plants form one of the important synoptic themes that links aspects of biology. The spider diagram shows some of these links. You can use the cross-referencing system in this book to add greater detail to the diagram.

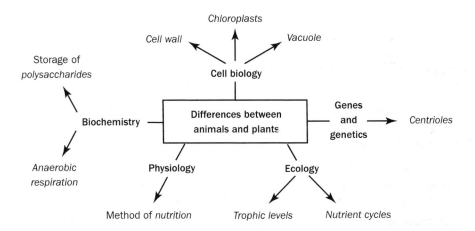

Annelida: the animal phylum which contains segmented worms such as earthworms and leeches. Members of the Annelida share the following features:

* they have worm-shaped bodies which are divided into distinct segments

* they possess small bristles or chaetae on their body surfaces which assist with loco-motion

* there is a fluid filled *coelom* which helps to form a skeleton against which the locomo-tary muscles can contract when the animal moves.

anorexia nervosa: a disorder in which patients starve, use laxatives or make themselves vomit in order to lose weight. The underlying causes are psychological and lead to a mis-conception of being overweight or a fear of becoming fat. The symptoms include severe loss of weight and, in female patients, stopping of *menstrual* periods.

antagonistic substance: substances which block the action of *neurotransmitters* at *synapses*. A synapse is a junction between two nerve cells. A small quantity of a chemi-cal neurotransmitter is released by the presynaptic nerve cell. This diffuses across the gap between the cells and fits into a receptor molecule in the plasma membrane of the postsynaptic nerve cell. This leads to the production of a nerve impulse. The molecules of the neurotransmitter have a specific shape which enables them to fit into the receptor molecule. There are other substances which have similar shapes and they can also fit into these receptors. If the receptor site is occupied by another different molecule, it will not be available to the neurotransmitter, so a nerve impulse cannot be produced in the postsynaptic nerve cell.

There are many examples of drugs and other substances which can act as antagonists. The South American arrow poison, curare, blocks the action of acetylcholine at the junction between nerve and muscle. This leads to paralysis. *Beta-blockers* are drugs which are important in the treatment of heart disease. They also have an antagonistic effect.

antagonistic muscles: a pair of muscles which, on contraction, produce opposite effects to each other. Muscles can only do work by shortening or contracting. Therefore, if a limb has to be moved, two muscles are required to do this: one will bend the limb when it contracts, the other will straighten it. The two muscles are said to be antagonistic. The term does not apply only to muscles which involve the movement of limbs. The earthworm has two sets of muscles in its body wall which are antagonistic. Circular muscles will make the animal longer and thinner when they contract; longitudinal muscles will make it shorter and fatter.

anther: the part of the *stamen* of a flower in which the *pollen* develops and from which it is released.

antibiotic: a substance which is produced by one type of *microorganism* which kills or stops the growth of another. There are a number of exceptions to this definition, however, so it is difficult to produce a general definition of the term. Antibiotics used in medicine are often altered chemically to make them more effective and some are made entirely by chemical means. Substances which have an antibiotic effect have now also been found in a wide range of different animals and plants, ranging from toads to snowdrops.

There are a number of small but important differences between the biochemical processes which take place in microorganisms and those that take place in mammals. Antibiotics work by affecting those processes which are found only in microorganisms. Penicillin, for example, prevents chemical bonds forming which strengthen bacterial cell walls. Since human cells do not have cell walls, penicillin has no effect on them. The diagram on page 19 shows the effects of some antibiotics on bacteria.

If, as a result of its action, bacteria are killed, the antibiotic is referred to as a *bactericidal* (biocidal) antibiotic. If the bacteria are simply prevented from reproducing, then it is called a *bacteriostatic* (biostatic) antibiotic.

hint Extensive use of antibiotics has led to the evolution of antibiotic-resistant strains of bacteria. Antibiotics do not cause these bacteria to mutate. What happens is that bacteria vary. Some will be resistant to a particular antibiotic while others will not. If a person is treated with a course of antibiotics, most of the bacteria causing the infection will die. The resistant ones will survive and multiply, passing on the gene for antibiotic resistance.

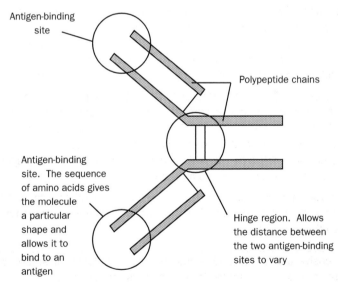

Ciprofloxacin
Inhibits DNA
replication

Streptomycin
Inhibits protein
synthesis

DNA ⟶ mRNA ⟶ proteins

Penicillin
Prevents cell wall
formation

The effects of some antibiotics on bacteria

antibody: a molecule secreted by certain lymphocytes in response to stimulation by the appropriate *antigen*. The term describes simply what the molecule does. It acts against foreign bodies. Antibodies are types of proteins known as immunoglobulins. The way in which these molecules work is closely related to their structure. This is explained in the diagram.

Antigen-binding
site

Polypeptide chains

Antigen-binding
site. The sequence
of amino acids gives
the molecule
a particular
shape and
allows it to
bind to an
antigen

Hinge region. Allows
the distance between
the two antigen-binding
sites to vary

The structure and function of an antibody

anticodon: the sequence of three bases on a molecule of transfer RNA (*tRNA*) which corresponds to a particular *codon* on an *mRNA* molecule. During *translation* a ribosome becomes attached to a molecule of mRNA. The ribosome is just the right size to hold two codons of mRNA. Suppose one of these codons has the sequence of bases GCU which codes for the amino acid alanine. Only a tRNA molecule with the complementary anticodon, CGA, will be able to bind to this codon. A tRNA molecule with this anticodon will always carry alanine.

19

antigen: a molecule which triggers an immune response by the *lymphocytes*. Although usually proteins, other types of molecule such as polysaccharides, lipids and even nucleic acids can act as antigens. In certain circumstances, inorganic substances such as nickel can bind to proteins in the body and can also act as antigens. This is what happens, for example, in a person who develops an allergy to metal earrings. The one feature that all antigens have in common is the large size of their molecules.

hint An *antigen* is a molecule which triggers an immune response. An *antibody* is a molecule secreted in response to stimulation by the appropriate antigen. Take care not to confuse these basic terms.

antiseptic: a substance which kills *microorganisms* but can be used safely on the skin or the mucous membranes of the body. Antiseptics are like *disinfectants* in that they kill most of the harmful microoganisms present. They differ, however, in that they are safe to use on the body surface itself.

aorta: the main artery of the body. In a mammal, it leaves the left ventricle of the heart and curves over to run through the abdomen. The arteries which supply all the other organs of the body (with the exception of the lungs) branch off the aorta. The wall of the aorta contains baroreceptors which monitor blood pressure and chemoreceptors which monitor changes in concentration of carbon dioxide and oxygen in the blood.

apoplastic pathway: the pathway by which substances go through the cell walls and the intercellular spaces of a plant. Water moves through the roots into the *xylem*. It also moves out of the xylem and through the leaf tissue. There are three separate pathways along which it can travel. The apoplastic pathway described here, the *symplastic* pathway through the cytoplasm of the cells, and the *vacuolar* pathway which involves passing from vacuole to vacuole. The apoplastic pathway is probably the most important of the routes by which water moves through plant tissues. There are no barriers to movement so it is able to diffuse freely.

aqueous humor: a watery solution found in the front part of the eye underneath the *cornea*. It is continually being formed by a structure known as the ciliary body and reabsorbed back into the blood. It has two important functions. The cornea and lens are both transparent structures but they are composed of living cells which require a supply of nutrients. The aqueous humor provides these nutrients. It also helps to maintain the shape of the front part of the eye.

arteriole: a vessel which takes blood from the smaller *arteries* to the *capillaries*. Arterioles are very small in diameter and, like all blood vessels, have a lining of epithelial cells. Their walls contain large numbers of muscle fibers. Many arterioles also have rings of muscle where they join with the capillaries. These are called *sphincter* muscles. By contraction of the muscle fibers in the walls and the sphincter muscles, the blood supply to particular capillary networks can be regulated to meet the needs of the part of the body concerned.

arteriosclerosis: a term which can apply to a number of conditions affecting arteries.

artery: a blood vessel which takes blood from the heart towards the *capillaries*. In mammals, arteries usually contain blood which is rich in oxygen but there is one important exception to this. The blood going to the lungs along the pulmonary artery has come from the right side of the heart. This is the blood which has returned from the tissues and is therefore low in oxygen. Like all blood vessels, arteries have a lining composed of epithelial cells. Their walls are very thick and contain a large amount of elastic tissue and muscle. When the ventricles of the heart contract, blood at high pressure is forced into the arteries and causes the walls to stretch. When ventricle contraction stops, the pressure of the blood falls and the elastic tissue contracts. This stretching and contraction of the elastic tissue in the artery walls helps to even out blood flow.

The activity of different organs in the body is always changing. During a period of strenuous exercise, for example, the muscles require as much blood as possible; less is required by the digestive system. The walls of smaller arteries contain a lot of muscle fibers. These may contract and make the diameter of the artery smaller and so help to regulate the blood supply to a particular organ.

Arthropoda: the animal phylum containing insects, spiders and crustaceans such as crabs and woodlice. There are more arthropods both in terms of individuals and species than any other organism. Members of the Arthropoda share the following features:

* they possess an *exoskeleton* made of *chitin*. This hard outside layer has to be molted before an arthropod can increase in size. Growth is therefore discontinuous

* they have jointed limbs or appendages which are used for a variety of purposes including locomotion and feeding.

artifact: something seen in a specimen looked at under a microscope which was not present in the living cell. In order to prepare specimens for microscopic examination, they have to be treated in a number of ways. The tissue usually has to be preserved, cut into sections and stained. These procedures involve substances which may alter the cell. Great care has to be taken, then, in interpreting what is seen under the microscope because of artifacts which may arise in preparation of the specimen. This is particularly important with electron microscopes because it is not possible to observe living cells.

artificial insemination: the introduction of sperms into the vagina of a female by artificial means. Artificial insemination is of considerable importance in the production of cattle and pigs. The process in cattle is summarized in the flowchart on page 22.

Artificial insemination in cattle has a number of advantages over natural insemination:

* it allows farmers to select the qualities that they want in a calf. By keeping extensive records, the characteristics of the calves fathered by a particular bull can be predicted with some accuracy. The farmer can then choose the particular bull from which the semen sample may be collected

* bulls are rather aggressive and difficult animals to keep. Farmers can save on the costs and risk of keeping a bull

* artificial insemination reduces the risks of passing on sexually transmitted diseases from one animal to another.

ascorbic acid: see *vitamin C*

asexual: reproduction by any means which does not involve the fusion of *gametes*. One of the most important features of asexual reproduction is that it involves processes which give rise to the production of genetically identical offspring. Many *microorganisms*, such as bacteria and fungi, reproduce in this way and asexual reproduction is also widespread in plants. Among animals it is less common and is generally found only in more primitive groups such as the *Cnidaria*. The production of buds by yeast cells, of spores by mosses

Semen is collected from suitable bulls at an artificial insemination center. The bull ejaculates into an artificial vagina and the semen is collected

↓

The semen is checked for quality by examining under a microscope. It is then diluted and stored in liquid nitrogen at a temperature of $-196\ °C$

↓

When required, the sample is thawed and placed in a special pipette

↓

A long tube is used to introduce the semen into the cervix of a cow which is in season, or *estrus*

Artificial insemination

and ferns and the splitting of an amoeba to form two separate individuals are all examples of asexual reproduction.

In flowering plants, asexual reproduction is also known as vegetative propagation, and frequently involves the development of fairly large structures such as bulbs in plants like daffodils and onions or special types of underground stem as in, for example, potatoes and irises. Not only are these structures involved in asexual reproduction but they may also help the plant to survive extreme climatic conditions. With the first frosts, for example, the aerial parts of a plant such as an iris die. The underground stem, however, contains the nutrients necessary for the plant to survive the winter and to produce a number of new plants the following spring.

The fact that asexual reproduction produces genetically identical offspring is used commercially where it has considerable economic importance. Suppose a new variety of rose has been developed. Letting it reproduce sexually by pollinating the flowers and growing the resulting seeds would result in a variety of genetically different offspring. This would be of little use to the plant breeder. If more plants are produced by asexual reproduction, however, the offspring will all have the desired characteristic as they will be genetically identical. The process by which large numbers of genetically identical plants can be produced from cells taken from the growth region of a single parent is known as *micropropagation*.

association area: part of the *brain* which is involved with analysing and interpreting sensory information in the light of experience. The association areas form part of the *cerebrum* in the forebrain. They:

❋ link sensory information with previous knowledge, allowing us to recognize information

* link this sensory information with information from other receptors

* interpret the information in its present context.

As an example, think about an image formed on the retina of the eye. You could get the same size image from a small object close to the eye or a much larger object at a distance. It is the visual association area that links all the sensory clues with information stored in the memory to interpret the image.

asthma: a condition in which the flow of air to the gas exchange surface of the lungs is restricted. Some of the cells in the lining of the bronchi become sensitive to substances such as smoke, pollen or other atmospheric pollutants. Further exposure to these substances brings about an immune response which triggers an asthmatic attack. The smooth muscle in the walls of the bronchi contracts. Together with the production of large amounts of mucus, this leads to a narrowing of the airways and the formation of a blockage which limits the amount of air that can reach the alveoli.

atheroma: fatty deposits formed in the walls of *arteries*. It is a common condition, seen in many adults, and is associated with a number of possible factors, ranging from diets high in *cholesterol* and refined sugars to smoking and lack of exercise. They may not produce any obvious signs of ill health but, later in life, atheroma might give rise to more serious diseases of the circulatory system.

> **hint** When you write about medical conditions such as atheroma, use appropriate biological language. Atheroma is the formation of fatty deposits in the walls of arteries, not the "furring up" of blood vessels.

atherosclerosis: a disease of the arteries caused by *atheroma* or fatty deposits in the walls. These can either block an artery directly or increase the chance of it being blocked by a blood clot. When atherosclerosis affects the *coronary arteries* it may lead to a heart attack or *myocardial infarction*.

atom: The smallest part of an *element* that cannot be broken down further by chemical means. An atom is made up of a nucleus which contains two sorts of particle, protons which have a positive charge, and neutrons, which as their name suggests, have no charge. Negatively charged *electrons* orbit round the nucleus. Atoms of the same or of different elements may combine to form *molecules*.

ATP (adenosine triphosphate): an important molecule, found in all living cells, which is involved in the transfer of energy. A molecule of ATP is made up from a molecule of *adenine* joined to the five-carbon sugar, *ribose* and to three phosphate groups. When ATP is broken down, the third phosphate group is lost and a considerable amount of energy is released. The reverse reaction can also take place. ADP can join with a phosphate group to produce ATP. In this case energy is required. These reactions are summarized below:

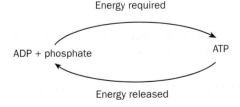

23

Most of the ATP within a cell is produced using energy released during the process of respiration. A small amount is also made in photosynthesis. Some of the ways in which ATP is used in living organisms are shown below.

The importance of ATP:

* ATP releases energy.

  This energy may be used in many different chemical reactions, for *active transport* and for movement.

* ATP can supply phosphate to various chemical compounds. This is important in some of the biochemical reactions in both *respiration* and *photosynthesis*.

* ATP is the molecule from which *cyclic AMP* is produced.

  Cyclic AMP is an important messenger molecule in cells.

The importance of ATP is one of the key synoptic themes that links together different aspects of biology. The spider diagram shows some of these links. You can use the cross-referencing system in this book to add greater detail to the diagram.

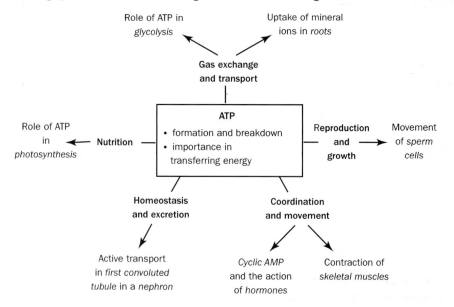

hint ATP consists of adenine, a five-carbon sugar, and three phosphate groups, so its molecules are larger than those of glucose.

atrial systole: the stage in the *cardiac cycle* where the walls of the atria contract and force blood into the ventricles.

atrioventricular node: a small area of muscle in the wall of the heart between the atria and the ventricles. It plays an important part in coordinating the beating of the heart. At the start of each heartbeat, a wave of electrical activity spreads from the *sinoatrial node* or pacemaker over the walls of the atria. This brings about contraction of the atria. The muscle fibers in the atria are completely separate from those in the ventricle, except in one small area, the sinoatrial node. It is only through this node that electrical activity can pass from the atria to the ventricles. There is a short delay here before the electrical activity spreads to the base of the ventricles. This is important in allowing emptying of the atria to be completed before the ventricles start to contract.

atrioventricular valve: one of the valves between the atria and the ventricles in the heart. These valves are made of fibrous tissue and are opened and closed by blood pressure. During the part of the *cardiac cycle* in which the pressure in the atrium is higher than that in the ventricle, the valve is open and blood is able to flow through into the ventricle. During *ventricular systole*, the muscle in the wall of the ventricle starts to contract and the pressure of the blood that it contains rises. As a result, the valve shuts. This enables blood to be pumped out through the arteries leaving the heart and prevents backflow into the atria. When a stethoscope is placed against the chest wall, distinctive heart sounds can be heard. The first of these is due to the atrioventricular valves closing.

The valve on the left side of the heart has two flaps of fibrous tissue and is known as the bicuspid valve; that on the right side has three flaps and is called the tricuspid valve.

hint Atrioventricular might seem to be a very complex term but it tells you precisely where the atrioventricular valves are situated. Call them right and left atrioventricular valves and you won't have to worry which is the bicuspid and which the tricuspid.

atrium: a chamber of the heart which receives blood returning from the organs of the body. Fish have a *single circulation* and consequently the heart of a fish has only one atrium. Mammals, on the other hand, have a *double circulation*. Their hearts have two atria. The right atrium receives blood returning from the organs of the body, while the left atrium receives oxygenated blood which the pulmonary veins bring back from the lungs.

hint It might seem very simple, but make sure that you do not confuse the right and left sides of the heart. The right atrium receives blood returning from the organs of the body; the left atrium receives oxygenated blood which the pulmonary veins bring back from the lungs.

attenuation: the growth of bacteria or viruses in a culture so that they lose their ability to cause disease. In order to complete its life cycle inside the cell of a host, a pathogen such as a bacterium needs a number of so-called virulence genes. These genes are not necessary, however, for the growth of the same bacterium in an artificial culture and, over a period of time, tend to become ineffective through the normal processes of mutation. Attenuated bacteria or viruses are often used for making *vaccines*.

auricle: see *atrium*. A term used to describe the chambers of the heart which receives the blood returning from the organs of the body.

australopithecine: a genus of extinct African *hominids*. Like members of the genus *Homo*, they could walk upright although there were slight differences in the shape of the pelvis and upper leg bones. However, in relation to their body size, they had smaller brains. There were a number of species of australopithecine, but these can be roughly divided into two distinct types. One group was rather lightly built, the other much more robust.

autoimmune disease: a disease which is caused by the body producing an *immune reaction* against its own tissues. Damage may either be caused by *antibodies* or by *T-cells*.

Autoimmune diseases are more common in older people. Some of these diseases affect specific organs or systems as in the case of pernicious anemia. In this disease, antibodies are made against a molecule in the gut necessary for the uptake of vitamin B_{12}. Other examples, such as rheumatoid arthritis which causes painful swelling of the joints in many elderly people, have a more general effect on the body.

autonomic nervous system: the part of the nervous system which supplies the muscles and glands which are not under conscious control. The autonomic nervous system is divided into two sections. The *sympathetic nervous system* plays an important part in enabling the body to react to stress, while the *parasympathetic nervous system* is more important when the body is at rest.

autosome: a chromosome which is not a *sex chromosome*. In a human body cell, there are 23 pairs of chromosomes. One pair of these are sex chromosomes; the other 22 pairs are autosomes.

autotroph: an organism whose method of nutrition is *autotrophic*.

autotrophic nutrition: a method of nutrition in which an organism builds up the organic molecules that it requires from simple inorganic molecules such as carbon dioxide and water. In order to do this, an energy source is necessary. For many autotrophs this is light energy, in which case the type of nutrition is referred to as *photoautotrophic nutrition*. In some organisms, however, the energy comes from another chemical reaction. This is *chemoautotrophic nutrition*. Autotrophs are *producers* and are at the base of all *food chains*.

auxin: a *plant growth substance* which has a number of different functions, one of the most important of which is cell enlargement. The flowchart explains how it is thought that cell enlargement is brought about.

Presence of auxin leads to the active transport of H^+ ions into the cell wall

$\downarrow$

The resulting change in pH leads to the breakdown of chemical bonds in the cell wall. As a result the cell wall softens

$\downarrow$

Cell can take in more water by osmosis. It can enlarge as the cell wall no longer restrains it

However, auxins have other effects and it is thought that they are also involved in switching on a number of different genes associated with growth.

Synthetic auxins have a number of important commercial applications:

* they may be used as weed killers. Since some of them are toxic to many weeds but nontoxic to grasses and cereals, they may be used to control weeds in cereal crops or on lawns

* they can be used to stimulate the formation of roots on cuttings of plants such as geraniums

26

- fruit drop is a common problem in orchards. Large amounts of fruit fall from trees before it is ripe. Spraying with auxin helps to prevent this from happening.

axil: the angle between the upper surface of a leaf and the stem to which it is joined. It contains a bud, known as an axillary bud. One of the features of a potato tuber which confirms that it is a modified stem and not a root is that it has tiny scale leaves on its surface, each of which has a bud in its axil.

axon: a long process in a *neurone* which carries *nerve impulses* away from the cell body.

bacillus: a rod-shaped bacterial cell. Examples of bacilli include *Lactobacillus bulgaricans* which is found in yogurt and the various species of *Salmonella* which produce food poisoning.

bacteria: an important group of microorganisms. Bacteria are *Prokaryotae* and their cells show the typical features of this kingdom. They are small in size, do not have nuclei or organelles such as *mitochondria*, and their *DNA* is in circular strands in the cytoplasm. Although some bacteria are harmful and are responsible for diseases such as cholera and plague, which have had enormous effects on human populations, others form the basis of much of modern *biotechnology*. In addition, they play a vital role in ecological processes. *Nutrient cycles* rely on the action of bacteria in making the elements locked up in complex organic molecules available to plants once again.

Bacilli or **rods**

Salmonella typhi causes typhoid fever
Bacillus subtilis is used in genetic engineering

Cocci

or in irregular colonies

May occur singly

or in chains

Staphylococcus aureus causes boils

Streptococcus lactus is used in making cheese

Spirilla

Treponema pallidum causes syphilis

Vibrios or **curved rods**

Vibrio cholerae causes cholera

Identifying bacteria by their shapes

28

A number of features are used to identify bacteria. The shape of an individual cell is important and enables it to be classified as a *bacillus*, a *coccus*, a *spirillum* or a *vibrio*.

With their small size and the enormous number of species involved, shape alone is inadequate as a means of identifying individual bacteria. The molecules which they contain and the biochemical processes which occur within their cells are also important and allow techniques such as *Gram's staining* to be used to distinguish between bacteria which look very similar under a microscope.

bacterial chromosome: see *chromosome*

bactericidal: used to describe a substance which kills bacteria. Killing is essential for sterilization so *disinfectants* and *antiseptics* are bactericidal compounds. There is a wide range of substances available, but their effectiveness depends on the microorganism involved. The spores formed by certain species of bacteria are extremely resistant to substances which would kill other forms of life. Similarly, relatively few substances are effective against *viruses*. Many drugs used to combat disease, however, are bacteriostatic and simply prevent bacterial reproduction.

bacteriostatic: used to describe a substance which prevents bacterial reproduction. Many *antibiotics* are bacteriostatic as they interfere with the mechanism of protein synthesis. They prevent bacteria multiplying while the body's natural defences destroy those already present.

balanced diet: a diet which contains all the basic nutrients in the proportions necessary to maintain health. The exact composition of a balanced diet will vary throughout a person's life. Such factors as age, sex, pregnancy and lactation will all have an influence. To be balanced, consideration needs to be given to:

* carbohydrates. Sufficient carbohydrate needs to be taken in order to meet the energy requirements of the body. Excess can lead to being overweight. A correlation exists between an excessive intake of carbohydrates and a number of conditions. These include heart disease, diabetes and dental caries

* lipids. It is necessary to eat lipids but there is concern over diets which contain high amounts of animal fats. There is a link between an excessive intake of animal fats and heart disease

* protein. Individuals who are actively growing, and pregnant or lactating women need increased protein in their diets. It is essential that the protein in the diet includes the necessary amounts of *essential amino acids*. This is particularly true of vegans, who eat no food of animal origin

* mineral ions. These are required in varying amounts and, again, the amount actually needed depends on the specific circumstances of the individual concerned. Women, for example, tend to require a higher intake of iron to meet the loss from the body through menstruation. Pregnancy and lactation demand a higher level of calcium

* vitamins. These are required in small amounts for a variety of purposes. There is no evidence that consumption of extra amounts of vitamins has a beneficial effect. Most suitably varied diets will contain sufficient vitamins to meet a person's requirements adequately.

basal metabolic rate (BMR): the *metabolic rate* of a person who is at complete rest. It is a measurement of the amount of energy required for vital activities such as the action of the heart and the muscles associated with breathing. It should be measured some time after a meal, so as to reduce as far as possible the effect of food in the stomach. It also needs to be determined at a comfortable environmental temperature since processes such as shivering have a considerable effect on metabolism. It is usual to give figures for basal metabolic rate in energy units per square meter of body surface per hour. This takes into account variation in body size.

There are a number of factors which affect basal metabolic rate.

* age. In humans, basal metabolic rate has a maximum value at about one year old. It then falls more or less continuously for the rest of a person's life

* sex. At all ages, females have a slightly lower basal metabolic rate than males

* state of nutrition. People who are undernourished tend to have lower basal metabolic rates

* health. A good example is provided by the *thyroid gland*. This gland produces a hormone which controls metabolic rate. A person with an underactive thyroid will therefore have a low metabolic rate while someone who has a tumor in the thyroid gland may produce large amounts of thyroid hormone and have a very high metabolic rate.

> **hint** Don't forget that basal metabolic rate is a rate, so it refers to the amount of energy used in a given time.

base: in biology, the term usually refers to one of the components of a nucleic acid molecule. A molecule of DNA or RNA is made up of a large number of *nucleotides*. Each of these consists of a five-carbon sugar, a phosphate and a nitrogen-containing base. There are five different bases and one of these is found in each nucleotide. Some of the properties of these bases are shown in the table.

Property	Adenine	Cytosine	Base Guanine	Thymine	Uracil
Type of base	Purine	Pyrimidine	Purine	Pyrimidine	Pyrimidine
Nucleic acid in which base is found	DNA & RNA	DNA & RNA	DNA & RNA	DNA only	RNA only
Base with which it pairs	Thymine (DNA) Uracil (RNA)	Guanine	Cytosine	Adenine	Adenine

> **hint** This is an area of biology where spelling really matters. There are lots of biological terms similar to adenine, for example: adenosine, alanine, even Anadin! So it is particularly important to spell the names of bases correctly.

basophil: see *granulocyte*

batch fermentation: a method of fermentation which involves separating the products at the end of the process. The *microorganisms* involved are put in a *fermenter* along with the necessary medium. Nothing is then added or removed during fermentation. At the end of the process, the required product is extracted and purified. This procedure may be contrasted with *continuous fermentation* where the substrate is added and the products removed while fermentation is going on. Batch fermentation has some advantages over continuous fermentation:

- if contamination occurs, it only results in the loss of a single batch. Because it is relatively easy to set up a batch culture, this results in minimal financial loss
- batch culture is a relatively short-term process and does not involve setting aside a fermenter for a long period of time. It is much easier to switch the use of fermenters and make different products at different times. This is more versatile
- it is much easier to maintain the correct environmental conditions over a single batch fermentation than over the long period of time required by continuous fermentation.

benign tumor: see *tumor*

B-cell: a lymphocyte which, when it comes into contact with the appropriate antigen, develops into an antibody-secreting cell. B-cells have receptors on their surfaces which will only bind to specific antigens. Once a B-cell has encountered this antigen, it is stimulated to divide rapidly to produce a large number of identical cells called a clone. This clone contains two sorts of cells, *plasma cells* and *memory cells*.

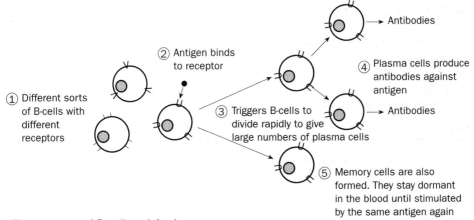

The response of B-cells to infection

Benedict's test: a biochemical test which can be used to show the presence of reducing sugars. If the specimen is hydrolyzed first by boiling with an acid, it can be used to test for a nonreducing sugar. For details, see *biochemical tests*.

beta-blocker: a drug which can be used to treat angina and high blood pressure. The membrane surrounding a cell, the *plasma membrane*, has a number of protein molecules in it. Some of these act as receptors. They have a particular shape into which another molecule such as a hormone fits. Heart muscle has beta-adrenergic receptors which are the receptors into which *adrenaline* molecules fit. A beta-blocker has a similar molecular shape to that of adrenaline. It will also fit into the beta-adrenergic receptor. When it does so, it blocks the receptor so adrenaline will not affect the heart.

Beta-blockers were first developed to combat angina, a condition which is characterized by severe chest pains which accompany exercise. When a person normally undertakes greater physical activity, adrenaline is secreted. This increases the activity of heart muscle with the result that more oxygen is required. In a person with heart disease, the supply of oxygen cannot be increased enough so angina results. By blocking the action of the adrenaline, this condition should not arise. As well as preventing angina, beta-blockers can also be used to treat high blood pressure or *hypertension*.

bicuspid valve: the valve between the left atrium and the left ventricle in the heart of a mammal. See *atrioventricular valve*.

bilateral symmetry: used to describe organisms which can only be cut in one plane to produce a mirror image. This is because their organs are arranged similarly on either side of a line which goes down the center. Most bilaterally symmetrical animals are motile, that is, they can move. Bilateral symmetry is thought to be an adaptation for movement allowing the equal contraction of muscles on either side of the body. The other important feature of bilaterally symmetrical animals is that they tend to have a distinct head region. It is the front end of a moving animal which is most likely to encounter food and the different stimuli which occur when it enters a new environment. The feeding apparatus and sense organs therefore tend to be concentrated at the front end. Organisms which have more than one plane of symmetry are said to show *radial symmetry*.

bile: a solution produced in the *liver* which empties into the small intestine through the bile duct. It is a slightly alkaline liquid containing, apart from water and various inorganic ions, *bile salts* which play an important part in the digestion of fat, and *bile pigments* which are excretory products formed from the breakdown of *hemoglobin*. Bile does not contain any digestive enzymes. Between meals, bile is stored in the gall bladder. When partly digested food from the stomach enters the first part of the small intestine, the hormone *cholecystokinin* is released. This results in the emptying of the gall bladder.

bile pigment: a substance formed by the breakdown of *hemoglobin*. *Red blood cells* do not live for very long. After about 120 days in the circulation, they die and the hemoglobin which they contain is broken down. The diagram shows that the heme group is split off from the globin part of the molecule. The iron from the heme group is used to make more hemoglobin while the rest of the group is converted into bile pigments which are excreted in the bile (see page 33).

hint There are no digestive enzymes in bile. Bile does contain salts, however, which help to break fats into smaller droplets. This provides a large surface area so that they can be broken down more rapidly by *lipase* enzymes.

bile salts: a component of bile involved in the digestion and absorption of fat. Without the presence of bile, digestion is much less efficient and there is a considerable increase in the amount of undigested fat found in the *feces*. Bile does not contain any digestive enzymes. It does, however, contain bile salts which emulsify fats, that is, they break the fat found in food into tiny droplets which increase the surface area for the action of *lipase* enzymes. As well as this, bile salts also activate lipases and play a part in the absorption of the products of fat digestion. Most of the bile salts are reabsorbed by the small intestine and transported back to the *liver* where they enter the bile again.

Hemoglobin
broken down to

Heme

Globin
(a protein)

Iron removed
and stored

Rest of molecule
converted to
bile pigments

Amino acids

The formation of bile pigments from hemoglobin

binary fission: a form of *asexual* reproduction taking place in single-celled organisms. It is the way that most bacteria reproduce and it is also found among members of the *Protoctista* such as amoeba. The parent cell divides to form two identical daughter cells. Given favorable conditions, such as a plentiful supply of nutrients, binary fission can lead to a rapid increase in numbers.

binocular vision: vision in which both eyes face forward. It enables distance to be judged extremely accurately. Binocular vision is found in humans and other *primates* where it is thought to be an adaptation associated with living in trees. In such a habitat it is clearly very important to be able to judge the distance between branches as accurately as possible.

bioaccumulation: the way in which certain substances are concentrated in the bodies of animals, the higher their position in a food chain. In the early 1950s, the substance, DDT, was widely used to control insect pests. It proved to be very effective but, after a while, it was noticed that there was a decline in the number of predators, such as sparrow hawks, peregrines and other birds of prey. What was happening was that DDT was present in the bodies of insects which were being eaten by small birds. These birds were not breaking the insecticide down but storing it in the body fat. A bird of prey, like a sparrow hawk, eats large numbers of smaller birds. Each time, a little more DDT was entering its body and accumulating in its tissues. Although eating a single contaminated bird would have had little effect, eating numbers of them led to the buildup of lethal concentrations of insecticide. The table shows the results of one study into the effect of bioaccumulation on the concentration of organochlorine insecticides like DDT.

Species	Food	Concentration of organochlorine insecticides in the body tissues (parts per million)
Sparrow hawk	Small birds	3.8
Wood pigeon	Plant material	0.3

DDT probably provides the most well-known example of a substance which became concentrated by bioaccumulation, but there are other examples. An incident which happened

33

biochemical oxygen demand (BOD)

in Japan in the 1950s involved the discharge of factory waste high in mercury. It became concentrated by the various organisms in the food chain until it reached levels in shellfish as high as 102 parts per million. When these shellfish were eaten by the local people, they suffered the effects of severe mercury poisoning.

biochemical oxygen demand (BOD): the amount of oxygen which is removed from a sample of water in a given time. It is a measure of how polluted the sample is. If a sample of water contains a lot of organic matter, it will support large numbers of bacteria. The more bacteria there are present, the more oxygen they will remove from the water and, therefore, the higher the biochemical oxygen demand. The BOD of a water sample is determined by incubating it at a temperature of 20 °C for five days and measuring the amount of oxygen used.

biochemical tests: a series of simple tests which can be used to identify proteins, lipids and carbohydrates. The tests required at this level are summarized in the table below.

Substance	Test	Brief details of test	Positive result
Protein	Biuret test	• Add sodium hydroxide solution to the test sample • Add a few drops of dilute copper sulphate solution	Solution turns mauve
Lipid	Emulsion test	• Dissolve the test sample by shaking with ethanol • Pour the resulting solution into water in a test tube	A white emulsion formed
Carbohydrates			
Reducing sugars	Benedict's test	• Heat test sample with Benedict's reagent	Orange-red precipitate formed
Nonreducing sugars		• Heat test sample with Benedict's reagent to confirm that there is no reducing sugar present • Hydrolyze by heating with dilute hydrochloric acid • Neutralize by adding sodium hydrogencarbonate • Heat test sample with Benedict's reagent	Orange-red precipitate formed
Starch	Iodine test	• Add iodine solution	Turns blue-black
Cellulose		• Add Schultze's solution	Turns purple

hint These tests are all very straightforward but you need to learn them. Questions asking you to describe one or more of these tests appear very often in examination papers. In view of this, it is surprising how many times examination candidates confuse them.

34

biodegradable: something may be described as biodegradable if it can be broken down by the bacteria and fungi normally responsible for the processes of decay. Biodegradable *pesticides*, for example, are converted by soil microorganisms into less harmful substances while biodegradable plastics can be disposed of more easily when they are no longer required. One advantage of using *enzymes* in industrial processes is that, since they are *proteins*, they are biodegradable. *Decomposers* involved in the *nitrogen cycle* are able to break them down and release the nitrogen they contain as ammonia, just as they would with naturally occurring proteins.

biodiversity: an abbreviation of biological diversity, this term refers to the variety of living organisms in a particular area. Biodiversity has three main components:

* Species diversity. This is a measure of both the number of individuals and the number of species in a community. See *diversity* for more detail.

* Genetic diversity. There is genetic variation between members of a species. In animals such as tigers which have been reduced to a number of small isolated populations, the genetic diversity is often very low and can result in the accumulation of harmful recessive *alleles* resulting from the breeding of closely related animals.

* Ecosystem diversity. A wide range of different habitats in an area will allow a much greater species diversity. Conservation often involves managing an area so as to maintain ecosystem diversity.

biological control: the use of a pest's natural predators or parasites to limit its population. In their natural *ecosystems*, populations of organisms are kept more or less constant in size by their many *predators* and *parasites*. When an organism is introduced into a completely new area, these predators and parasites are often left behind and they can multiply unchecked to become serious pests. This has happened with animals such as the rabbit in Australia and plants such as ragwort in New Zealand. By finding and introducing a suitable predator or parasite, numbers can be brought under control again. This is biological control and has been used to reduce the numbers of many pests. The virus disease, myxomatosis was used against rabbits, the cinnabar moth against ragwort and, in the United Kingdom, the parasitic wasp, *Encarsia*, has been used successfully in controlling whitefly in large greenhouses. Biological control has a number of advantages over chemical control. Although not wiping out the pest completely, it reduces its numbers to a level where economic damage is no longer significant and, if the parasite or predator has been chosen carefully:

* it is very specific, affecting only the pest species

* it only needs to be introduced once rather than used repeatedly like pesticides. It is therefore usually much cheaper

* it does not pollute the environment

* pests do not usually become resistant to organisms which are used for biological control.

hint If biological control is in your specification, learn a real example. Don't simply write about keeping a cat to control mice or introducing ladybirds into your garden to feed on the aphids. This will not gain you any credit.

biomass: a measure of the amount of living material present such as the biomass of plants in a rainforest or of worms in the soil. Biomass is expressed in units such as $g\ m^{-2}$ which reflect both the mass and the size of the sample. Because the amount of water in living organisms is very variable, samples are often dried, in which case, the term dry biomass is used. It is important to note, however, that it is not essential to use dry biomass. There are some circumstances where it is far more sensible and convenient to consider fresh or live biomass. Ecological *pyramids* may be expressed in terms of the biomass at each *trophic level*.

biosensor: a device which uses biological molecules to detect and measure the levels of certain chemicals. It usually consists of a membrane to which a layer of *enzymes, antibodies* or receptor molecules is attached. Binding of the biological molecule and the test substance produces a small electrical signal which can be read by a suitable means. Biosensors have a number of advantages over chemical tests:

- because these molecules are very specific, they will only bind to the specific test substance. Chemical tests often respond to a range of similar substances
- they can provide direct measurements of quantities, such as blood glucose concentration, without the need for time-consuming laboratory procedures
- they provide quantitative results and can easily monitor changes over a period of time
- they can be used to measure the concentration of chemically unstable molecules.

biosphere: this encompasses all the living organisms found on Earth together with the nonliving part of their environment. The biosphere includes the land, the oceans and the lower part of the atmosphere.

biotechnology: making use of microorganisms or biochemical reactions to produce useful products or to carry out useful processes. Although biotechnology was used in the making of beer and bread by the ancient Egyptians, there has been an enormous increase in the number and variety of processes in which it is now involved. All of us are familiar with some of the wide range of products now made by these techniques. Medically important substances such as insulin, factor VIII and growth hormone are produced with the aid of *genetic engineering*; enzymes are involved in a variety of applications ranging from the manufacture of "stone-washed" jeans to soft-centered chocolates, and *microorganisms* play an important role in the manufacture of many food products.

biotic: an ecological factor which is concerned with the activities of living organisms. Predation, food availability and competition are all examples of biotic factors which will affect the distribution of various organisms. Rainfall, soil pH and temperature, on the other hand, are concerned with the nonliving part of the environment and are *abiotic* factors. Biotic factors are usually *density-dependent*. That is, they are factors whose effects on a population are relatively greater at higher population densities than at lower ones.

bipedal: able to walk on two legs. With the exception of birds, most land-dwelling vertebrates are quadrupeds – they use all four legs to walk. Humans, however, are fully bipedal. One of the most important ways of determining whether fossil bones are human is to look for evidence of bipedalism. This is reflected in the structure of many parts of the skeleton.

- The foramen magnum is the opening in the skull where the spinal cord leaves the brain. It is where the vertebral column meets the skull. In bipedal animals it lies underneath the skull; in quadrupeds it is farther towards the back.

- The pelvis is short and broad instead of long and narrow.

- The femur or upper leg bone has a distinctive twist which brings the knee joint directly under the pelvic girdle.

There are a number of theories which have been advanced to explain why humans developed bipedal locomotion. These include ideas such as increasing the ability to see over the top of vegetation in the grassland which probably formed the habitat of early humans and enabling large quantities of food to be brought back to provide for the needs of females and young. All this, however, is speculation. We will never know for certain.

bipolar cell: a type of *neurone* or nerve cell found in the *retina* of the *eye*. A bipolar cell has a cell body containing the nucleus. From this cell body there are two long, thin branches. One of these is a dendron which brings impulses from rod or cone cells to the cell body. The other is an axon which transmits impulses away from the cell body to other nerve cells. The *acuity* of the eye (its ability to distinguish between objects close together) depends on each cone cell having a connection with a single bipolar cell. Stimulation of cone cells, even if they are very close together, therefore results in separate impulses in individual bipolar cells and ultimately leads to impulses in different neurones in the optic nerve. The sensitivity of the eye, however, depends on the fact that several rod cells connect to a single bipolar cell. In this way, a number of weak stimuli can lead to the generation of a nerve impulse enabling a person to see a very faint image.

birth rate: the number of children born each year per thousand of the population.

Biuret reaction: a biochemical test which can be used to show the presence of a protein. For details, see *biochemical tests*.

bivalent: the paired homologous chromosomes. During the first stage of *meiosis*, homologous chromosomes come together so that they lie side by side. They shorten in length and get thicker and it becomes apparent that each of the chromosomes forming the bivalent consists of a pair of chromatids. The formation of bivalents only occurs in meiosis; it does not happen in *mitosis*.

blastocyst: a stage in the development of humans and other mammals. The zygote which results from fertilization of the egg cell divides and develops into a hollow ball of cells. The wall in one part of this ball is thicker and forms a mass of cells. These cells will eventually become the *embryo* while the *placenta* will develop from the remaining ones. The blastocyst is really equivalent to the *blastula* which is a characteristic stage in the development of all animals.

blastula: a stage in the development of an animal. When an egg cell has been fertilized by a sperm, a *zygote* is formed. The zygote is a single cell. It divides and divides again to produce a hollow ball of cells called a blastula. The blastula will eventually develop into an *embryo*. The presence of a blastula is a characteristic shared by all members of the kingdom *Animalia*.

blood cells: One of a number of different types of cell found in the blood. In a mammal, there are two main types of blood cell. These are *red blood cells*, or erythrocytes, whose main function is the transport of respiratory gases. White blood cells, or *leucocytes*, are larger than red blood cells and there are fewer of them. They are formed from cells in the bone marrow and they are concerned with *immunity*.

blood clotting: The mechanism by which a blood clot forms when tissue is damaged. A protein called *fibrinogen* is present in blood plasma. When the body is wounded, soluble fibrinogen is converted to insoluble *fibrin*. This forms a mesh over the surface of the wound which traps red blood cells and forms a clot. Clots stop further blood from escaping and also help to prevent the entry of pathogenic bacteria. A complex mechanism controls the process of blood clotting. Fibrinogen can only be converted into fibrin in the presence of the enzyme, *thrombin*. Thrombin is normally present in the blood in an inactive form known as prothrombin. Prothrombin can be converted to thrombin in the presence of a number of substances or factors, some of which are only released at the site of tissue damage. This is summarized in the diagram.

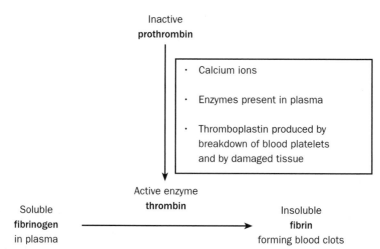

The blood clotting process

In addition to this mechanism, blood is normally prevented from clotting inside the body by a number of anticoagulants which are produced in the body.

blood glucose pool: the total amount of glucose in the blood at any one time. Accurate control of blood glucose is very important. If the concentration falls too low, the central nervous system ceases to function correctly. If it rises too high, then there will be a loss of glucose from the body in the urine. Hormones including *insulin* and *glucagon* play an important part in maintaining a constant level of glucose in the blood. Although the overall level stays within narrow limits, it is important to realize that glucose is always being added to and removed from the blood glucose pool. Digestion and absorption of carbohydrates and conversion from the body's stores of glycogen and fats tend to increase the blood glucose level, while such processes as respiration decrease it. This is summarized in the diagram opposite.

blood groups: one of the types into which a person's blood may be separated. *Red blood cells* have protein molecules in their plasma membranes. Some of these proteins act as *antigens* and it is the presence of these which determines blood group. There are many different systems of grouping blood but the ABO system is probably the best known. There are four blood groups in this system, A, B, AB and O and these are determined by the presence of the relevant antigen. Individuals with blood group A, therefore, have antigen A

The blood glucose pool

on their red blood cell plasma membranes; those who are group B have antigen B; group AB individuals have both antigen A and antigen B while group O have neither of these antigens. It is the presence of these antigens which determine whether or not one person's blood may be safely given to another during *blood transfusion*. The antigens are inherited and their presence is determined by a single gene with three alleles, I^A, I^B and I^O. The alleles I^A and I^B are *codominant* while the allele I^O is *recessive* to both I^A and I^B. The table shows how these blood groups are genetically determined.

Blood group	Antigens on plasma membrane of red blood cells	Possible genotypes
A	A	$I^A I^A$ or $I^A I^O$
B	B	$I^B I^B$ or $I^B I^O$
AB	A and B	$I^A I^B$
O	None	$I^O I^O$

blood system: the transport system found in many animals. Blood systems have a number of characteristics:

- they are *mass flow* systems which provide a rapid means of transport from one part of the body to another
- they contain blood, a fluid whose composition varies from animal to animal. Blood transports a number of different substances. It often contains a respiratory pigment such as *hemoglobin* which makes it very effective in transporting oxygen

- molecules and ions are exchanged between the blood and the surrounding cells by means of *diffusion* and *active transport*

- a pump is necessary to move the blood within the system. This may be a blood vessel with contractile walls or a specialized structure known as a *heart*.

Animals such as insects have an open system in which blood is pumped out by the heart into large blood spaces. It flows slowly through these bathing the surrounding tissues directly. Fish and mammals, on the other hand, have a closed circulatory system in which the blood flows through and stays in a series of *blood vessels*. This has a major advantage in that it allows for much better control of the supply of blood to particular organs or regions of the body. The blood system in fish is a *single circulation*, passing through the heart only once in its passage through the body. A mammal has a *double circulation*. The blood passes through the heart twice in its passage through the body.

blood transfusion: the transfer of blood from one person to another. The ability to do this safely depends on the *blood groups* of the individuals concerned. *Red blood cells* have protein molecules in their plasma membranes. Some of these proteins act as *antigens* and it is the presence of these which determines blood group. For example, the red blood cells belonging to group A individuals have antigen A on their surfaces while those belonging to people with group B blood have antigen B. In addition to this, these individuals also have *antibodies* present in their plasma. Someone with group A will have a specific antibody against antigen B. It is convenient to call this antibody b. Similarly, a person with group B blood will have antibody a.

Because of this, group A blood can obviously be safely given to someone else with blood group A. Both have antigen A but neither has antibody a. However, if blood of group B is given to a person with group A blood, there is a problem. Group B individuals have blood with antigen B. It is given to someone with antibody b. The antigen combines with the antibody causing the red blood cells to stick to each other. This is called *agglutination* and can lead to blockage of blood vessels and possibly the death of the person concerned.

hint Distinguish carefully between the *agglutination* and blood clotting. Agglutination is the sticking together of red blood cells which occurs when antigens on the cell plasma membrane combine with antibodies in the plasma. Clotting is entirely different. This happens at the site of an injury when red blood cells become trapped in a mesh formed from the protein, *fibrin*.

blood vessel: part of the system which transports blood through the body. In a mammal, blood is pumped away from the heart along large *arteries*. These branch into smaller arteries and eventually lead into still smaller *arterioles*. Blood passes from the arterioles into the *capillaries* which are very small in diameter. Capillaries form a network of fine tubes in almost every organ in the body. They are vital to the blood system as it is here that exchange of substances between the blood and the cells of the body takes place. From the capillaries, blood returns to the heart via the venules and, finally, the *veins*.

There are differences in the pressure of the blood and the way in which it flows in the various parts of the blood system. The structures of the different sorts of blood vessel are related to this. The table opposite shows some of these features.

Feature	Artery	Capillary	Vein
Pressure of blood	High	Lower than in the arteries	Very low
Blood flow	Pulsatile. The flow is not even; it surges	Even blood flow	Even blood flow
Structure of walls	Thick wall containing a lot of elastic tissue and muscle	Wall made up of lining cells only. No muscular or elastic tissue	Wall not as thick as that of an artery. Little elastic tissue or muscle
Presence of valves in wall	Not present	Not present	Present

B-lymphocyte: see *B-cell*

body mass index (BMI): a method of relating a person's body mass to his or her height. If body mass is measured in kilograms and height in meters, it is calculated from the formula:

$$BMI = \frac{mass}{height^2}$$

Worked example: A man is 1.8 m tall and weighs 80 kg. Calculate his body mass index.

$$Body\ mass\ index = \frac{mass}{height^2} = \frac{80}{1.8^2} = \frac{80}{3.24} = 24.7$$

The importance of this index is that it allows us to compare the body mass of people of different height. A person who has had or is recovering from a serious illness may be underweight and will have a low body mass index while someone who is obese will have a high body mass index. The table shows some typical values.

BMI	Interpretation
under 20.0	underweight
20.0–24.9	normal
25.0–29.9	overweight
over 30.0	obese

Bohr effect: this is what happens to the *oxygen-hemoglobin dissociation curve* when there is an increase in the amount of carbon dioxide present. The greater the amount of carbon dioxide, the further the curve is pushed to the right. This means that, in the presence of carbon dioxide, hemoglobin will give up a greater amount of the oxygen that it is carrying. This is particularly important during exercise. As the level of activity increases, so does the rate of respiration of the muscles. They will, therefore, produce greater amounts of carbon dioxide and this will result in the release of greater amounts of oxygen from the blood.

bone: a supporting tissue in the skeletons of mammals and other members of the phylum *Chordata*. Bone is made up of cells embedded in a matrix of *connective tissue*. This matrix contains organic material which is largely the protein, *collagen*. This is impregnated with calcium phosphate and other inorganic salts. The arrangement of the materials within the bone matrix gives it great strength and enables it to withstand the considerable forces which act on it. Bone cells are found throughout this matrix in small spaces called lacunae.

Tiny canals filled with cytoplasm connect these lacunae to each other, while blood vessels supply the living bone tissue with the oxygen and nutrients it needs and remove waste products. The activity of the various types of bone cell enables bone to be reabsorbed or added to. This property enables an individual bone to be repaired, to grow or to change to meet different forces which act on it at various stages in an animal's development.

Bowman's capsule: see *renal capsule*

brain: the organ which is responsible for coordinating the activities of the nervous system. The brain of a mammal is extremely complex. It is divided into three main regions, each of which is made up of a number of parts with their own specific functions. The main functions of some of the major parts of the brain are given below.

⊛ The forebrain

 – The *cerebrum*, or cerebral hemispheres, controls the body's voluntary behavior. It is responsible for analyzing and interpreting much of the information from sense organs such as the eyes, as well as for controlling many motor activities, learning and memory.
 – The *hypothalamus*. This is an area on the floor of the forebrain which has a number of important functions. These include the control of many aspects of *homeostasis*, integration of the nervous and hormonal systems and the coordination of the *autonomic nervous system*.

⊛ The midbrain

⊛ The hindbrain

 – The *cerebellum* controls posture and balance as well as coordinating the overall smooth movement of the body.
 – The *medulla oblongata* contains centers which coordinate and control breathing, heart rate and blood pressure.

bronchiole: a small airway in the lung which connects the larger bronchi with the alveoli or airsacs. The structure of bronchioles differs from that of *bronchi* as there is no cartilage in the walls.

bronchitis: a respiratory disease involving inflammation of the *bronchi*, the main airways to the lungs. The onset of acute bronchitis, which is caused by bacterial or viral infection, is quite sudden. Chronic bronchitis, however, is a long-term condition which is associated with atmospheric pollution and cigarette smoking. The early stages of chronic bronchitis involve the production of considerable amounts of *mucus* from the enlarged mucus-producing glands which line the bronchi. This is accompanied by frequent bouts of coughing. As the disease progresses, there is a progressive narrowing of these airways. The resulting breathlessness may condemn patients to increasing disability. The condition is often associated with *emphysema*.

bronchus: one of the larger tubes carrying air in and out of the lungs. Air passes from the mouth and nose into the lungs by way of the *trachea* or windpipe. The trachea divides into two main bronchi which then divide into other small bronchi. The walls of the bronchi, like those of the trachea, are lined with ciliated epithelial cells. Particles in the air which is breathed in are trapped in *mucus* secreted by *goblet cells* in the airways of the lung. The *cilia* beat continuously and the mucus and the particles which have been trapped are carried upwards into the back of the throat. There are rings of cartilage in the bronchial walls which prevent the tubes from collapsing as air is breathed in.

Bryophyta: the plant phylum which contains mosses and liverworts. Members of the Bryophyta share the following features:

* like all plants, members of this phylum show an *alternation of generations*. In the Bryophyta, however, the dominant generation is the *gametophyte*

* they do not have true roots or leaves

* they do not have vascular tissue. There is no *xylem* or *phloem* present so bryophytes rely on diffusion to move substances from one place to another.

buffer: A solution or substance which can absorb hydrogen ions. As a result of this, the *pH* of a buffer solution will change very little if small amounts of either an acid or an alkali are added to it. Buffer solutions are used in investigations where it is necessary to maintain a solution at a particular pH value. They also occur naturally. *Hemoglobin*, *plasma proteins* and phosphates are all important buffers which help to keep the pH of mammalian blood constant.

hint If you want to investigate the various factors which affect the rate of an enzyme-controlled reaction, make sure that you use an appropriate buffer either to alter the pH or, if you are investigating another factor, to keep the pH constant.

bundle of His: specialized fibers of heart muscle which run from the *atrioventricular node* to the base of the heart. They play an important part in coordinating the beating of the heart. A wave of electrical activity passes rapidly down them to the base of the ventricles. It then spreads upwards to all parts of both ventricles along finer branches known as *Purkyne tissue*. This electrical activity stimulates the heart muscle to contract. Because of the arrangement described, the ventricle muscle starts contracting from the bottom. This squeezes blood upwards into the arteries leaving the heart.

C_3 **pathway:** part of the biochemical pathway of photosynthesis. In the *light-independent reaction*, carbon dioxide is taken into the plant and a molecule is formed which contains three carbon atoms. This is part of a cycle of chemical reactions which is known as the *Calvin cycle* or, because the first molecule produced contains three carbon atoms, the C_3 cycle or pathway. The reaction in which carbon dioxide combines with ribulose bisphosphate to produce glycerate 3-phosphate is not very efficient. Some plants such as maize and sugar cane use a different method of trapping carbon dioxide. This is known as the C_4 *pathway*.

C_4 **pathway:** a photosynthetic pathway in which the carbon dioxide taken into the plant is first used to form a molecule which contains four carbon atoms. In the C_3 pathway, the reaction in which carbon dioxide combines with ribulose bisphosphate to produce the three-carbon molecule glycerate 3-phosphate is very slow particularly if the concentration of carbon dioxide falls much below its normal level. In hot, dry conditions the *stomata* of plants often close during the daylight hours and the carbon dioxide level in the leaf will fall as it is used for photosynthesis. The C_4 pathway involves a carbon dioxide-concentrating system which helps to avoid this problem. Carbon dioxide is taken into the mesophyll cells of the plant and converted to a molecule which contains four carbon atoms. This reaction can take place in very low concentrations of carbon dioxide. It can then be released into the relatively few cells in which the C_3 pathway takes place producing high concentrations of carbon dioxide in these cells. The disadvantage of this pathway is that it requires rather more energy in the form of ATP. It is therefore only effective in conditions of high light intensity. Plants which have the C_4 pathway include a number of tropical species such as sugar cane and maize.

calcium: an important nutrient required by both animals and plants. Calcium has a number of functions:

* it is an important component of the *cell walls* of plant cells (calcium pectate). Calcium salts are also important in skeletal structures such as the bones of mammals and the shells of molluscs

* calcium is essential for a variety of processes in a mammal, including *blood clotting*, muscle contraction and transmission across *synapses* in the nervous system

* a *second messenger* is a molecule found inside a cell which responds to the presence of hormones outside the cell. It works by activating a particular enzyme inside the cell. Calcium is an important second messenger.

Calvin cycle: a biochemical cycle which forms part of the *light-independent reaction* of *photosynthesis*. The main stages in process are summarized in the diagram.

44

- Carbon dioxide enters the cycle, combining with the 5-carbon sugar, ribulose bisphosphate (RuBP) to form two three-carbon molecules of glycerate 3-phosphate (GP).

- GP is converted to triose phosphate or glyceraldehyde 3-phosphate (GALP). This reaction requires *ATP* and reduced NADP, two substances which are formed in the *light-dependent reaction*.

- Some of the triose phosphate is used to produce glucose and other molecules. This is not part of the Calvin cycle.

- The rest of the triose phosphate goes through a series of reactions in which RuBP is again produced. ATP is also required for this process.

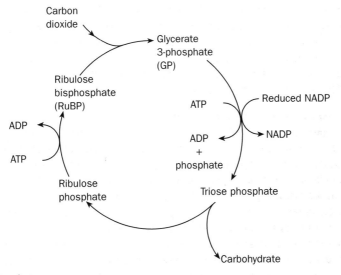

The Calvin cycle

hint The Calvin cycle is part of photosynthesis; the *Krebs cycle* is part of *respiration*.

calyx: the outer part of a *flower* made up of a ring of sepals. It surrounds the flower bud, protecting the structures that are developing inside it.

cancer: see *tumor*

capillarity: the ability of water to move up a fine tube. A xylem vessel has a diameter in the range of 20 to 400 μm. *Xylem* can therefore be considered as forming a series of fine tubes which extend up the plant from the roots to the leaves, and water will, therefore, rise up through the xylem by capillary action. Although capillarity does help to explain how water gets to the leaves of a plant, the maximum height which could be achieved by this process is only about a meter and this is certainly not enough to get the water to the top of even the smallest tree. Therefore there must be other mechanisms that are responsible for the movement of water. The most likely explanation of how this is done is the *cohesion–tension theory*.

capillary: a very small blood vessel through the walls of which substances are exchanged with the cells of the body. As there are very large numbers of capillaries, it means that most cells in the body are no more than about fifty micrometers from one of these blood vessels. The diameter of an individual capillary is somewhere between about five and twenty micrometers. This is about the same as an individual red blood cell, so these cells are often squeezed out of shape as they pass along the capillaries. Capillary walls consist of a single layer of epithelial cells. There is no muscle nor any of the elastic tissue which is found in the walls of larger *arteries* and *veins*.

The blood pressure at the arterial end of a capillary is high and forces water and small soluble molecules out through the walls forming the *tissue fluid* which surrounds the cells of the body. Much of this water flows back into the capillary at its venous end as the *water potential* of the tissue fluid is higher than that of the blood plasma at this point. There is, therefore, a continuous circulation of fluid out of the capillaries and back into them. This takes useful substances to the cells and returns waste products to the blood.

Blood flows into capillaries from the *arterioles*. Arteriole walls contain muscle fibers which may contract and reduce the diameter of the vessel. In this way, the blood supply to the capillary system in a particular organ is always being adjusted to meet its needs.

hint Capillaries in the skin are involved in controlling temperature. In hot conditions, more blood flows through these capillaries so more heat is lost by physical processes such as radiation. Capillaries do not respond to a change in body temperature by moving up or down in the skin!

capsid: the protein coat which surrounds a *virus* particle. The main function of the capsid is to protect the nucleic acid molecule which is contained inside the virus. As well as this, some viruses have specific receptor molecules in their protein coats. These enable recognition of, and attachment to, host cells.

capsomere: one of the protein units that makes up the *capsid* of a *virus*. When seen with an *electron microscope*, many viruses have a regular geometric shape. They are polyhedrons. Each face of the polyhedron is made up a number of protein units or capsomeres each looking rather like one of a tightly packed group of billiard balls.

capsule: a layer which may be found outside the *cell wall* in a bacterial cell. Capsules are not found in all *bacteria*, or even in all bacteria of the same species. Where they occur, they are usually made of polysaccharides although some species of bacteria have protein capsules. The presence of a capsule is one of the features which helps to decide whether a particular strain of bacterium is harmful or not. Bacteria which are enclosed in capsules are often able to resist the action of *phagocytes* in an animal's blood and are, therefore, not as easily destroyed.

capture-recapture method: a method of estimating population size by marking a number of animals, releasing them and then observing the number of marked animals in a second sample. The method is particularly suitable for estimating the population of animals which move around a considerable amount. It makes a number of assumptions. These are:

* the number of organisms in the population does not change significantly between the time when they were captured and marked and the time when the second sample was taken. This applies to births, deaths and migration

* the marked animals mix randomly in the population

* marking does not affect the animal in any way such as, for example, by making it more conspicuous to predators.

The following example shows how population size can be estimated using this method.

Worked example: Pitfall traps were set and 60 beetles were trapped. These were marked and released. In a second sample of 66 beetles trapped four days later, 20 were marked. Estimate the population of beetles.

The calculation relies on the simple relationship:

Proportion of marked beetles in population $=$ Proportion of marked beetles in the sample

or

$$\frac{\text{Number of marked beetles in population}}{\text{Total number of beetles in population}} = \frac{\text{Number of marked beetles in sample}}{\text{Total number of beetles in sample}}$$

$$\frac{60}{\text{Total population}} = \frac{20}{66}$$

$$\text{Total population} = \frac{60 \times 66}{20}$$

$$= 198$$

hint If you have to calculate the size of a population using the capture-recapture method, don't try to learn a formula – it is far too easy to get it wrong. Just remember that the proportion of marked animals in the sample will be the same as the proportion of marked animals in the population as a whole.

carbohydrate: a compound which contains the elements carbon, hydrogen and oxygen. Carbohydrates get their name because the hydrogen and oxygen are in a two to one ratio, the same as in water. They are either sugar molecules or are built up from sugar molecules joined by means of *condensation* reactions. Carbohydrates whose molecules contain a single sugar unit are *monosaccharides*; those with two sugar units are *disaccharides* and those with many units are *polysaccharides*. The table below shows some features of a number of biologically important carbohydrates.

Example	Type of carbohydrate	Role in living organisms
Glucose	Monosaccharide	Important as a source of energy in *respiration*
Ribose		A five-carbon sugar found in *nucleic acids*
Sucrose	Disaccharide	The form in which sugars are transported in the *phloem* of plants
Lactose		The main carbohydrate in milk
Maltose		A product of starch digestion
Glycogen	Polysaccharide	The storage carbohydrate in animals
Cellulose		Important structural molecule in plant *cell walls*
Starch		The storage carbohydrate in plants

carbon cycle: the way in which the element carbon circulates or cycles in an *ecosystem*. The basic principles are exactly the same as for other *nutrient cycles* and the diagram overleaf is based on this.

47

The particular points to note in the biological part of the cycle are:

* the inorganic form of carbon taken up by the plants is carbon dioxide. This is absorbed by the leaves for use in photosynthesis. It is not taken up through the roots like other elements

* carbon dioxide is released directly as a result of respiration by all the other organisms involved in this cycle as well as by the decomposers.

Much carbon has been locked up in coal, oil and limestone formed many millions of years ago and in forests which have existed for hundreds, if not thousands, of years. Processes such as the burning of fossil fuels and the clearing of forests have led to the release of this carbon dioxide and an increase in its level in the atmosphere. This is regarded as one of the main contributors to the *greenhouse effect* and possible *global warming*.

 hint The only form in which carbon is taken up by plants is carbon dioxide. Plants do not absorb carbon-containing ions through their roots.

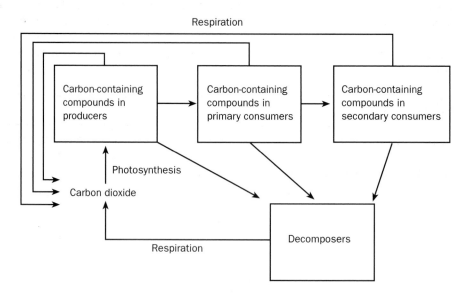

The carbon cycle

carbon sink: a stage in the *carbon cycle* involving a net removal of carbon dioxide from the atmosphere. One of the most important carbon sinks is provided by the huge number of tiny algae which form the plankton floating in the surface waters of the oceans. As these algae photosynthesize, they convert carbon dioxide dissolved in the water into various organic compounds. The seas in many parts of the world are heavily polluted and there are real concerns that pollution could have an adverse effect on planktonic algae. Contrary to what many people think, tropical rainforests are not carbon sinks. The total amount of carbon dioxide they remove from the atmosphere in photosynthesis is more or less equal to the total amount added by the combined respiratory activity of the producers, the consumers and the decomposers.

carbonic anhydrase: an enzyme which catalyzes the reaction between carbon dioxide and water to produce carbonic acid. The carbonic acid then dissociates to form hydrogen ions and hydrogencarbonate ions.

$$CO_2 + H_2O \rightarrow H_2CO_3 \rightarrow H^+ + HCO_3^-$$

In the blood plasma, there is no carbonic anhydrase and this reaction takes place very slowly. However, there are considerable amounts present in the *red blood cells* so the reaction takes place about ten thousand times quicker here. As a result there is a higher concentration of hydrogencarbonate ions in the red blood cells than in the blood plasma. The hydrogencarbonate ions diffuse into the plasma and are transported to the lungs. The reverse series of reactions occur in the lungs and the carbonic anhydrase catalyzes the breakdown of carbonic acid to form carbon dioxide and water.

carcinogen: a substance or other agent which will cause *cancer*. Examples of carcinogens include ultraviolet light and chemicals such as those found in the tar in cigarette smoke. Carcinogens cause damage to the *DNA*. Cells in which the DNA has been damaged, particularly in people who have inherited a tendency to develop cancer, may go on to become cancerous.

cardiac accelerator nerve: a *nerve* which increases the rate at which the heart beats. The *cardiovascular center* in the medulla of the brain can control the rate at which the heart beats. The cardiac accelerator nerve is a branch of the *sympathetic nervous system*. It takes impulses from the cardiovascular center to the *sinoatrial node*, or pacemaker, stimulating it to beat faster. This leads to an increase in heart rate.

 hint Like all nerves, the cardiac accelerator nerve conveys impulses – not messages!

cardiac acceleratory center: part of the *cardiovascular center* in the brain. It sends impulses down the sympathetic nerve (cardiac accelerator nerve) to the sinoatrial node. This makes it beat faster and speeds up the heart rate.

cardiac cycle: the sequence of events which make up a heartbeat. It is a continuous cycle but it may be conveniently divided into three main parts.

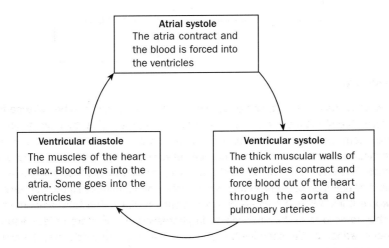

Atrial systole
The atria contract and the blood is forced into the ventricles

Ventricular diastole
The muscles of the heart relax. Blood flows into the atria. Some goes into the ventricles

Ventricular systole
The thick muscular walls of the ventricles contract and force blood out of the heart through the aorta and pulmonary arteries

The cardiac cycle

49

cardiac inhibitory center

The heartbeat is controlled by a small area of muscle in the wall of the right atrium which is called the *sinoatrial node* or pacemaker. From here, a wave of electrical activity spreads over the surface of the heart bringing about the events described in the diagram. Two nerves from the brain, the *cardiac accelerator nerve* and the *vagus nerve* are able to send impulses to the sinoatrial node and alter the rate of the heartbeat.

cardiac inhibitory center: part of the *cardiovascular center* in the brain. It sends impulses down the parasympathetic nerve (vagus nerve) to the sinoatrial node. This slows down the heart rate.

cardiac muscle: the type of muscle which is found in the heart. Its microscopic structure is similar to *skeletal muscle* and it also has the same characteristic striations. The individual muscle fibers, however, branch frequently to form a network which spreads through the walls of the atria and ventricles. The heart is able to beat continuously throughout the life of the animal and, to do this, it requires a lot of energy. Normally, a very high proportion of this is supplied by *aerobic respiration*. Anaerobic respiration does not release enough *ATP*. The muscle of the heart, therefore, has an abundant blood supply and large numbers of *mitochondria*.

If the nerves supplying a skeletal muscle are cut, the muscle will no longer contract. Cardiac muscle shows an important difference. It will continue to beat rhythmically even after all the nerves supplying it have been cut. Because it can beat without being stimulated either by nerves or hormones, cardiac muscle is said to be myogenic.

cardiac output: the total amount of blood pumped out by the heart per minute. There are two things which contribute to cardiac output. The first of these is the stroke volume. This is the volume of blood pumped out each time the heart beats. The second is the heart rate, the number of beats per minute. The relationship between cardiac output, stroke volume and heart rate is given by the equation:

cardiac output $=$ stroke volume $\times$ heart rate

An increase in physical activity requires an increase in cardiac output. More oxygen has to be supplied and more waste products removed. This increase in cardiac output is achieved by increasing both the stroke volume and the heart rate. In some animals, fish for example, it is accomplished mainly by increasing the stroke volume. The heart rate increases very little. In mammals and birds, however, the increase in cardiac output is mainly due to a faster heart rate.

> **hint** If it is on your examination, learn (carefully!) the simple equation:
> cardiac output $=$ stroke volume $\times$ heart rate
> Don't confuse cardiac output with pulmonary *ventilation*. Cardiac output is a measure of the amount of blood pumped out by the heart. Pulmonary ventilation concerns the volume of air taken into the lungs.

cardiovascular center: part of the hindbrain which is responsible for regulating heart rate. There are two centers concerned. These are the cardiac acceleratory center which is responsible for increasing the heart rate and the cardiac inhibitory center which slows it. Both centers are found in the part of the hindbrain known as the medulla oblongata. The flowchart opposite shows how these centers can influence the rate at which the heart beats.

50

These two centers do not work independently. Their activities are integrated and they work together to match the heart rate to the needs of the body.

Nerve impulses convey sensory information to cardiac inhibitory center

↓

Impulses pass down parasympathetic nerve (vagus) to sino-atrial node

↓

Acetylcholine released by parasympathetic nerve endings cause impulses from sinoatrial node to decrease, slowing down heart rate.

Nerve impulses convey sensory information to cardiac acceleratory center

↓

Impulses pass down sympathetic nerve (cardiac accelerator nerve) to sino-atrial node.

↓

Noradrenaline released by the sympathetic nerve endings cause impulses from sinoatrial node to increase, speeding up heart rate.

Cardiovascular center

carotenoid: yellow, red or brown pigments found in plants. Carotenoids occur in flowers such as wallflowers and in fruits such as tomatoes where they contribute to the bright colors responsible for attracting insects for pollination and/or birds for dispersal. Other carotenoids are found in leaves where they play a role in trapping light used in *photosynthesis*. One very widespread and biologically important carotenoid is β-carotene which gives carrots their distinctive orange color. Mammals are able to break down β-carotene during digestion to form vitamin A.

carotid body: a group of chemoreceptors found in the walls of the carotid arteries. They detect changes in the carbon dioxide concentration in the blood but they are also stimulated by changes in blood pH or oxygen concentration. Sensory information from the carotid bodies goes to the centers in the brain which regulate breathing.

carrier molecule: a protein found in the *plasma membrane* which helps in the transport of molecules into or out of a cell. Substances can pass through plasma membranes in a variety of ways. *Facilitated diffusion* is one of these. It is a process that requires protein carrier molecules. Like all proteins, a carrier protein has a specific shape. In particular, it has a receptor site into which the substance which is being transported fits. This brings about a change in the shape of the protein carrier which results in the transported molecule being passed through the membrane. A similar process occurs with *active transport*, the main difference being that, in this case, the change in shape of the carrier molecule which is responsible for transporting the substance across the membrane requires energy in the form of ATP.

hint Use the term "receptor site" for the part of the carrier molecule to which the transported substance binds. Use "active site" only when referring to *enzymes*.

51

carrying capacity: the maximum stable *population* that a particular environment can support. The wildebeest is an antelope which lives in the Serengeti National Park in Tanzania. In 1960, an outbreak of disease reduced its numbers to approximately 10 animals per square kilometer. Its population then rose and levelled out at between 50 and 60 animals per square kilometer. This figure represents the carrying capacity.

cartilage: a hard, flexible supporting tissue important in skeletons. In animals such as sharks, the skeleton is made entirely of cartilage. There is no *bone* present at all. In humans and other mammals, there is much more cartilage present in the young animal but this is gradually replaced by bone as it grows. Cartilage has various functions. It is more compressible than bone so the cartilage found at the ends of bones and between the vertebrae enables the body to withstand the shocks and jarring which accompany movement. Its flexibility is ideally adapted to supporting such structures as the nose, the larynx and the trachea.

Like bone it consists of cells which are embedded in a matrix. This matrix contains various materials including a large number of fibrils of the protein, *collagen*, but it is not impregnated with inorganic salts. The cartilage cells or chondrocytes are found in little groups enclosed in spaces known as lacunae. Unlike bone, substances can reach these cells by diffusing through the matrix. There are no blood vessels found within it.

cascade effect: the way in which small amounts of a *hormone* can cause a target cell to produce very large amounts of a particular product. In simple terms, a hormone molecule binds to a receptor molecule on the *plasma membrane*. This activates messenger molecules such as *cyclic AMP* inside the cell. In turn, each of these messenger molecules activates many enzyme molecules and each enzyme molecule affects large numbers of substrate molecules. In this way, a single molecule of the hormone *adrenaline*, for example, can lead to the production of many thousands of glucose molecules in a liver cell.

Casparian strip: a band of a waterproof material called suberin which runs around the walls of each of the endodermal cells. The *endodermis* in a plant is a ring of cells between the outer part of the root and the *vascular tissue* in the center. The Casparian strip prevents water and dissolved substances from getting into the vascular tissue by going through the cell walls and intercellular spaces. As a result, these substances have to pass through the cytoplasm of the endodermal cells which can therefore control their movement into the xylem.

catabolic reaction: a chemical reaction in living organisms which involves the breakdown of larger molecules and the release of energy. Examples of catabolic reactions include *respiration* and the breakdown of storage molecules such as *starch* and glycogen. Metabolism involves a combination of catabolic reactions, and *anabolic reactions* in which large molecules are synthesized from smaller ones.

cataract: the clouding of the lens of the eye; it results in blurred vision. The commonest type of cataract is that associated with old age but cataracts may also arise as a result of metabolic diseases such as *diabetes* or from injury or prolonged exposure to infrared radiation. Cataracts may be treated surgically and glasses, contact lenses, or an artificial lens inserted in the eye used to compensate for the loss of the original lens.

cell: the basic unit from which living organisms are built up consisting of a mass of cytoplasm surrounded by a *plasma membrane*. The body of an individual plant or animal contains many different types of cell, each type being specialized and adapted for a particular function. The process by which cells become specialized in this way is called *differentiation*. A group of similar cells which carry out a particular function is known as a *tissue*. Different tissues make up *organs* while several organs combine to form a *system*.

The development of electron microscopes has allowed cell structure to be investigated in detail. The cytoplasm that makes up most of the cell has been shown to contain a large number of different *organelles*, each of which has a specific function.

There are two basic types of cell structure found in living organisms. Prokaryotic cells are found among bacteria and other members of the kingdom *Prokaryotae*. They are characterized by the absence of a nucleus and have very few organelles. *Eukaryotic* cells are found in the organisms belonging to the other four kingdoms. They have nuclei as well as large numbers of different organelles. Some, like mitochondria, are surrounded by membranes while others, like ribosomes, are not.

> **hint** Do make sure that you can identify a cell. Hemoglobin and enzymes are molecules. Mitochondria, chloroplasts and nuclei are organelles. Molecules and organelles are not cells!

cell cycle: the continuous cycle of cell growth and cell division. The cell cycle consists of three main stages which are summarized below:

* interphase – cell grows and increases in size; new proteins are synthesized and new cell organelles are made; DNA replicates

* mitosis – the genetic material divides

* cytokinesis – the cytoplasm of the cell and its organelles divide more or less equally between the two daughter cells.

The length of an individual cell cycle, even in the same organism, is variable. Many things combine to determine its precise length. It will depend, for example, on temperature and nutrient supply.

The cells of many higher organisms are generally only able to go through a limited number of cell cycles. Once they have become specialized, they are unable to divide any more. Cancer cells however can carry on dividing indefinitely. Knowledge of the mechanism which controls the cell cycle may lead to the discovery of ways in which tumor growth may be limited.

> **hint** DNA replication takes place during interphase, not during mitosis.

cell fractionation: the process in which cells are broken up and the different types of *organelle* separated from each other. The flowchart on the next page summarizes the main steps in the procedure.

Using techniques like this, it is possible to get a suspension containing a single type of organelle. Various techniques can then be used to investigate the structure and functions of the organelle concerned.

hint When organelles are separated from each other in the processes of cell fractionation, tissue is first chopped up and homogenized in a salt solution. Organelles such as mitochondria are delicate and are surrounded by partially permeable membranes. The salt solution is of such a concentration that the organelles will not be damaged by the osmotic movement of water. The purpose of the salt solution is to prevent the organelles, not the cells, from being damaged. The object of chopping and homogenising the tissue is to smash the cells up anyway!

> Tissue homogenized. It is broken up and suspended in a buffer solution to keep the pH constant. This solution has the same water potential as the original tissue and is kept cold

↓

> The mixture is filtered to remove any material which has not been properly broken up

↓

> The filtrate is put in a centrifuge and spun at a low speed. The larger organelles such as nuclei and chloroplasts fall to the bottom to form a pellet. These can be removed and suspended in a fresh solution if they are required

↓

> The fluid supernatant can be put back in the centrifuge and spun at a higher speed. Smaller organelles such as mitochondria now separate out in the pellet

Cell fractionation

cell membrane: a membrane found either on the outside of a cell or within it. Each cell in a eukaryotic organism is surrounded by a *plasma membrane* and its cytoplasm also contains membranes and membrane-bounded organelles such as *mitochondria* and *chloroplasts*. Cell membranes are extremely thin. They are only about 7 nm thick and so cannot be seen with a light microscope. A transmission *electron microscope*, however, shows a cell membrane as consisting of three lines forming a sandwich. The two outer lines are dark in color while there is a lighter one in between. Cell membranes are made mainly of *lipids* and *proteins* but the actual amount of these substances differs from cell to cell. As it is impossible, even with an electron microscope, to see how the actual molecules are arranged in a cell membrane, it is necessary to produce a model to explain the membrane's properties. The most accurate model of membrane structure that has been developed is the fluid mosaic model which describes most of the properties of a cell membrane.

Cell membranes play a very important part in the biology of cells. Not only are they involved in regulating the movement of substances into and within cells but they also have important receptor molecules on their surfaces which enable cells to respond to chemicals such as *hormones*. Because of this, the adaptation of cell membranes forms one of the important synoptic themes that links aspects of biology. The spider diagram shows some of these links. You can use the cross-referencing system in this book to add greater detail to the diagram.

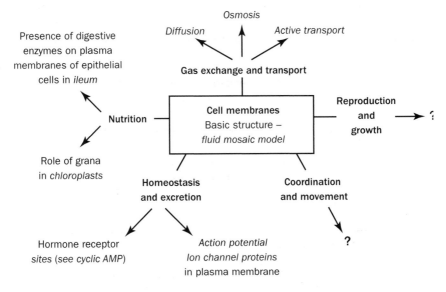

Cell membrane structure

cell sap: the liquid contained inside the *vacuole* of a cell. It contains water in which there are dissolved substances such as sugars and mineral ions. These substances are usually at a much higher concentration than they are in the surrounding cytoplasm. This means that there is a *water potential* gradient and water moves into the vacuole by *osmosis*. This is important in maintaining the cell in a turgid state and providing a plant with support. Cell sap may also contain *pigments* such as the anthocyanins which are responsible for producing the blue or red colors of many flowers.

cell-surface membrane: the *cell membrane* found on the outside of a cell. It is also known as the plasma membrane.

cell wall: a rigid layer which surrounds individual cells of plants, bacteria and fungi. Animal cells do not have cell walls. In plants, the most abundant component is *cellulose*. Cellulose molecules form long parallel chains which are linked into bundles called microfibrils. These cellulose microfibrils are cemented together in a matrix containing a variety of other substances. The resulting structure is extremely strong and resistant to both compression and tensile (pulling) forces. It is, however, freely permeable, allowing water and other substances to pass readily through it. In the cell walls of *sclerenchyma* and *xylem*, this wall is further strengthened by the addition of *lignin* which cements and binds the cellulose microfibrils together.

The cell wall provides support for the cell. In addition, it plays a vital part in allowing turgor pressure to build up in the cell, contributing to the cell's ability to help in the support of the plant as a whole. The cell walls of bacteria and fungi share many of the functions of the cell walls of plants although there are differences in their chemical structure. The main molecule

55

in bacterial cell walls is peptidoglycan, while in those of fungi it is the nitrogen-containing polysaccharide, *chitin*.

> **hint** A cell wall is not a special sort of cell membrane found in plants. When you are asked to write about membranes, don't even think about mentioning cell walls!

cellulose: a *polysaccharide* which is one of the main components of plant *cell walls*. The diagram shows how molecules of a form of *glucose* called β-glucose are joined by condensation reactions to give long, straight chains of cellulose.

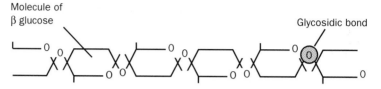

The structure of cellulose

These chains are further linked into bundles by *hydrogen bonds*. It is because they are arranged in these bundles that they can form strong but permeable cell walls in plants. An additional feature of cellulose is that, although it is a very common carbohydrate, it is difficult to digest. Mammals which live to a large extent on plant material usually have partnerships with microorganisms. The mammal provides the microorganisms with large amounts of cellulose from its diet. In turn, the microorganisms produce various enzymes with which they can break the cellulose down into smaller molecules which can be absorbed by the mammal concerned. This is an example of *mutualism*.

> **hint** You may be asked to write about the ways in which cellulose molecules are adapted for their functions. Make sure that you describe the adaptations of cellulose and not the adaptations of cell walls.

central nervous system: this is made up of the *brain* and *spinal cord* and is responsible for coordinating and controlling the nervous system.

centrifuge: an instrument which is able to separate out particles of different size and density by spinning them at high speed. *Cell fractionation* is a process in which cells are broken up and the resulting suspension centrifuged. The different organelles have different densities and settle to the bottom of the centrifuge tube at particular spinning speeds. A centrifuge of the type normally found in school or college laboratories can generate a force about one thousand times that of gravity (1000 g). This is enough to separate larger organelles such as nuclei and chloroplasts. Ultracentrifuges which can generate forces up to 10 000 g can separate much smaller organelles.

centriole: structures found in animal cells which are associated with the separation of chromosomes during mitosis. They are small, hollow cylinders each containing a ring of microtubules. They separate from each other during the early stages of *mitosis*, one going to one pole of the cell and one to the other. They were thought to be responsible for organizing the spindle, a system of protein fibers which are involved in the separation of the chromatids, but this is not now considered to be the case.

centromere: the region on a chromosome which holds the daughter chromatids together during the early stages of cell division. The centromere is also the region to which the spindle fibers attach. Contraction of the spindle fibers pulls the daughter chromatids apart during anaphase.

cerebellum: part of the hindbrain which controls posture and balance as well as coordinating the overall smooth movement of the body.

cerebral hemispheres: see *cerebrum*

cerebrovascular accident: this is a stroke. The term refers to the effects produced by a serious interruption to the blood supply of the brain. It may be caused by a blood clot or by the rupture of an artery wall. The actual effects vary according to the part of the brain involved. Damage to the right side of the *cerebellum*, for example, may produce loss of feeling in or paralysis of the left side of the body.

cerebrum: part of the forebrain, made up of the two cerebral hemispheres, responsible for control of voluntary behavior. In humans, this part of the brain is very large and covers most of the midbrain and hindbrain. Studies of the brain suggest that different parts of the cerebrum have different functions. These may be grouped together and involve:

- sensory areas – those parts of the cerebrum which are associated with the receipt of sensory information. They include areas involving specific senses, such as vision and hearing, as well as those concerned with sensory information from the general body surface. The size of the area is related to the number of receptors involved

- association areas – the parts of the cerebrum which are responsible for interpreting sensory information in the light of experience. They are obviously associated with memory and learning

- motor areas. These are the areas from where the impulses which go to voluntary muscles originate.

CFC (chlorofluorocarbon): a substance used as an aerosol propellant, in refrigeration and in containers used in fast-food packaging. It is biologically important as it is mainly responsible for the thinning of the ozone layer.

chance: this term effectively means the same as luck. Suppose measurements were made of the lengths of two sets of broad bean seeds and the mean values calculated. The figures would probably differ slightly from each other. There could be two possible explanations of this difference. It might be that it resulted from the beans having been obtained from different genetic varieties or from plants grown in different environments. On the other hand, the difference might simply be due to chance, a matter of luck that more small beans were found in one sample than in the other. In analyzing the results of investigations like this, we can use statistical tests to determine the *probability* of any differences being due to chance. On the basis of this, we can decide whether the results are biologically significant or not.

chemoautotroph: an organism which makes use of the energy which it obtains from chemical reactions to produce its organic compounds. All autotrophs make the organic substances which they require from simple inorganic ones such as carbon dioxide and water. In order to do this they require a source of energy. *Photoautotrophs* obtain their energy from light but chemoautotrophs obtain it from chemical reactions. Some nitrifying bacteria can convert ammonia into nitrites, while others produce nitrates from these nitrites. Both of these reactions are oxidation reactions which release the energy that enables the organism to build up its carbohydrates and other organic compounds from carbon dioxide and water. Other chemoautotrophic organisms are found in the ocean depths. They are associated with volcanic vents and again use the energy released from chemical reactions to synthesize organic molecules. Since no light can penetrate this far, these chemoautotrophs are the *producers* on which a variety of marine worms and other organisms depend.

chiasma (plural **chiasmata**)**:** the place where chromatids are seen to cross over each other in a cell which is undergoing meiosis. During the first stage of *meiosis*, the *chromosomes* come together in their homologous pairs, each chromosome consisting of a pair of *chromatids*. There are, therefore, four strands, which represent the two pairs of sister chromatids. Chiasmata occur at the points where these strands, or chromatids, break and rejoin in the process known as *crossing over*. The number of chiasmata that occur on a pair of chromosomes is variable. Generally speaking, the longer the chromosome, the greater the number of chiasmata that are likely to form.

chitin: a nitrogen-containing *polysaccharide*. Chitin is found in the *exoskeletons* of insects and other *arthropods* and in the *cell walls* of *fungi*. The basic units of the molecule are linked together by *condensation* reactions to make up long chains. *Hydrogen bonds* link the chains together and help to make chitin rigid and strong.

chlorophyll: the green pigment found in organisms that photosynthesize which is responsible for the capture of light energy. There are a number of slightly different sorts of chlorophyll molecule which differ from each other in the details of their chemical structure but they all share certain features:

* they have a head region with a magnesium atom at its center. This is the part of the molecule which absorbs light energy

* slight differences in the chemical groups attached to the head region produce different chlorophyll molecules which are able to absorb light with slightly different wavelengths

* they have a tail region which is soluble in lipids. Because of this, it is the part that is anchored in the membranes of the chloroplast.

chloroplast: a chlorophyll-containing *organelle* found in the cells of plants and algae. It is where photosynthesis takes place. The chloroplasts found in plant cells are small, flattened discs, each about five micrometers in diameter. They are found in the cytoplasm but can change their position in response to differences in light intensity. Chloroplasts are surrounded by an outer envelope consisting of two cell membranes. Inside there are further membranes. These run through the chloroplast and are stacked in places to form structures rather like piles of coins. These membrane stacks are the grana. The membranes inside the chloroplast contain molecules of *chlorophyll*, and they are where the *light-dependent reaction* of photosynthesis takes place. Light energy is trapped by the chlorophyll molecules and used to produce chemical energy in the form of *ATP* and also to reduce the coenzyme NADP. Surrounding these membranes is the rest of the chloroplast. This is called the stroma. It contains the enzymes associated with the *light-independent reaction* of photosynthesis. In this stage, sugars are produced from carbon dioxide with the ATP and reduced NADP formed in the light-dependent reaction.

cholecystokinin: a *hormone* produced by the wall of the upper part of the small intestine. It stimulates the contraction of the gall bladder, releasing *bile*. It used to be thought that another hormone, called *pancreozymin*, increased the secretion of digestive enzymes by the *pancreas*. It is now known that only one hormone is involved. Because of this, this hormone is now usually called cholecystokinin-pancreozymin or CCK-PZ.

cholesterol: a lipid which plays an important part in living organisms. Some of the cholesterol required by the body is taken in in the diet and some is formed in the liver. Cholesterol is an important component of *cell membranes* and is a precursor of *bile salts* and steroid hormones such as *testosterone* and *progesterone*. High levels of cholesterol in the blood are associated with *atheroma*. Its level is therefore often monitored in older people and drugs may be given to reduce blood cholesterol.

cholinergic: used to describe a nerve cell or *synapse* in which the *neurotransmitter* is *acetylcholine*. In mammals and other vertebrate animals, cholinergic synapses are very common. They are found throughout the central nervous system. Neuromuscular junctions between motor neurones and the muscles they supply are also examples of cholinergic synapses. In the autonomic nervous system, some of the synapses in the *sympathetic* part and all of those in the *parasympathetic* part are cholinergic.

Chordata: the animal phylum to which humans and other mammals belong. Fish, amphibia, reptiles and birds are also chordates. Some of the features shared by members of this phylum are:

* at some stage in their lifecycles they have gill slits. These are very obvious in fish and can also be seen clearly in the tadpoles of frogs and toads. In the other groups they are only visible while the animal is an embryo

* there is a post-anal tail. An earthworm is not a chordate; its anus is right at the end of the body and there is no tail behind it. A snake may look worm-like but a careful examination will show that the anus is about two-thirds of the way back along the body. The rest of the animal is tail. Snakes are members of the phylum, Chordata

* the nerve cord forms a hollow tube situated just under the dorsal surface. In animals which are not chordates, the nerve cord is solid and on the ventral side

* below the nerve cord is a cartilaginous rod-like structure called a notochord. This is the feature which gives the phylum its name of Chordata.

chorionic villus sampling: a method of obtaining cells from an *embryo* for *genetic screening*. The technique can be carried out as early as eight weeks into the pregnancy. *Ultrasound* is used to determine the position of the developing embryo and, by means of a fine tube inserted through the vagina and cervix, some cells are obtained. These are taken from the chorionic villi, structures which form part of the developing *placenta*.

chromatid: one of the two strands of genetic material that make up a *chromosome*. When chromosomes become apparent during the early stages of *mitosis* or *meiosis*, each can be seen to consist of a pair of chromatids. These are held together by a *centromere*. The two chromatids are then pulled apart to the opposite poles of the cell during the later stages of cell division. In mitosis, the daughter chromatids are genetically identical. In meiosis, they often differ genetically because of *crossing over*.

hint Distinguish carefully between *chromatids* and *chromosomes*, particularly when you are describing what happens during cell division.

59

chromatin: the *DNA* present in the nucleus of a nondividing cell. It is only when a cell is actually dividing that the DNA contained in its nucleus can be seen to be packaged into *chromosomes*. When the cell is in interphase, the DNA is much more spread out. In this condition it is known as chromatin.

chromatography: a technique used to separate individual substances from a mixture. In its simplest form, a solution is made. This is loaded onto a strip of absorbent chromatography paper by applying it at a point just above the bottom. The paper is then suspended so that its end dips into a suitable solvent. The solvent moves up the paper by capillary action taking the substances which were loaded on as it goes. These substances will then separate out at various levels. The places where they end up depends on their solubility in the moving solvent and the ease with which they are absorbed by the paper. It is then possible to identify the individual components, either by comparing their positions on the chromatogram with standard solutions, or by calculating their R_f *values*.

hint In answering questions about chromatography, do not confuse the terms "solute", "solvent" and "solution."

chromosome: one of the thread-like structures found in the *nucleus* on which the genetic material of the cell is organized. It may be thought of as a package of genes. A chromosome consists of a tightly coiled length of *DNA*, closely associated with a small amount of *RNA* and a number of proteins. These protein molecules carry out a variety of different functions. One group, the histones, are involved in the packaging of the DNA.

In nondividing cells, the chromosomes are very long and thin and cannot be seen as separate structures. The material from which they are made, however, can be stained and is known as chromatin. During cell division, the chromosomes, become shorter and thicker and can then be identified as distinct structures. In a human body cell, there are 23 *homologous* pairs of chromosomes, the members of each pair having the same distinctive appearance. This enables them to be identified and the chromosomes of one individual compared with another. Twenty-two of these pairs are found in the body cells of all humans and these are known as *autosomes*. In addition there are two *sex chromosomes* which are usually referred to by the letters X and Y.

In *prokaryotic* organisms like bacteria, there is less DNA present. It forms a loop which is sometimes called a bacterial chromosome. Since this structure is not associated with histone packaging proteins, it is not a true chromosome.

cilia: tiny, hair-like organelles found on the surface of certain cells. They have a distinctive internal structure. Each has a ring of nine pairs of tubules towards the outside and a single pair in the center, the so-called 9 + 2 structure. These enable the cilia to beat. Cilia are found in a number of places:

⊛   single-celled organisms like paramecium have rows of cilia covering the cell surface. The animal moves through the water by beating these cilia in a rhythmic, coordinated way

⊛   filter-feeding worms and molluscs have tentacles or gills covered in ciliated cells. These maintain a constant flow of mucus which traps small food particles and draws them towards the mouth

⊛   ciliated epithelial cells line much of the respiratory pathway in mammals. Cilia continuously waft a mixture of mucus and trapped particles away from the gas-exchange surface.

ciliary muscles: the muscles in the eye which are responsible for altering the shape of the lens.

The way in which these muscles work is shown in the diagram.

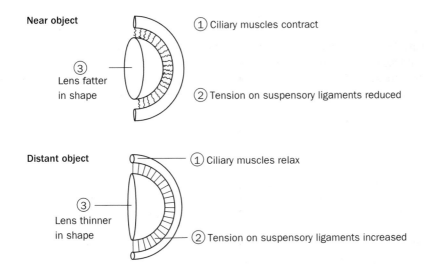

Focusing on near and distant objects

While you are reading this book, you are focusing on the print which is very near to the eye, so your ciliary muscles have contracted. This reduces the tension in the suspensory ligaments attaching the muscle to the lens. As a result, the lens becomes fatter in shape. When you look at an object in the distance, the ciliary muscles relax. This increases tension in the suspensory ligaments and the lens is pulled into a thinner shape. Change in shape of the lens allows both close and distant objects to be seen clearly, a process known as *accommodation*.

circadian rhythm: the internal 24-hour rhythm which governs the behavior of many organisms. The changes which take place during the course of a single day have a considerable effect on animals. In hot, dry climates, for example, it may be a considerable advantage to be nocturnal, hunting for food during the cooler hours of the night and resting in the shade during the day. Organisms have a natural biological clock that enables them to adapt to conditions like this. Experimental evidence suggests that this is an internal mechanism but that it has to be maintained by making constant reference to slight changes in environmental conditions. Lizards, for example, when kept under constant environmental conditions, still show a 24-hour pattern of activity. However, unless there are slight differences in the environmental conditions, this pattern starts to drift. Circadian rhythms are found widely in living organisms. Examples include leaf movements and growth patterns in plants, the vertical migration of small planktonic organisms in the sea and patterns of hormone secretion in animals.

class: a level of *classification* below that of *kingdom* and phylum.

classical conditioning: a form of learning in which an unrelated, neutral stimulus produces the response which was originally associated with an appropriate stimulus. The early work on classical conditioning was carried out by the Russian physiologist, Pavlov. Dogs were fed and the amount of saliva that they produced was collected and measured. The stimulus provided by the presence of food in the mouth produced the response of an increase in the production of saliva. Pavlov found that if a neutral stimulus such as a flashing light was presented at the same time as the original stimulus, eventually, the animal would respond to the neutral stimulus alone.

classification: the way in which living organisms are divided into groups. No one knows for certain how many different species of living organisms there are but it is probably around fifteen million. For a biologist to find patterns of similarities and differences among these organisms, they need to be sorted out and classified. The system that is used has two particular characteristics:

* it is a hierarchical system. This means that we can keep breaking everything down into smaller and smaller groups. All living organisms belong to one of five *kingdoms*; each kingdom is divided into a number of phyla; each phylum is broken down into classes and so on down to genera which are divided into individual species

* it is phylogenetic. There are all sorts of ways of classifying a group of living organisms – whether they are edible or not, their color and size – but none of these would help us to understand the evolutionary relationships between the different members. The system used is based on important biological differences so it shows these relationships clearly.

The table below shows how one organism, a brown rat, "*Rattus norvegicus*", is classified.

Classification of the brown rat, *Rattus norvegicus*

Kingdom	Animalia	Rats are clearly members of the animal kingdom
Phylum	Chordata	They possess the characteristics of the phylum, Chordata
Class	Mammalia	They have hair, sweat glands and suckle their young
Order	Rodentia	Gnawing mammals
Family	Muridae	The group containing rats and mice
Genus	*Rattus*	The scientific name of a species has two words. The first
Species	*norvegicus*	word gives the genus and the second, the species to which the organism belongs.

> **hint** The correct order in which the various taxonomic groups are placed is:
> kingdom, phylum, class, order, family, genus, species
> Devise a suitable mnemonic to help you to learn this sequence. You might try:
> King Philip Claimed Our Family Gold and Silver
> but you ought to be able to do better than this!

climax community: the different species of organisms that make up the final stage in an ecological succession. Over most of Britain, for example, the climax community is some form of deciduous woodland.

clomiphene: a drug which stimulates ovulation and is used in the treatment of infertility in women. In a fertile woman, at the beginning of each menstrual cycle the hormone *FSH* stimulates a follicle to develop into a mature *ovarian follicle*. The follicle cells produce another hormone, *estrogen*, which has a number of different effects on the reproductive system. One of these is to inhibit the release of more FSH. It does this by *negative feedback*. Some women who are infertile produce small amounts of FSH, not enough to stimulate a follicle to develop. If these women are given clomiphene early in a menstrual cycle, the drug blocks the negative feedback action of estrogen so that larger amounts of FSH are secreted. This may be enough to stimulate follicle development leading to ovulation and the prospect of pregnancy.

clone: a group of genetically identical organisms all produced from the same parent individual by *asexual reproduction*. The ability to produce clones has important applications in plant breeding. Varieties that have arisen by *mutation* or as the result of deliberate crosses may be multiplied to give large numbers of identical individuals for commercial purposes. All named varieties of roses and apples, for example, are clones. The technique is also important in genetic engineering.

close season: a period of the year in which a particular species of animal may not be caught or harvested. The stocks of almost all the world's commercially important fish have been seriously depleted by overfishing. In order to maintain existing stocks and allow them to recover, a variety of conservation measures need to be put in place. One of these is to make use of close seasons which do not allow fishing during the main breeding season of the species concerned. This measure is extremely difficult to enforce but is used with freshwater species.

Cnidaria: the animal phylum containing hydra, jellyfish, sea anemones and corals. Members of the Cnidaria share the following features:

* they are *diploblastic* and *radially symmetrical*
* they catch their prey with special stinging cells called nematocysts
* their digestive cavity or enteron has only one opening
* their nerve cells are not grouped together to form distinct nerves. Activities are coordinated by a simple mesh-like arrangement of nerve cells called a nerve net.

coccus: a bacterium which is more or less spherical in shape. Cocci range from approximately 0.5 µm to 1.2 µm in diameter. They include a number of species of *Streptococcus* which are important in cheese making as well as the bacteria responsible for boils and some of the infections which may follow surgical operations. Cocci divide in a number of different ways and these give rise to characteristically shaped clumps of cells. Division once, for example, gives rise to a pair of cells, diplococci. Irregular division gives the clumps which are characteristic of *Staphylococcus*. Some different cocci are shown in the diagram below.

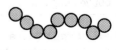

Diplococci   Staphylococci   Streptococci

Different types of cocci

codominant: two *alleles* are said to be codominant if both are expressed in the *phenotype* of the heterozygote. A different system is used to represent codominant alleles from that used to represent a pair of alleles in which one is *dominant* and one *recessive*. In cattle, one of the genes responsible for coat color has two codominant alleles. The gene may be represented by the letter C. The two alleles are C^R which is responsible for producing red hairs and C^W which is responsible for producing white hairs. The heterozygote will have the genotype $C^R C^W$. Since these alleles are codominant, both will be expressed and the animal will have a mixture of red and white hairs giving an overall color called roan.

codon: a sequence of three bases on an *mRNA* molecule which codes for an amino acid (see genetic code). There are also three codons which do not code for an amino acid. Their presence indicates the end of a particular protein. They act as stop signals rather like the presence of a full stop at the end of a sentence.

coelom: a cavity which develops in the mesoderm of *triploblastic* animals. The coelom is thought to have been an advantage in the development of movement in animals. It separates the muscles of the body wall from the various internal organs and so allows locomotion to be independent of movements of internal organs. The coelom can also transport molecules and cells from some parts of an animal to others. In soft-bodied animals, such as worms belonging to the phylum *Annelida*, the coelom is full of liquid and acts as a sort of fluid-filled or hydrostatic skeleton.

coelomate: an animal which possesses a *coelom*.

coenzyme: a molecule, other than the substrate, which is needed for an *enzyme* to work. It acts rather like a shuttle, transporting atoms or molecules from one enzyme controlled reaction to another. Some important coenzymes are involved in the processes of respiration and photosynthesis. In many of the stages of *aerobic respiration* hydrogen is released. This is picked up by the coenzyme, NAD, which is reduced in the process:

NAD + hydrogen $\rightarrow$ reduced NAD

The reduced NAD is then fed into the *electron transport system* or respiratory chain. Many of the coenzymes important in respiration are made from *vitamins*. NAD, for example, is produced from nicotinic acid, one of the B group vitamins.

cohesion–tension theory: a theory used to explain the movement of water from the roots up to the leaves of a plant. The main steps in the argument are:

* water is lost from the leaves during *transpiration*. It moves from the saturated air inside the leaf to the drier air outside by the process of evaporation

* water molecules stick to each other. This property is called cohesion. As the water is lost by evaporation from the leaf, more is pulled up because of this capacity of water molecules to cling to each other

* since transpiration is pulling the water column in the *xylem* upwards and gravity is tending to pull it down, the water column is stretched. It is under tension

* water molecules are also attracted to and cling to the walls of the xylem. This is known as adhesion and helps in pulling the water upwards.

See hint box at top of next page.

64

hint An elastic band makes a very useful model with which to explain evidence which supports the cohesion–tension theory. Pull on it and the band is under tension. So how would you explain the following observation?

The trunk of an actively transpiring tree changes in diameter over a 24-hour period. During the daytime its diameter is less than at night.

The answer … When the tree is actively transpiring, water is being pulled up through the xylem. The water column will therefore be under tension, just as when you pull on an elastic band. Pulling on an elastic band puts the elastic under tension and it gets thinner. So does the tree trunk.

collagen: a fibrous *protein* found in animals. Almost a third of all the protein found in the human body is collagen. Each collagen molecule consists of three *polypeptide* chains which are wound tightly around each other to give a helix. This gives the molecule a structure which is very resistant to stretching, a vital property, for example, for its functions in tendons and as a component of the matrix of *bone* and *cartilage*. If collagen is boiled it can be converted to gelatin.

collecting duct: the last part of a *nephron* or kidney tubule. The part of the nephron known as the *loop of Henle* is responsible for setting up a gradient of salt concentration in the medulla of the kidney. This means that the *water potential* of the fluid in the collecting duct is always higher than that in the medulla and water can be removed by *osmosis* all the way down the collecting duct. This is important in bringing about the production of concentrated urine.

The hormone *ADH* influences the permeability of the cells which line the collecting duct. When the body is dehydrated and needs to conserve water, this hormone is secreted. It makes the cells lining the collecting duct more permeable to water by increasing the number of channels in the plasma membrane of the cell through which water passes. More water is thus reabsorbed and smaller quantities of more concentrated urine are produced.

collenchyma: a type of specialized supporting tissue found in plants. The main characteristics of the cells found in collenchyma are that they are rather elongated and have *cell walls* with extra *cellulose* thickening at the corners. Collenchyma tissue is found either in strands or in a continuous cylinder around the outside of young stems. This enables it to resist the compression and stretching that occur as the plant is blown from side to side by the wind. Like *parenchyma* cells, the cells of collenchyma have living contents. Their walls can still be stretched, so they are particularly important in the support of young, growing stems.

colon: the part of the alimentary canal between the small intestine and the rectum. This is the part in which *feces* are produced and stored before being removed from the body. Its main function is the absorption of water. The colon contains many bacteria. Some of these are *mutualistic*. These bacteria receive a continuous supply of nutrients from the material in the intestine. In turn, they synthesize a number of *vitamins* which may be absorbed by the host and, in some herbivorous mammals such as rabbits and horses, they play an important part in the digestion of cellulose.

colostrum: a secretion produced by the mammary glands immediately after birth. It differs from milk in that it contains more protein and much less fat and sugar. It provides the young mammal with *antibodies*. These are protein molecules which are too large to pass across the *placenta*. They provide passive immunity against some diseases, and are thought to be particularly important in preventing gut infections which might otherwise prove dangerous to a newly born animal.

commensalism: a nutritional or *symbiotic* relationship between two different species of organisms where one gains but the other suffers no direct harm. There is no physiological link between the organisms concerned. House sparrows may be described as having a commensal relationship with humans. The sparrow consumes considerable amounts of human food waste. This causes no direct harm to humans, however. More different organisms have probably established commensal relationships with humans than with any other species.

community: all the living organisms present in an *ecosystem*. These organisms have different feeding habits and occupy different *trophic levels*. Organisms within these trophic levels are related to each other by *food chains* and *food webs*. A community consists of different species of organism and a measure of the number of species present is the *diversity*.

competition: a relationship between different organisms requiring the same resources. In most ecological situations, resources are limited. There is, for example, only so much nitrate in the soil or light energy striking the surface of the ground. Some organisms can gain these resources at the expense of others. These will be successful in terms of competition and will be more likely to survive. Competition can be between members of the same species when it is known as *intraspecific competition* or between members of different species when it is called *interspecific competition*.

hint When you write about competition, use the right language. Organisms compete for resources. They do not fight for them.

competitive inhibition: when the rate of reaction of an enzyme is slowed down or stopped by a molecule which is similar in shape to its normal substrate. An *enzyme* molecule has an *active site* formed by a group of amino acid molecules. The active site has a particular shape into which the *substrate* molecule normally fits to form an enzyme-substrate complex. This complex then breaks down to form the products and release the enzyme molecule. A competitive inhibitor is characterized by having molecules which are very similar in shape to those of the substrate. They will also fit into the enzyme's active site and will prevent it from being occupied by the normal substrate. Because there are fewer available enzyme molecules, the rate of reaction will slow down. Competitive inhibitors, therefore, can be said to compete with the normal substrate for the active site of the enzyme.

condensation: a chemical reaction involving the joining together of two molecules with the removal of a molecule of water. Condensation reactions are very important in the formation of biologically important *polymers*. This is the way in which amino acids join to produce *polypeptides*, glucose molecules form *starch* and *glycogen* and individual nucleotides combine to produce the *polynucleotides* that make up DNA and RNA. Molecules which are formed in this way can usually be broken down by the addition of water molecules, a process known as *hydrolysis*.

hint Sort out the meaning of the terms condensation and hydrolysis. Condensation is the joining of smaller molecules to form larger ones and involves the removal of a molecule of water. Hydrolysis is the breakdown of larger molecules into smaller ones and involves the addition of water.

conditioned reflex: see *classical conditioning*

contractile vacuole: an organelle found in the cytoplasm of some single-celled organisms which is involved in the removal of excess water from the cell. Amoeba is a small organism which lives in freshwater. Its cytoplasm contains a higher concentration of dissolved molecules and mineral ions than is present in the surrounding water. As a result, the cytoplasm has a lower *water potential* and water moves into the organism by *osmosis*. If this water is not expelled, the cell will eventually burst. The contractile vacuole is involved in removing the excess. It gradually fills with water. This is an active process and involves *ATP*. Eventually, it is full and discharges its contents out through the plasma membrane. It is thought that the contractile vacuole may also be involved in the excretion of the organism's metabolic waste products.

cone cell: a color-sensitive receptor cell found in the retina of the eye. Cone cells contain pigments which are sensitive to light of different colors. In the cone cells of the human eye, there are three different pigments, one most sensitive to blue light, one to green and one to red. Blue light only stimulates cones containing the blue-sensitive pigment. This means that the brain will interpret the object concerned as being blue. Orange objects appear orange because they reflect both green and red light. If an orange object is viewed, the light reflected from this will cause the cones containing the green-sensitive pigment as well as those containing the red-sensitive pigment to be stimulated. The brain will interpret this object as being orange. The cone cells are very tightly packed into the fovea. There are about fifty thousand of them to each square millimeter. It is this distribution of cone cells and the fact that each one is connected individually to a sensory neurone in the retina which give the eye its ability to see fine detail. It is an arrangement which means that each cone cell is able to give rise to a nerve impulse in a separate sensory neurone.

Coniferophyta: the plant phylum which contains the conifers. Many conifers are large trees which are able to grow in cold regions such as in the far north or high in mountains. Members of the Coniferophyta share the following features:

* they are trees with thin, needle-like leaves
* the *seeds* which are formed as a result of fertilization are not enclosed in fruits
* the reproductive structures are associated with cones.

connective tissue: a type of *tissue* found in mammals and other vertebrates which originates from specific cells called mesenchyme cells. Connective tissue often contains cells called fibroblasts which produce fibers of proteins such as elastin and *collagen*. These cells lie in a surrounding matrix of non-cellular material. This is made of a variety of proteins and polysaccharides in which are also found the protein fibers produced by the fibroblasts. Connective tissue has a variety of different functions. It can bind other tissues together or it may form *tendons*, *ligaments*, *bone* or blood. Many of these tissues have functions associated with support. In addition, connective tissue plays an important part in the defensive mechanisms in the body.

conservation: in biological terms, this is a way of maintaining the *diversity* of ecosystems and the living organisms which inhabit them. There are two extreme views of conservation. On the one hand, there are those who feel that nature should be left to take its course. They argue that natural environments should be preserved and protected from all human influence. At the other extreme is the idea that wildlife should be actively managed so that it can be exploited on a sustainable basis. An example of these two approaches is the conservation of African elephants. Before poaching became a serious problem in Kenya and some other East African countries, total protection of relatively small areas meant that the elephant population exceeded the *carrying capacity* of the land. As a result, serious damage was done by the animals to their environment and large numbers died. In parts of southern Africa, however, surplus animals were deliberately killed in such a way that a sustainable yield of meat, skins and ivory could be produced. See also *genetic conservation*.

consumer: a *heterotroph*. All *food webs* and *food chains* ultimately depend on *producers*, organisms such as plants that can produce organic molecules from simple inorganic ones. These are eaten by consumers. Primary consumers feed directly on producers. In turn the primary consumers are fed on by secondary consumers and the secondary consumers, by tertiary consumers. In the simple food chain:

nettle plant $\rightarrow$ large nettle aphid $\rightarrow$ two-spot ladybird

the large nettle aphid is a primary consumer and the two-spot ladybird, a secondary consumer. In practice, it is not always easy to describe the *trophic level* to which an organism belongs as precisely as this. Many animals feed at more than one trophic level. The bush cricket, for example, feeds directly on nettle plants so it is a primary consumer. It also feeds on other animals which makes it a secondary or even a tertiary consumer.

continuous variation: *variation* in which there is a complete range of measurements from one extreme to the other. In other words, individuals do not fall into discrete categories. The graph below shows figures for the milk yields of a large number of cows.

Because it is based on measurable features, continuous variation is sometimes referred to as quantitative variation. There is a complete range from the cow with the lowest yield to the one with the highest. Where there is a pattern of continuous variation like the one shown, it often reflects a characteristic which is controlled by a number of different genes and where the environment plays a significant part in determining the final pattern of variation.

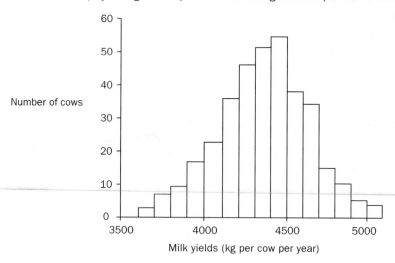

Milk yields in cattle

cornea: the transparent area at the front of the eye. The cornea plays an important part in focusing the light onto the *retina*. Light rays are bent when they pass from one medium to another which has a different density. This is called refraction and, in the eye, occurs mainly when the light rays pass from the air into the cornea. However, the angle through which rays of light are bent by the cornea is always the same. This creates a potential problem because light rays from objects which are close to the eye need to be bent more if they are to be focused. The *lens* is able to change shape so it is this structure which is able to change the amount of refraction, allowing both close and distant objects to be brought into focus on the retina.

The cornea consists of living cells which must be supplied with nutrients such as glucose. This is done by the liquid which lies behind it, the *aqueous humor*. Certain medical conditions result in the cornea becoming cloudy and sight being lost, which can be overcome by a corneal transplant. A piece of cornea can be transplanted from one person to another without being rejected as would happen, for example, with skin. This is because it does not have a blood supply.

corolla: the part of a *flower* which is inside the *calyx* and consists of a number of petals. The corolla is of particular importance in insect-pollinated flowers where it serves two important functions. Its bright colors attract insects and it also serves as a suitable landing platform for them. Many different insects visit flowers and the color and shape of the corolla is often related to the particular species which are responsible for their pollination. Generally speaking, bees are attracted to flowers which are blue or purple, flies to those which are yellow or white and butterflies to red- and pink-colored flowers. Flowers which have an open structure are more likely to attract flies and small beetles while those which are closed or tubular are frequently pollinated by bees. The corolla is very much reduced in wind-pollinated plants.

correlation: two variables are said to be correlated if a change in one of them is reflected by a change in the other. A simple method by which we can find out if a correlation occurs is to plot the two variables on a graph and draw the line of best fit. These graphs are called scatter diagrams. Three types of scatter diagram are shown on the next page.

Care must be taken in interpreting scatter diagrams. Just because two things are correlated, it does not mean that one causes the other. The lower scatter diagram shows the relationship between mean daily intake of fat and the number of cases of breast cancer in various countries.

It shows a clear positive correlation between the two. We must not jump to conclusions, however. This does not mean that eating fat causes breast cancer. There may be any number of other possible explanations. For example, it might be argued that women who eat more fat come from more affluent areas. They may therefore drink more alcohol or smoke more cigarettes.

coronary heart disease: disease affecting the coronary arteries which supply the muscle of the heart. When one of these arteries becomes blocked, the area of heart muscle that it supplies is deprived of oxygen. It therefore dies and gives rise to what is known as a heart attack or a myocardial infarction. There are three main reasons why blockages occur in the coronary arteries:

* *atherosclerosis.* This is due to the buildup of fatty material or *atheroma* in the lining of the artery wall. Eventually this material along with fibrous tissue and calcium salts will form hard plaques which lead to the narrowing of the artery lumen

69

* thrombosis. The presence of a blood clot in one of the coronary arteries. This is often associated with atheroma and it is thought that the blood clot often forms when the surface of one of the plaques breaks away. The clot blocks the lumen

* spasm. The muscle in the wall of the coronary artery contracts and goes into a spasm. The reasons for this are not really understood but it clearly produces a narrowing of the lumen.

corpus luteum: a group of cells in the ovary of a mammal which develops from an *ovarian follicle* after *ovulation* has taken place. It is mainly concerned with the production of the hormone *progesterone*. During the *menstrual cycle*, progesterone is responsible for maintaining the lining of the uterus. If fertilization has not taken place, after about two weeks the corpus luteum starts to break down and the amount of progesterone falls. As a result of this, the lining of the uterus starts to break down and is finally lost from the body during menstruation. If pregnancy does occur, however, the corpus luteum remains active and continues to produce progesterone for about three months. After this, progesterone production is taken over by the *placenta*.

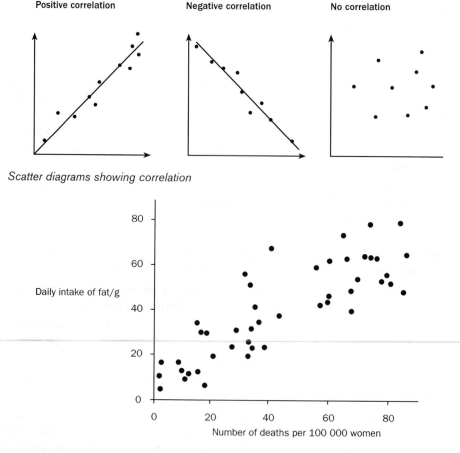

Positive correlation   Negative correlation   No correlation

Scatter diagrams showing correlation

Scatter diagram showing correlation between mean dietary fat intake and breast cancer

70

cotyledon: a leaf from a plant embryo found inside a *seed*. When seeds germinate, the cotyledons may stay below the surface and supply nutrients to the developing plant. In other plants they emerge above ground level, turn green and photosynthesize. However, they are much simpler structures than the leaves which form later and still act mainly as a store of food for the young plant. The number of cotyledons present is an important feature in classifying flowering plants. The seeds of *monocotyledons* have a single cotyledon in each seed while those of the *dicotyledons* have two.

counter-current system: a system in which two fluids flow in opposite directions increasing the efficiency with which substances are exchanged between them. In a fish, water rich in oxygen flows over the surface of the *gills*. Oxygen diffuses from this water into the blood which is flowing through the capillaries. If both water and blood flowed in the same direction, equilibrium would rapidly be reached and no more oxygen would be taken up by the blood. By having the two fluids flowing in opposite directions, however, the water always has a higher oxygen concentration than the blood in the nearby capillary. Thus a concentration gradient is maintained across the entire exchange surface making the uptake of oxygen much more efficient. A similar system is found in the *placenta* where the blood of the mother and the blood of the fetus generally flow in opposite directions. As well as being involved with gas exchange, counter-current systems between blood entering and leaving limbs and organs are involved in maintaining body temperature.

courtship behavior: the behavior which comes before and is associated with mating. A great variety of different patterns of courtship behavior exists. In some species there is little or no preliminary interaction between the individuals involved; in others courtship involves long, complex sequences of behavior. Courtship has a number of different functions which include:

- the attraction of a mate, possibly from a considerable distance. Female silk moths, for example, produce substances called *pheromones*. Male silk moths respond to these and can use them to locate females from a distance of several miles

- courtship behavior enables an animal to identify a member of the correct species. Different species of firefly have different patterns of flashing lights. Females will only respond to the correct pattern

- it enables male and female to synchronize their reproductive behavior. With sticklebacks, the behavior of one individual stimulates a response from the other. This ultimately leads to fertilization of eggs laid by the female in the nest built by the male

- it may keep male and female together while the young are reared. Storks greet each other when food is brought to the nest with a noisy bill-clattering display.

cristae: the folded inner membranes of mitochondria. The final stage in aerobic *respiration*, oxidative phosphorylation, is associated with these *cell membranes* which therefore contain the enzymes and carrier molecules associated with this process.

crop rotation: the process consists of changing the crop grown in a particular field each year. This practice has a number of advantages:

- Many important agricultural pests are specific and only affect certain crops. Their eggs and spores often remain in the soil after the crop has been harvested and infect next year's crop if the same plant is grown. Changing the crop regularly limits the buildup of pests.

71

- Different crops remove different proportions of mineral ions from the soil. By varying the crop, the mineral resources in the soil may be exploited to the full.

- The bacteria present in the root nodules of leguminous plants (such as clover) fix nitrogen. When these plants decay, their nitrogen containing compounds are broken down by soil bacteria. This increases the soil nitrate content for the following crop.

cross-fertilization: fertilization where male and female *gametes* come from different organisms. This term should be compared with *self-fertilization* where both gametes come from the same organism.

cross-sectional study: a method of collecting information about growth by measuring individuals of different ages. Each individual is only measured once. A *longitudinal study* is where the same individual is measured at different ages. Cross-sectional studies have some advantages over longitudinal studies:

- they can be carried out more rapidly as there is no need to wait for individuals to age. This, of course, makes them less expensive

- very large samples can be measured as there is no need to keep track of individuals. Because of this, they provide a lot of information about the range of variation at a particular age

However, there are some disadvantages as well:

- cross-sectional studies cannot show individual differences in growth patterns. Study of variation between individuals can produce evidence about the influence of genetic and environmental factors on growth

- because the data obtained from a cross-sectional study is presented as the average size at a particular age, the growth curve which results can be misleading. If data is collected for humans, for example, it is very difficult to see the growth spurt associated with puberty. This shows clearly in longitudinal studies.

crossing over: in cell division, this is where chromatids break and rejoin so that genes from one chromatid become attached to genes on another chromatid. During the first stage of *meiosis*, the *chromosomes* come together in their homologous pairs, each chromosome consisting of a pair of *chromatids*. There are, therefore, four strands, representing the two pairs of sister chromatids. These chromatids may break and rejoin with other chromatids. This results in the genes on one part of a chromatid becoming attached to the genes which were present on part of another chromatid. If one of the chromatids concerned originally came from the mother while the other came from the father, there would be a new mixture of genetic material. Crossing over is very important as it produces genetic variation.

cultural evolution: the changes which have taken place in human societies over the course of time. The large brains and upright posture that are features of modern humans have evolved gradually. Evidence from fossils can reveal much about these physical changes. At the same time, however, there have been changes in the structure of human communities. With the development of skills, such as tool making and basic agriculture, human communities have progressed from small groups that depended on hunting and scavenging for meat and gathering wild plants, to the complex societies that exist today. There were a number of significant steps in our cultural evolution. Some of the more important ones are summarized in the table on the facing page.

Step	Evidence first found	Evidence
Making of tools	About 2.5 million years ago	• Presence of marks and scars on the surface of the tool where it was deliberately shaped • Marks on accompanying bones suggesting deliberate cutting or hitting • Presence of stones of different material from surrounding rocks
Control of fire	Good evidence only from about 0.5 million years ago	• Ash deposits found associated with signs of human occupation
Domestication of plants	Between 12000 years ago and 9000 years ago	• Evidence poor • Presence of cereals in which the grains do not fall easily from the stalks
Domestication of animals	Between 9000 years ago and 8000 years ago	• Discovery of relatively large numbers of bones from young and female animals • Evidence from animal bones showing a reduction in body size • Much more variation in features such as horn length and shape. These are features by which individual animals may be recognized

cuticle: a noncellular layer found on the surface of leaves and on the outside of many animals. In plants it consists of a waxy layer which is impermeable to water and helps to prevent water loss. In insects, the cuticle is a more complex structure consisting of three layers:

- a very thin outer waxy layer known as the epicuticle; this layer is waterproof and, like the cuticle on a leaf, helps to limit water loss

- the exocuticle which is extremely hard and strong; it is made up of chitin (a nitrogen-containing *polysaccharide*) and *protein*

- a slightly softer inner layer, the endocuticle; this consists mainly of chitin.

The cuticle completely surrounds the insect and is unable to stretch. Because of this, insect growth takes place in a series of steps. An insect molts its cuticle and growth only takes place before the new cuticle has fully hardened.

cyanobacteria (blue-green algae): a group of prokaryotic organisms related to *bacteria*. They are interesting to biologists in a number of ways:

- they are among the oldest organisms known. Stromatolites are a mixture of sediment and cyanobacteria. Some fossil stromatolites have been estimated to be over 2 000 000 000 years old

- some cyanobacteria are important nitrogen-fixing organisms (see *nitrogen fixation*)

- in many nutrient-rich lakes and rivers, the numbers of cyanobacteria increase enormously in what is called an algal bloom. The death of these organisms leads to severe depletion in the amount of dissolved oxygen (see *eutrophication*). Some of the cyanobacteria involved produce toxins which can be poisonous to an animal drinking the water.

cyclic AMP: a messenger molecule found inside cells. It is formed when a *hormone* binds with a receptor site on the *plasma membrane*. This causes an enzyme to be released which converts *ATP* to cyclic AMP. Cyclic AMP then activates the enzymes which control the biochemical pathways associated with the action of the hormone.

cystic fibrosis: an inherited condition in which affected individuals lack a protein which is responsible for the transport of chloride ions. Molecules of this protein are found in the *plasma membranes* and normally allow chloride ions to pass into cells. Absence of them results in the production of thick, sticky mucus. This has a number of effects on the body. It can accumulate in the lungs making the person very susceptible to lung infections and it can also prevent the secretion of pancreatic enzymes. At present the condition is treated by physiotherapy to get rid of the mucus which collects in the lungs and by providing extra pancreatic enzymes so that food can be digested efficiently. In the future, it is hoped that *gene therapy* may be used. Researchers are investigating ways of introducing the gene for making the chloride-ion transporting protein into the cells of affected people.

Cystic fibrosis is a *recessive* condition and the gene responsible for producing the protein concerned has been identified as being on chromosome number 7. Because there is a one in four chance of a second child with cystic fibrosis being born to parents who have already had one affected child, it is important that such parents should receive *genetic counseling*.

cytochrome: a molecule which forms part of the *electron transport chain*. A cytochrome is a protein which is combined with another chemical group containing iron or copper. They are found on the membranes inside *mitochondria* and *chloroplasts*.

cytokinin: a *plant growth substance* which has a number of different functions, one of the most important of which is that it stimulates cell division. Cytokinins are produced in the growing root tips of plants and are transported upwards to the stems and leaves.

Micropropagation of plants requires the isolation of plant cells and their growth in a suitable medium to produce a mass of similar cells called callus. If cytokinin and another plant growth substance, *auxin*, are added to the callus, roots and shoots will form. Roots form if a lot more cytokinin is added than auxin; shoots form if the amount of cytokinin is only a little more than the amount of auxin. Use of these two plant growth substances in the correct ratios is important in producing new plants by this technique.

cytoplasm: the contents of a cell outside its *nucleus*. The cytoplasm consists of a fluid matrix called the cytosol. The cytosol is a solution containing ions and smaller molecules such as simple sugars and amino acids in which larger insoluble molecules are suspended. It also contains a network of protein filaments which provide the cell with its shape and make movement possible. Within this matrix, *organelles* such as *mitochondria* and *ribosomes* are suspended.

cytosine: one of the *nucleotide bases* found in nucleic acid molecules. Cytosine is a pyrimidine which means that it has a single ring of atoms in each of its molecules. When two polynucleotide chains come together, cytosine always bonds with guanine. The atoms of these two bases are arranged in such a way that three hydrogen bonds are able to form between them.

cytosol: the fluid matrix which forms part of the *cytoplasm* of a cell. It consists of a solution of ions and small molecules as well as various larger, insoluble molecules.

dark reaction: the process in which carbon dioxide is reduced to form carbohydrate during photosynthesis. The term "dark reaction" is misleading since the biochemical reactions involved take place in the light rather than in the dark. This is because they need ATP and reduced NADP which are both produced by the action of light energy. The term "light-independent" is therefore preferred by most biologists. See *light-independent reaction*.

deamination: the removal of the amino (NH_2) group from an *amino acid*. If more protein is eaten than is required, the body is unable to store the excess amino acids which result. They are broken down in the *liver* where the amino group is removed from the molecule and converted into ammonia. Unfortunately ammonia is very toxic. Only in animals that live in water can it be diluted sufficiently to be excreted safely. In mammals, it is converted into a much less poisonous compound, *urea*, in a cycle of biochemical reactions called the ornithine cycle. The remainder of the molecule can then either be respired or converted into carbohydrate or fat.

decline phase: the final stage in a *population growth curve* in which there is a fall in the number of living organisms present. The reasons for this are usually associated either with a lack of nutrients or with a buildup of toxic waste products.

decomposer: a *microorganism* that breaks down the organic compounds in dead material and waste products into carbon dioxide, water and simple inorganic ions. As soon as an animal or plant dies, its dead tissue is colonized by bacteria and fungi. Any soluble sugars present are rapidly broken down by the first microorganisms to arrive. Structural substances such as those that make up plant cell walls take much longer to decompose.

deficiency disease: a disease caused by the lack of particular nutrients in the diet. Some examples of deficiency diseases are shown in the table.

Nutrient lacking	Deficiency disease	Explanation
Iron	Anemia	Iron is an essential component of hemoglobin. Insufficient iron in the diet lowers the amount of hemoglobin present and therefore reduces the blood's efficiency at transporting oxygen.
Iodine	Goiter	Iodine is a component of hormones produced by the thyroid gland. Inadequate iodine causes swelling of the neck produced by enlargement of this gland.
Vitamin A	"Night blindness"	Vitamin A is essential for formation of rhodopsin, the pigment found in blindness rod cells in the eye. Lack of rhodopsin results in an inability to see in dim light.

Vitamin C	Scurvy	Essential for the formation of collagen, a protein important in connective tissue. The symptoms of scurvy reflect this and involve bleeding from the gums and the failure of wounds to heal.
Vitamin D	Rickets	Vitamin D controls the absorption of calcium from the gut and its metabolism. Without adequate calcium, bones fail to calcify and skeletal deformities result.

deflected succession: a type of ecological *succession* in which human activity prevents a *climax community* from being established. An area of grassland, for example, would normally give way to scrub and finally, if left long enough, to woodland. The woodland would be the climax community. If, however, grazing animals such as sheep are kept on the area, they will eat the young woody plants which will be unable to establish themselves. As long as the land is grazed by the sheep, succession will not progress beyond the grassland stage. This is a deflected succession.

degenerative disease: a disease brought about by decline in the efficiency of the body systems as a result of aging. Examples of degenerative diseases include *senile dementia* resulting from degeneration of the brain, *cataracts* which affect the lens in the eye and *osteoarthritis* which affects joints. As people live longer, the importance of degenerative disease increases.

dendron: a long, thin process in a *neurone* that carries nerve impulses towards the cell body.

denitrification: the conversion of nitrates to nitrogen gas by certain bacteria. This is part of the *nitrogen cycle*. It is particularly important in soils such as those that are waterlogged or poorly aerated. In these soils the oxygen content is low and denitrifying bacteria thrive. Economically, these bacteria are very important, since they reduce the amount of nitrate that is available to plants by changing it to nitrogen gas. This cannot be used directly.

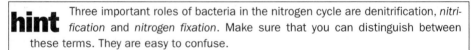 **hint** Three important roles of bacteria in the nitrogen cycle are denitrification, *nitrification* and *nitrogen fixation*. Make sure that you can distinguish between these terms. They are easy to confuse.

density-dependent factor: a factor whose effect on a population is relatively greater at higher population densities than at lower ones. Take, as an example, caterpillars feeding on cabbage plants. If there are large numbers of caterpillars and they have a high population density, they will eat a lot of the cabbage. The resulting lack of food will lead to the death of large numbers of caterpillars. On the other hand, if there is a low population density of caterpillars, food supply will have very little effect on the caterpillar population. Food supply is said to be a density-dependent factor regulating population size. The higher the population density of caterpillars, the greater the effect on the food supply and the greater the number of caterpillars that will die. Most *biotic* factors operate in this way. The incidence of disease, predation and parasitism will all have a relatively greater effect on populations at high population densities.

density-independent factor: a factor whose effect on a population is more or less the same, whatever the population density of the organism concerned. Many *abiotic* factors are density-independent. Heavy rain may flood the soil around a pond. It is likely to have a similar effect on the populations of small animals living there whatever their population densities.

deoxyribonucleic acid: see *DNA*

detritivore: an animal which feeds on the pieces of partly broken down plant or animal tissue known as detritus. Many soil organisms such as earthworms and woodlice are detritivores. Together with *decomposers*, they play an important part in the cycling of soil nutrients. Investigations have shown that breakdown by decomposers is much faster if detritivores are present. Since the detritivores break up the tissue, they expose a much greater surface area to attack and break down by decomposers.

diabetes (diabetes mellitus): a condition in which the level of glucose in the blood cannot be properly controlled. There are two forms of diabetes. In the insulin-dependent form, the pancreas fails to secrete enough *insulin* to control the blood glucose level. After a meal, glucose is absorbed from the small intestine. There is not enough insulin produced to convert the excess glucose to *glycogen* for storage in the liver. It rises to a level that is too high to be reabsorbed from the filtrate in the kidneys and, because of this, it appears in the urine. In addition, the high concentration of glucose in the kidney filtrate means reabsorption of water is not as effective as normal and large quantities of dilute urine are produced. This can result in severe dehydration. As there is very little glycogen stored, the body starts to break down fats and proteins to use as *respiratory substrates* and there is a rapid loss in weight. By carefully regulating the diet and regular injection of insulin, the condition can be controlled effectively. The second form of diabetes is non-insulin dependent. This usually affects older people. They continue to produce insulin, but it is no longer effective. This results in cells failing to take up glucose and a consequent rise in its level in the blood. There is no point in giving such people extra insulin. The condition can only be controlled by careful regulation of dietary carbohydrate.

dialysis: a method of separating small molecules from larger ones. Suppose it is necessary to separate a mixture of glucose and starch. The solution containing both substances is put in a partially permeable bag and suspended in distilled water. The small glucose molecules are able to diffuse through the pores in the membrane of the bag into the surrounding distilled water, while the starch molecules are too large and remain behind. Artificial kidneys rely on this principle. In one of these machines, blood is kept apart from a suitable solution of salts by a partially permeable dialysing membrane. The small soluble molecules of *urea* pass out of the blood and through the membrane, while the larger protein molecules and blood cells are unable to do so.

diastole: the stage in the *cardiac cycle* which involves relaxation of the heart muscle. During this stage, the ventricles fill with blood.

diastolic blood pressure: the blood pressure in the main arteries measured when the *ventricles* are filling (*diastole*). At this stage in the *cardiac cycle*, blood pressure in the arteries will be at its lowest. It is the pressure to which blood falls between heartbeats.

dietary fiber: the food which cannot be digested by enzymes in the course of its passage through the human gut. The most important components of dietary fiber are the non-starch *polysaccharides*. These are substances such as *cellulose* which make up plant cell walls. Not surprisingly then, cereal and vegetable foods are particularly good sources of dietary fiber. It is impossible to say whether a lack of fiber in the diet can cause a particular disease but low-fiber diets have been linked with a variety of conditions including diabetes, heart disease and cancer of the colon. One of the main problems in interpreting the available evidence is that low-fiber diets are often associated with other risk factors such as a high intake of fats.

dietary reference values: a set of figures which relate to the intake of particular nutrients in the human diet. The amount of a given nutrient will differ from person to person. Some people will require larger than average amounts, some less than average. The graph shows the estimated requirements for a particular nutrient in a large sample of people. It is a bell-shaped or normal distribution curve. It allows us to fix several other points. These are:

- the Estimated Average Requirement (EAR). This is the mean value
- the Reference Nutrient Intake (RNI). This is the point which is defined mathematically as two *standard deviations* above the Estimated Average Requirement. Intakes above this will be adequate to meet the needs of most people
- the Lower Reference Nutrient Intake (LRNI) is defined as the point two standard deviations below the Estimated Average Requirement. An amount less than this might be adequate for some individuals in the population but, for many, it will almost certainly not be enough.

Unfortunately, the data on which such values are based are not very reliable, and because of this, care has to be taken in both in interpreting them and in using them in practical situations.

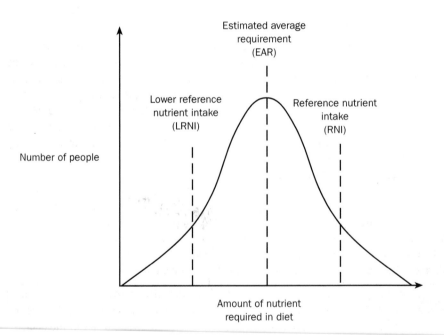

dicotyledon: a member of the group of flowering plants characterized by having two *cotyledons* in each of their seeds. There are two main groups of flowering plants – the Monocotyledones or monocotyledons and the Dicotyledones or dicotyledons. They differ from each other in a number of important ways in addition to the number of cotyledons contained in their seeds. Some of these are summarized in the table on the facing page.

78

Monocotyledones	Dicotyledones
(e.g. grasses and cereals, orchids, lilies)	(e.g. buttercups, dandelion, oak tree)
Seeds contain a single cotyledon or seed leaf.	Seeds contain two cotyledons or seed leaves.
Flower parts such as petals usually in threes or multiples of three.	Flower parts such as petals usually in fours or fives.
Leaves with parallel veins.	Leaves with veins which spread out in a net-like pattern.
Do not possess a cambium and rarely have woody stems.	Many of the larger members possess a cambium and have woody stems.

differentiation: the process in which a cell becomes specialized for a particular function. In a plant root, for example, only the cells immediately behind the tip are able to divide. If a longitudinal section of a root is examined with a microscope, these cells can be identified quite easily, as various stages of *mitosis* can be observed. Having divided, the cells then elongate and finally differentiate into tissues such as *xylem* and *phloem*. Differentiated cells, such as those in the xylem and phloem no longer have the ability to divide.

Undifferentiated cells like those in the root and shoot tips are used when large numbers of plants are produced from a single parent by *micropropagation*.

diffusion: the movement of the *molecules* or *ions* from where they are in a high concentration to where they are in a lower concentration. In liquids and gases, molecules possess kinetic energy and are continually moving about. As this movement is at random, an equilibrium will eventually be reached where the molecules are spread out evenly. This is diffusion and it is very important in the movement of substances through the *plasma membrane*. Fat-soluble molecules dissolve in the *phospholipid* part of the membrane and diffuse very readily. Molecules such as water, oxygen and carbon dioxide diffuse readily because of their very small size. Ions, which are charged particles, and larger molecules such as glucose cannot pass through the phospholipid layer. Instead they go through special channel proteins. Since all of the molecules and ions involved are moving anyway, the cell does not require the use of energy for this process. It can therefore be described as a form of passive transport. The rate at which substances diffuse into and out of cells is governed by a number of different factors including the area of the membrane over which they diffuse, the difference in concentration on either side of the membrane and the membrane thickness. *Fick's law* is a simple way of expressing this relationship.

hint In living organisms, surfaces over which diffusion takes place are always moist. The reason for this is that water has very small molecules. The body fluids of most terrestrial organisms have a higher concentration of water molecules than the surroundings. Water will, therefore, always diffuse out where oxygen can diffuse in.

digestion: the process in which the large insoluble molecules which make up the food of an organism are broken down to smaller, soluble molecules. The place where digestion takes place differs from organism to organism. Mammals, for example, have specialized guts, while amoeba engulfs its food by *phagocytosis* and then digests it once it is inside the cell. Fungi are *saprophytes*. They secrete enzymes onto the surface of their food, digest it and then absorb the soluble molecules which result.

79

Digestion in a mammal involves both physical and chemical processes. The chewing action of the teeth, the churning by the muscles in the wall of the stomach and the action of bile, which breaks fats down into an emulsion of tiny droplets, are purely physical. *Enzymes* complete the process of digestion by chemically breaking down the individual food molecules. All digestive enzymes work in the same way. They are hydrolases which means that they are able to break chemical bonds by the addition of water. This is hydrolysis and it is involved in the breaking down of proteins to amino acids, polysaccharides such as starch to monosaccharides like glucose, and lipids to fatty acids and glycerol.

hint Distinguish between the four processes which take place in the alimentary canal: ⬦ ingestion ⬦ digestion ⬦ *absorption* ⬦ *egestion* Digestion only involves the breakdown of large, insoluble molecules into smaller, soluble ones.

dihybrid cross: a genetic cross between individuals in which genes at two different loci (see *locus*) are considered. It is important in explaining a cross such as this to set out the explanation clearly and logically. The example below shows how to do this.

In peas, height of the plant and pod color are determined by two genes which are on different pairs of chromosomes. Pure-breeding, tall plants with green pods were crossed with pure-breeding dwarf plants with yellow pods. The F_1 plants which resulted from this cross were all tall with green pods. Show by means of a suitable genetic diagram, the expected results of a cross between two of these F_1 plants.

If all the F_1 plants were tall and had green pods, then the allele for tall would be dominant to that for dwarf, and the allele for green pods would be dominant to that for yellow pods.

These alleles may be represented by the following letters:

T	tall	**G**	green pods
t	dwarf	**g**	yellow pods

Parental phenotypes: Tall, green pods   Tall, green pods

Parental genotypes: TtGg   TtGg

Gametes: (TG)(Tg)(tG)(tg)   (TG)(Tg)(tG)(tg)

Offspring genotypes:

		Male gametes			
		TG	Tg	tG	tg
	TG	TTGG	TTGg	TtGG	TtGg
Female	Tg	TTGg	TTgg	TtGg	Ttgg
gametes	tG	TtGG	TtGg	ttGG	ttGg
	tg	TtGg	Ttgg	ttGg	ttgg

Offspring phenotypes:

Tall, green pods:	9
Tall, yellow pods:	3
Dwarf, green pods:	3
Dwarf, yellow pods:	1

80

hint Here are some important points to remember in setting out a dihybrid cross:

- Make sure that you use the word "allele" when you mean allele and "gene" when you mean gene.
- In an examination, it is very easy to confuse capital and small letters. Try to avoid using letters such as C and c or S and s to represent alleles.
- Set out your method in a clear, logical way. The approach used with this example has been recommended and is used by all examination boards. Use it as a model for your answers.
- Draw a ring round your gametes and remember that, since gametes are haploid, you can only ever have one of a pair of alleles in each ring.
- Set out your cross as a checkerboard or Punnett square. You are not as likely to make mistakes this way. Avoid lines linking gametes to the offspring genotypes. This gets very confusing.
- Cross off each of the offspring genotypes as you count it. It will save the embarrassment that comes from having a total of 15 or 17!

dilution plating: a method of counting the number of live bacteria in a culture. Suppose a sample of the culture was poured on an agar plate and incubated for a suitable time. Colonies of bacteria would appear on the surface of the agar. Each of the colonies would have come from a single bacterium. By counting the colonies, the number of live bacteria in the original sample could be determined.

In practice, there would be so many bacteria present in the culture, that it would not be possible to do this. It would be necessary to dilute the sample first. A known volume, say 1 cm^3, is taken from the culture and diluted by adding 9 cm^3 of distilled water. This is a 10^{-1} dilution. After mixing thoroughly, 1 cm^3 of the resulting suspension is removed and further diluted with distilled water. This process is continued, resulting in a series of dilutions, each one-tenth of the concentration of the previous one. A sample is taken from each dilution and poured on a separate agar plate. The plate on which the number of colonies is small enough to count but large enough to give accurate results is taken. The colonies are counted and the number obtained multiplied by the dilution factor to give the number of live bacteria in the original sample.

dioecious plants: plants in which male and female flowers are found on separate plants. Holly has separate male and female trees. For *pollination* to occur pollen must be transferred from a male tree to one with female flowers. The biological advantage of this arrangement is that self-pollination is impossible; cross-pollination always takes place.

diploblastic: an animal body pattern in which there are two layers of cells. The outer layer is called the ectoderm and the inner layer the endoderm. Together, the two layers surround a gut cavity or enteron. Animals that belong to the phylum *Cnidaria*, which includes jellyfish, sea anemones and corals, have this basic body structure. All other animal phyla which you are likely to study are *triploblastic* and have a body pattern which is based on three layers of cells (see the diagram on page 82).

diploid: a term which can refer to cells, organisms or stages in the life cycle in which the nuclei contain two copies of each *chromosome*. The diploid number of chromosomes is variable and differs from one species to another. In humans, this number is 46 or 23

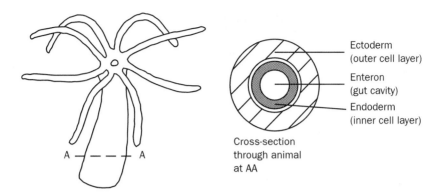

Ectoderm
(outer cell layer)

Enteron
(gut cavity)

Endoderm
(inner cell layer)

Cross-section
through animal
at AA

A diploblastic animal

pairs, in pea plants, 14, and in the fruitfly, *Drosophila melanogaster*, it is 8. Of the 23 pairs of chromosomes that are found in each body cell in humans, 22 pairs are found in both males and females and are called *autosomes*. The remaining pair, the X and Y chromosomes are the *sex chromosomes*. When gametes are formed, the process of *meiosis* results in male and female gametes each having half the number of chromosomes that are present in a body cell. In other words, they are *haploid*.

hint Different organisms have different diploid numbers of chromosomes. If you get an examination question asking you about chromosome number, look at the evidence very carefully before you give your answer. Don't make the mistake of thinking that all organisms are like humans in having 23 pairs of chromosomes in each body cell.

directional selection: *selection* which operates against one extreme in a range of variation. The graph opposite shows variation in hair length in house mice.

House mice are very common animals which live in a variety of places. Some populations of house mice are able to live in cold stores. Obviously, in such places there will be selection against mice with short hair. Mice with long hair will be at an advantage and will be selected for. As hair length is an inherited characteristic, over a period of time directional selection should result in an evolutionary change, with the mean hair length of house mice living in cold stores becoming longer.

disaccharide: a carbohydrate whose molecules contain two sugar units. *Maltose* is an example of a disaccharide. It is made from two *glucose* molecules joined together by a reaction in which a molecule of water is removed. This is known as a *condensation* reaction and produces a bond between the two glucose molecules called a glycosidic bond. The reaction will also go in the reverse direction. If a molecule of water is added to a maltose molecule, two molecules of glucose are produced. This is *hydrolysis* and is what happens during the final stages in the digestion of maltose.

disease: a disorder which affects an organism. It has a specific cause and can be recognized by particular signs and symptoms.

discontinuous variation: *variation* in which individuals fall into distinct categories. This is in contrast to *continuous variation* in which there is a complete range of measurements from one extreme to the other. The graph on page 84 shows figures relating to the number of petals in a sample of celandine flowers.

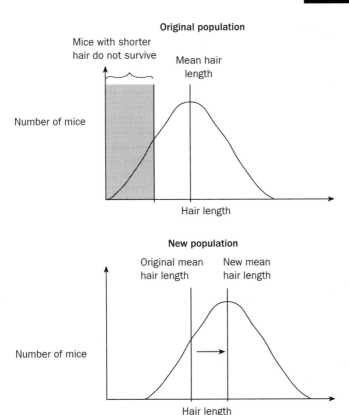

Original population

Mice with shorter hair do not survive

Mean hair length

Number of mice

Hair length

New population

Original mean hair length

New mean hair length

Number of mice

Hair length

Directional selection

Discontinuous variation is based on features which are not measured. Because they are either present or not, it is sometimes referred to as qualitative variation. In this example, flowers may have any number of petals from five to eleven, but they can have either one number or another. There are no intermediate categories. Where there is a pattern of discontinuous variation like the one shown, it often reflects a characteristic which is controlled by a small number of genes or even a single gene and where the environment plays an insignificant part in determining the final pattern of variation.

disinfectant: a substance which is used to kill *microorganisms* that are outside the body. Because they kill bacteria and other organisms, disinfectants are described as being *bactericidal* or biocidal. Disinfection kills most of the harmful bacteria but it does not necessarily result in the killing of all microorganisms. It is enough to disinfect a thermometer before taking a person's temperature. Surgical instruments, however, need to be sterilized before an operation. This involves the killing of all microorganisms.

disruptive selection: selection which favors the extremes in a range of variation at the expense of the mean. It is easier to find good examples of *stabilizing selection* and *directional selection* than of disruptive selection. The graph on page 84 shows the general principle involved.

There is a range of variation. Organisms whose *phenotypes* correspond to the mean in this range are at a disadvantage. They will be selected against, while selection will favor both of the extreme varieties. Over time this will result in the evolution of two distinct forms.

83

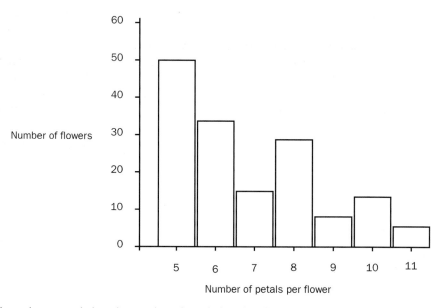

Discontinuous variation: the number of petals in celandine flowers

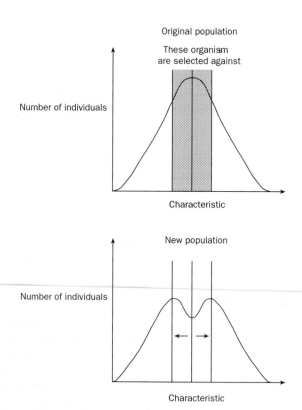

Disruptive selection

diversity: a way of describing the relationship between the number of individuals and the number of species in a community. A sand dune with large numbers of marram grass plants and very little else would have a low diversity. On the other hand, tropical rainforests with their great variety of different species would have a much higher diversity. In general terms, the harsher the environment, the fewer the species present and the lower the diversity. As with many ideas in ecology, it is useful to be able to express diversity in numerical terms. There are many different ways of doing this but one relatively simple one is shown in the example below.

Worked example: The table shows the number of birds of different species encountered on a walk across a park.

Species	Number of birds of this species encountered
Magpie	11
Black-headed gull	4
Carrion crow	4
Blackbird	1
Starling	37
House sparrow	7
All species	64

An index of diversity may be calculated from the formula:

$$d = \frac{N(N-1)}{\Sigma n(n-1)}$$

where N = total number of organisms of all species and n = total number of organisms of a particular species

$$d = \frac{(64 \times 63)}{(11 \times 10) + (4 \times 3) + (4 \times 3) + (1 \times 0) + (37 \times 36) + (7 \times 6)}$$

$$d = \frac{4032}{110 + 12 + 12 + 0 + 1332 + 42}$$

$$d = \frac{4032}{1508} = 2.7$$

On its own this figure does not mean very much, but it does allow the diversity to be compared with other areas. In this particular case, the diversity is fairly low, but it is higher than the value for the city center.

hint The key to calculating the index of diversity correctly is to remember that Σ means "the sum of."

division of labor: the adaptation of different parts of an organism to carry out different functions. Even within a simple one-celled organism like amoeba, different *organelles* have different roles. Regulation of the water content of the cell is carried out by a *contractile vacuole*; the metabolic pathways associated with respiration take place in the *mitochondria*, and *lysosomes* release the enzymes necessary for the digestion of food particles which

85

have been taken into the cell. Larger organisms have many cells. These cells are grouped together to form *tissues*, while the tissues form *organs*, and the organs form *systems*. Each system is adapted to carry out a specific function within the organism.

One of the evolutionary trends which is apparent in the animal kingdom is that the more advanced the animal, the greater the division of labor. Hydra is a small animal belonging to the phylum *Cnidaria*. It has a sac-like body consisting of just two layers of cells which enclose a cavity, the enteron. The enteron does several things. It functions as a simple gut as food is taken in and the process of digestion is started here. Oxygen diffuses from the water in the enteron into the surrounding cells, so the enteron is also associated with gas exchange. Finally, it acts as a simple hydrostatic skeleton, providing support for the organism. Thus the enteron is a single part of the hydra, with a number of different functions. In a mouse, on the other hand, these functions are all carried out by different systems. There are separate systems for digestion, gas exchange and support.

DNA: the molecule which forms the genetic material of all living organisms. In animals and plants, it is an important component of *chromosomes*, but is also found in mitochondria and chloroplasts. Chemically, DNA consists of two *polynucleotide* chains, one running in one direction, the other running in the opposite direction. Each chain has a sugar-phosphate backbone. One of four bases, adenine, cytosine, guanine or thymine is attached to each sugar in this backbone. These bases are joined by hydrogen bonds which are formed between the adenine in one chain and the thymine in the other and between the cytosine in one chain and the guanine in the other. The whole molecule is twisted to form a double helix with one complete turn of the spiral for every ten base pairs.

The process of *DNA replication* means that the molecule can be copied accurately and an exact copy passed on to other cells.

Genes are sections of a DNA molecule which code for particular proteins or polypeptides. During *protein synthesis*, another nucleic acid, *mRNA*, is produced which takes the gene code from the DNA in the nucleus of the cell to the cytoplasm where amino acids are assembled in the right order to produce the required protein.

> **hint** DNA is not a protein. A DNA molecule is built up from nucleotides, each of which consists of a molecule of deoxyribose, a phosphate group and an organic base. Proteins are built up from *amino acids*.

DNA probe: a single strand of *DNA* which is used to identify a particular gene. A DNA probe has a base sequence which is complementary to part of one of the DNA chains in the *gene* to be identified. The probe is usually made using *nucleotides* which contain radioactive phosphorus, ^{32}P. The DNA being tested is treated to separate its two chains. This DNA is then mixed with the probe which will bind to the complementary bases on one of the chains. The site at which it binds may then be identified by its radioactivity.

DNA replication: the process by which a DNA molecule can produce two exact copies of itself. DNA replication is an important part of the *cell cycle*. It takes place in the period of interphase between the times when a cell is dividing by *mitosis*. The flowchart summarizes the basic steps involved in DNA replication.

DNA normally replicates very accurately and the cell has a number of repair mechanisms which correct errors in replication. Nevertheless, uncorrected errors do occasionally happen and these are the causes of *gene mutations*.

The old DNA molecule unwinds, the hydrogen bonds break and its two chains separate. This process involves enzymes and other proteins

↓

Each of the old chains acts as a template for the formation of a new chain. The bases of free nucleotides in the cell line up against the complementary bases on each of the old DNA chains

↓

The nucleotides are now joined together by enzymes forming two new DNA molecules each formed from one of the existing chains which provided a template for the formation of a new one. This is known as semiconservative replication

There are a number of important differences between DNA replication and the process of *transcription* which takes place during *protein synthesis*. There are summarized in the table below.

Feature	DNA replication	Transcription
Molecule or molecules produced	Two molecules of DNA	Single molecule of mRNA
Strands of DNA involved	Both strands of DNA, each producing a new DNA molecule	One strand of DNA, the coding or sense strand, producing an mRNA molecule
Amount of DNA involved	Whole DNA molecule unwinds and separates	Only part of the DNA molecule unwinds and separates

hint DNA replication is completely different from transcription. Replication is the process by which a DNA molecule produces two exact copies of itself. Transcription is the first stage in protein synthesis and involves the formation of an mRNA molecule from one of the DNA strands.

dominant: an *allele* is said to be dominant if it is always expressed in the *phenotype* of an organism. In peas, the allele for green pods, G, is dominant to that for yellow pods, g. Since the allele for green pods is dominant, pea plants with either the genotype Gg or GG will have green pods.

hint You cannot tell whether an allele is dominant or recessive by simply counting the particular phenotypes. If you are given a family tree and 7 out of 12 individuals, for example, have green eyes, it doesn't prove that the allele for green eyes is dominant; it may simply be commonest. You must look at individual crosses for specific evidence.

dorsal root: part of a spinal nerve which joins with the *spinal cord*. The spinal nerves contain both sensory and motor *neurones*. Just before each nerve joins with the spinal cord, it splits in two producing a dorsal root and a *ventral root*. The dorsal root contains the sensory neurones which carry nerve impulses from the receptors to the spinal cord. There is a small swelling called a ganglion present on the dorsal root. This contains the cell bodies of the sensory neurones.

double circulation: a blood system in which the blood passes through the heart twice in its passage round the body. In mammals, for example, blood flows from the right side of the heart to the lungs. After it has become oxygenated, it returns to the left side of the heart from where it is pumped to the rest of the body. A complete circuit therefore involves passing through the right side and through the left side of the heart. The advantages of a circulation like this are that:

* oxygenated and deoxygenated blood are kept completely separate. They are not mixed. Because of this, the blood supplying the tissues is always saturated with oxygen

* blood is supplied to the tissues at a much higher pressure than is the case in a single circulatory system. This makes for much more efficient circulation.

Down's syndrome: a condition shown in humans which results from three copies of chromosome 21. During the first division of *meiosis*, the chromosomes belonging to pair number 21 fail to separate. This results in gametes containing an extra chromosome. When fertilization takes place, an individual will be produced with 47 chromosomes instead of the usual 46. There will be three copies of chromosome 21, one more than normal. Abnormalities arising from the failure of chromosomes to separate properly during meiosis become increasingly common with age and there is a clear correlation between the age of the mother and the probability of having a child with Down's syndrome.

Down's syndrome is an example of a chromosomal mutation. *Chromosome mutations* involve changes in large amounts of genetic material and many of them are either lethal or produce individuals who are severely handicapped. However, although people with Down's syndrome show a number of distinctive physical features, such as a characteristic facial shape and stature and have limited mental development, they are able to lead relatively normal lives.

downstream processing: the name given to a variety of procedures which follow the biological part of a *biotechnological* process. There are three basic steps involved. These are:

* separation of the crude product from the various other products, substrates, enzymes or microorganisms which may be present in the mixture

* concentration of the product to remove the large amount of water that is usually present

* purification of the crude extract to produce a single product.

ductus arteriosus: a blood vessel found in the fetus, it links the pulmonary artery directly to the aorta. While a fetus is developing inside its mother's uterus, its oxygen supply comes from its mother via the *placenta*. Its lungs do not function. Blood entering the right atrium of the heart either passes through a hole called the foramen ovale into the left atrium and on into the aorta or goes from the pulmonary artery through the ductus arteriosus into the aorta. At birth, the foramen ovale and the ductus arteriosus both close so that blood then flows from the right side of the heart to the lungs.

duodenum: the first loop of the small intestine. Partly digested food passes into the duodenum when a ring of muscle called the pyloric sphincter relaxes and allows the *stomach* contents to flow out. Alkaline mucus is secreted by glands in the wall of the duodenum. This neutralizes the very acid contents of the stomach and provides an environment of the correct pH for the enzymes secreted into this part to function. The duodenum receives *bile* which flows into it along the bile duct from the gall bladder. It also receives the digestive enzymes secreted by the *pancreas*. Both of these secretions are controlled by hormones produced by cells in the wall of the duodenum. The inner surface of the duodenum, like that of the rest of the small intestine, has many villi on its surface. These provide a very large surface area for the absorption of the products of digestion.

ECG: see *electrocardiogram*

Echinodermata: the animal phylum containing sea urchins and starfish. Members of the Echinodermata share the following features:

* adult members of the phylum show a pentamerous *radial symmetry*. This is a five-way symmetry. A starfish, for example, has five arms so there are five ways in which it can be cut to produce pieces which are a mirror image of each other

* they have a unique water-filled vascular system. Sea water enters the animal and flows along a series of canals to the tube feet. These tube feet have a number of functions including gas exchange, excretion and locomotion

* they have a thin epidermis, immediately under which is a layer of spines and plates made of calcium carbonate.

ecological pyramids: see *pyramids of number, biomass, energy*

ecosystem: an ecological unit which includes all the organisms living in a particular area as well as the *abiotic* features of their environment. Ecosystems can be of very different sizes. The desert ecosystem of North Africa, for example, stretches for many thousands of kilometers. A bird's nest could just as well be considered as an ecosystem inhabited by large numbers of insects and *microorganisms*. Although ecosystems are usually thought of as being self-contained, they do influence each other considerably. A migrating bird, for instance, may fly between ecosystems in different continents, while pollution of farmland in Europe and North America has even resulted in increased amounts of pesticides in the bodies of Antarctic penguins.

hint There are two elements to the definition of an ecosystem. The organisms that are present in a particular area and the abiotic aspects of their environment. That is why questions which ask you for the meaning of this term are often worth two points.

ectotherm: an animal which makes use of the environment to regulate its body temperature. A crocodile, for example, regulates its temperature by moving between water and land. During the night, when the air temperatures are low, the animal stays in the warmer water. Nevertheless its body temperature gradually falls. With daylight, it moves up onto the shore and basks in the sun. As a result of this, its temperature rises again. During the middle of the day, when it can be very hot, it returns to the now cooler water. The animals generally come onto the shore again in the late afternoon, warming up in the sun before returning to the water for the night. Despite being traditionally thought of as a "cold-blooded animal" whose temperature fluctuates considerably, the crocodile is able to maintain its own temperature between remarkably narrow limits by making use of fluctuations in the air and water temperatures.

effector: an organ which responds to stimulation by a *nerve impulse* and brings about a response or a change. In animals, *muscles* and *glands* are both effectors.

efferent: means coming from. In a *reflex arc*, for example, the efferent neurone transmits a nerve impulse from the spinal cord. In the *kidney*, the efferent arteriole takes blood from the *glomerulus*.

egestion: the removal of waste products from the body as *feces*. The *lumen* of the gut forms a cavity, which runs through but is effectively outside the body. Only when food has been digested and absorbed does it enter the body itself. Much of the material present in the feces of a mammal is undigested food and has, therefore, never actually been inside the body. Because of this, it is not correct to refer to the process of defecation as *excretion*; the word "egestion" should be used instead.

elastin: a protein which forms long, flexible fibers. These fibers are found in the skin, in *ligaments* and in the walls of arteries where their elastic properties are important.

electrocardiogram (ECG): a record of the electrical events associated with the beating of the heart. Electrodes are attached to the skin on the chest and the limbs. These are connected to the recording apparatus. When the heart beats, a wave of electrical activity passes over its surface, producing an electrocardiogram such as that shown in the diagram.

Various features can be recognized. The part labeled P represents the spread of electrical activity from the *sinoatrial node* over the surface of the atria, that labeled QRS represents its passage over the ventricles, and the part labeled T represents the electrical changes which occur during the filling of the heart.

Although electrocardiograms can be used to investigate heart function in healthy individuals, the information they contain is also used to monitor the heart in patients recovering from heart attacks and to investigate other forms of heart disease.

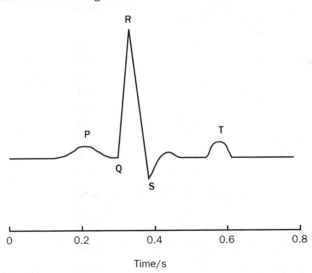

Time/s

An ECG from a normal heart

electron: A negatively charged particle which orbits the nucleus of an *atom*. The possession of electrons is very important in allowing atoms to form chemical bonds.

electron microscope

electron microscope: a type of microscope which makes use of a beam of electrons rather than visible light. Because the wavelength of electrons is much smaller than the wavelength of visible light, an electron microscope not only gives a high magnification but it also has high *resolution*. This means that details can be seen clearly. An electron microscope is very similar to a light microscope in the way in which it works but, instead of using glass lenses to focus a beam of light, it uses magnets to focus a beam of electrons. Electrons are very small so they are scattered if they hit molecules in the air. Because of this they must travel through a vacuum. Specimens that are to be examined must be cut into very thin sections, and these must be treated so that they can be examined in these conditions. The variety of chemical processes that are carried out in this preparation may change the appearance considerably. Features which have been introduced in this way are known as *artifacts* and care needs to be taken in interpreting electron micrographs because of the possible presence of artifacts. The main differences between a light microscope and an electron microscope are shown in the table.

Feature	Light microscope	Electron microscope
Source of illumination	Visible light	Beam of electrons
Method of focusing	Glass lenses	Magnets
Specimens which can be examined	Live or dead	Dead
Magnification	Student microscopes magnify up to about 400 times but the maximum magnification that can be achieved with a light microscope is about 1500 times	Up to 500 000 times
Resolution	About 0.2 µm	About 1 nm or 0.001 µm

There are two main types of electron microscope, the *transmission electron microscope* and the *scanning electron microscope*.

> **hint** You can tell whether a photograph of a cell has been taken with a transmission electron microscope rather than an optical microscope by looking for evidence of superior resolution.

electron transport system: a chain of carrier molecules in which energy released in the passage of electrons from one molecule to another is used to produce ATP. In *respiration*, a number of oxidation reactions take place in *Krebs cycle*. Hydrogen is released in this process and acts as a source of protons and electrons, as each hydrogen atom consists of one proton and one electron. The electrons are passed from molecule to molecule along the electron transport system. At each transfer a small amount of energy is released. This is used to pump the protons through the mitochondrial membrane in which the carrier molecules are found. This produces a proton gradient with more protons on one side of the membrane than on the other. The proton gradient acts as a store of potential energy. When the protons return through the membrane they release this energy which is then used to produce ATP. The protons and electrons finally combine with oxygen to produce water. The carrier molecules which make up the electron transport system in respiration are found on the inner membranes of the *mitochondria*.

A similar system is used to produce ATP in the *light-dependent reaction* of photosynthesis. Here, chlorophyll provides the electrons, while the protons come from water molecules.

electrophoresis: a method of separating molecules based on differences in their electrical charge. The substances to be separated are placed in a buffer solution on a layer of gel or a sheet of filter paper. Electrodes are arranged so that a direct current is passed through this medium. As a result, each of the substances to be separated will have a different charge and will move at a different rate. The separated substances will be found in bands whose position can be shown up by using a suitable stain.

Genetic fingerprinting relies on separating DNA fragments of different size. This is done by gel electrophoresis and relies on the fact that the rate at which a particular fragment of DNA moves is proportional to its mass.

ELISA technique: a sensitive method used by biochemists for measuring the concentration of a particular substance. It stands for enzyme-linked immunosorbent assay. The term is probably best explained by looking at these words in reverse order. An assay is simply a test used to measure the amount of a particular substance. Immunosorbent refers to the fact that it is based on an immune reaction which involves *antibodies*. Antibodies are very specific and a particular antibody will only bind with one particular molecule. In this technique, an antibody is used which is specific to the molecules of the substance being measured. The antibodies form complexes with these molecules. The more molecules of the substance present, the more antibody complexes that are formed. The number of antibody complexes can then be determined by using an *enzyme* which is easily measured and which will bind to the antibody complex. Hence the process is described as being enzyme-linked.

emulsion test: a biochemical test which can be used to show the presence of a lipid. For details, see *biochemical tests*.

embryo sac: a large cell found in the ovary of a flowering plant which contains a number of nuclei including the female gamete or egg nucleus. Inside the ovule of an immature plant is a single cell which divides by *meiosis* to produce four haploid daughter cells. Three of these cells disappear. The other one undergoes a number of mitotic divisions to produce a single cell, the embryo sac, containing eight nuclei. At fertilization, two of these nuclei, the polar nuclei, will fuse with one of the male gametes to become the *endosperm*. Another one, referred to as the egg nucleus, fuses with the other male gamete to become the zygote. This is summarized in the diagram opposite.

emphysema: a respiratory disease in which the walls of the *alveoli* or air sacs in the lungs break down. This causes these air sacs to enlarge with the result that the total area of their walls gets much less and gas exchange becomes less efficient. As a consequence, severe emphysema results in breathlessness and the only way to treat a patient may be by providing oxygen. Although it is not understood how emphysema occurs, it is known that it is often associated with cigarette smoking and chronic *bronchitis*.

endemic: a term which is used to describe diseases which occur frequently in particular areas or among the members of a particular population. Malaria for example can be described as being endemic to many tropical countries. Although cases of malaria do occur among people returning to the US from the tropics, the disease is not endemic to the US. People living in areas where a disease is endemic have often evolved adaptations which make them less susceptible to its effects. One such example is the occurrence of the sickle-cell *allele* in parts of Africa where malaria is endemic. In contrast, diseases such as

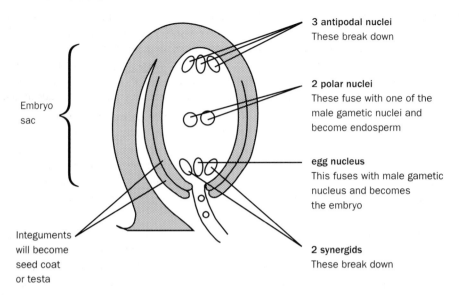

3 antipodal nuclei
These break down

2 polar nuclei
These fuse with one of the
male gametic nuclei and
become endosperm

Embryo sac

egg nucleus
This fuses with male gametic
nucleus and becomes
the embryo

Integuments
will become
seed coat
or testa

2 synergids
These break down

The embryo sac and its development

measles can have devastating effects, if introduced to isolated populations, such as native Americans living in the Amazon forests. The term can also be used to refer to other organisms whose distribution is restricted to specific areas.

endocrine gland: a *gland* which secretes a *hormone*. Many substances that are produced by glands leave the structure through a tube or duct. Endocrine glands are different, however, in that they produce hormones which are secreted into the blood system. They do not, therefore, have ducts and are often called ductless glands. Many endocrine glands have several different functions. The pancreas, for example, acts as an endocrine gland in producing insulin and glucagon, but also secretes digestive juices into the small intestine.

endocrine system: the system in the body responsible for the production of *hormones*. Endocrine glands are distributed throughout the body but, in addition, many organs which have other important functions are also part of the endocrine system. The kidneys, the stomach, the ovaries and the testes all produce hormones and are, therefore, part of the endocrine system.

endocytosis: the transport of large particles or fluids through the *plasma membrane* into the cytoplasm of a cell. The plasma membrane surrounds the particles concerned. A *vesicle* or *vacuole* is formed which is pinched off and moves into the cytoplasm. The term phagocytosis is generally used where solid particles are involved while pinocytosis refers to the cell taking in small droplets of fluid.

endodermis: a ring of cells between the outer part of the root or the cortex and the vascular tissue in the center. A band of waterproof material called the Casparian strip runs round the walls of each of these endodermal cells. It prevents water and dissolved substances from getting into the vascular tissue by going through the cell walls and intercellular spaces. As a result, these substances have to pass through the cytoplasm of the endodermal cells which can therefore control their movement into the xylem.

endogenous: coming from inside. Many organisms have a natural 24-hour *circadian rhythm* which governs their behavior. This has a considerable effect on animals. In hot, dry climates, for example, it may be an advantage to be nocturnal, hunting for food during the cooler hours of the night and resting in the shade during the day. Organisms have a natural biological clock that enables them to adapt to conditions like this. Experimental evidence suggests that this is an internal mechanism so it may be described as being endogenous.

Plant growth substances are chemicals produced by plants which are involved in the control of various aspects of their growth and development. Much early experimental work was done by treating plants with these substances and observing the consequences. However, there was one major difficulty with this approach. It involved making the assumption that the response of a plant to this treatment would be the same as that to the smaller concentration produced by the plant itself. In other words, the response would be the same as that produced by endogenous plant growth substances.

endometrium: the inner lining layer of the uterus. During the *menstrual cycle*, the endometrium goes through a series of changes. During the first part of the cycle, the hormone *estrogen* stimulates this lining layer to develop and thicken, preparing it for a possible pregnancy. Following ovulation, a second hormone, *progesterone*, is secreted by the corpus luteum. This maintains the endometrium. If pregnancy does not occur, the progesterone level falls and the lining is lost during menstruation.

endopeptidase: a protein-digesting enzyme, which breaks peptide bonds in the middle of a polypeptide chain, rather than peptide bonds at the ends. As a result, they produce smaller polypeptides rather than individual amino acids. This is shown in the diagram.

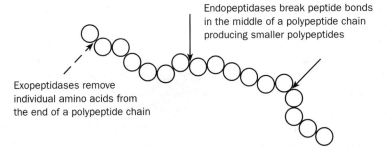

Endopeptidases break peptide bonds in the middle of a polypeptide chain producing smaller polypeptides

Exopeptidases remove individual amino acids from the end of a polypeptide chain

The action of endopeptidases

Pepsin and *trypsin* are examples of endopeptidases found in the guts of mammals. From the point of view of digestion, it is much more efficient to break a long polypeptide chain into shorter lengths before starting to remove individual amino acid molecules from the ends. There will be more ends for *exopeptidases* to work on. Endopeptidases are therefore secreted and act early in the digestive process.

endoplasmic reticulum: a network of membranes found in the cytoplasm of a cell. It consists of a complex system of pairs of membranes arranged parallel to each other enclosing flattened, fluid-filled spaces. These membranes may be covered with *ribosomes*, in which case they form rough endoplasmic reticulum. The ribosomes produce proteins which are transported through the spaces between the membranes. Smooth endoplasmic reticulum does not have ribosomes on its surface. Its main function is the production and transport of *lipids*.

endoscopy: a technique which involves the medical examination of the inside of body organs using an instrument called an endoscope. At its simplest, an endoscope is a long tube with a light at one end and a means of displaying an image at the other. Many modern endoscopes rely on fiber optics, so they are small in diameter and flexible. They enable detailed examination of the inside of an organ without the need for extensive surgery. Endoscopes are used for a variety of purposes, such as inspecting the external ear and eardrum for signs of damage or infection or for looking at the inside of the gut.

endoskeleton: a skeleton which forms an internal supporting system for the body. It is the type of skeleton found in mammals. In contrast to an *exoskeleton*, such as that in arthropods, the muscles are attached to its outer surfaces. The skeleton of a mammal is composed of two materials, *cartilage* and *bone*. These are living tissues and are capable of growth as the animal gets older and increases in size. There is no need for the complex process of molting which is associated with the possession of an exoskeleton.

endosperm: the tissue which supplies nutrients for the developing embryo in a seed. During fertilization in a flowering plant, one of the male nuclei from the pollen grain fuses with two of the nuclei in the ovule. The resulting nucleus contains three sets of chromosomes. It is triploid. This nucleus divides very rapidly and gives rise to a mass of large cells which surround and nourish the developing embryo. In the seeds of dicotyledonous plants, such as peas and beans, the endosperm has been entirely absorbed by the time the seed has developed fully. In monocotyledons such as maize and onion, there is still a lot of endosperm present in the mature seed. In these species, it acts as a food store for the young plant during germination.

endotherm: an animal that maintains its body temperature by using physiological mechanisms. Humans are good examples of endotherms. The diagram below summarizes the way in which body temperature is controlled.

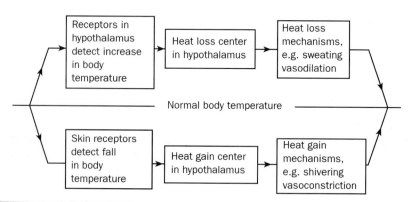

Control of body temperature in humans

If the temperature falls too low, usually as the result of being in cold conditions, this is detected by receptors in the skin. Information is sent by nerve impulses to the temperature control center in the *hypothalamus* of the brain. This coordinates the responses which lead to the body both producing more heat and limiting the amount of heat that it loses. On the other hand, if the body temperature rises, usually as a result of increased activity, there is an increase in blood temperature. This is detected by receptors in the hypothalamus itself and stimulates a range of responses such as increased sweating and *vasodilation* which lead to the loss of heat. Temperature control in an endothermic animal provides a good example of *negative feedback*. A change in body temperature leads to the operation of various mechanisms which return the temperature to its norm or resting value.

endotoxin: a poisonous substance or toxin contained inside the cells of certain bacteria. When the bacterium dies and its cell wall breaks down, the endotoxin is released and affects the tissues of the host. Endotoxins are produced by many bacteria including species of *Salmonella* and *Escherichia*. Whichever species is involved, the symptoms are similar and include fever, diarrhea and damage to small blood vessels.

enterokinase: an enzyme secreted by cells in the upper part of the small intestine. *Trypsin* is an enzyme which digests proteins in the small intestine. It is secreted in an inactive form called *trypsinogen*. Enterokinase is responsible for converting trypsinogen to trypsin.

enteron: the cavity enclosed by the body wall of an animal belonging to the phylum *Cnidaria*. It has a single opening which acts as both a mouth and an anus. The enteron has a number of functions other than acting as a digestive organ. Oxygen is exchanged between the water in the enteron and the cells which make up the inner layer of the body wall. In addition, as it is filled with water, it can act as a *hydrostatic skeleton*.

environment: the conditions surrounding an organism. These conditions can be divided into two main groups. *Abiotic* factors are concerned with the nonliving part of the environment. Rainfall, soil pH, temperature and humidity are all examples of abiotic factors which will affect the distribution of various organisms. *Competition* and *predation*, on the other hand, are concerned with the living part of the environment and are called *biotic* factors.

enzyme: a protein which is able to catalyze biochemical reactions. Many of the reactions which take place in living organisms would normally be very slow indeed. Enzymes are vital in allowing these reactions to take place rapidly in the conditions which are found inside living cells. There is a very large number of different enzymes in existence. A single mammalian cell, for example, may have as many as 2000 different chemical reactions taking place inside it. Each of these will be catalyzed by a different enzyme. They all work in a similar way, however, by lowering the *activation energy* necessary to start the reaction. As less energy is necessary, biochemical reactions can take place at the temperatures and pressures found in living cells. The way in which an individual enzyme molecule works may be explained by referring to a simple model, the *lock and key hypothesis*.

hint Some important things to remember about enzyme biology:
* Heating denatures enzymes and other proteins. It does not kill them!
* When an enzyme is heated, the bonds which give it its *tertiary structure* break. This leads to the molecule being denatured. Denaturation is not the result of the breaking of *peptide bonds*.
* High temperatures lead to enzymes being denatured. At low temperatures they work more slowly because molecules have less kinetic energy, therefore there are fewer collisions between enzyme and substrate. Low temperatures do not denature enzymes.

eosinophil: see *granulocyte*

epidemic: an outbreak of a disease that spreads rapidly through a population, affecting a large proportion of vulnerable members. Influenza is the commonest cause of epidemics. The influenza virus spreads easily from person to person and, in addition, it mutates frequently. As

97

a result the strain of influenza virus infecting people one year is likely to differ slightly from that most commonly found in the previous year. People who have had influenza in the past will only be immune to one particular strain; they will be susceptible to the new strain. Since only the more vulnerable members of the community (such as the elderly and those with chest disease) are usually vaccinated against the current strain of influenza a large proportion of the population always likely to be susceptible.

epidermis: the outer layer or layers of cells in a multicellular organism. In plants, the epidermis consists of a single layer of cells surrounding the tissues of the roots, stems and leaves. In the stems and leaves it is covered by a thin, waxy cuticle which helps to limit water loss. Pores in the cuticle, *lenticels* and *stomata*, allow the exchange of gases between the cells and the external environment.

In invertebrate animals, the epidermis is again made up of a single layer of cells which secrete a cuticle. This cuticle may consist largely of mucus as in earthworms or it may be more complex as in members of the phylum *Arthropoda*. Here the cuticle forms an *exoskeleton*. In mammals and other vertebrates, there is no cuticle over the surface. The epidermis instead consists of several layers of cells, the outer ones of which are cornified and dead.

epinephrine: *adrenaline*.

epistasis: a genetic cross in which a single characteristic is affected by two or more different genes which interact with each other. The diagram shows how two different genes may affect the color of sweet pea flowers.

Allele **A**   Allele **B**

Enzyme **A**   Enzyme **B**

Colorless precursor   ⟶   Colored intermediate   ⟶   Colored final
molecule   molecule   molecule
(White flowers)   ↓   (Red flowers)   ↓   (Purple flowers)

In the flowers, a colorless precursor molecule is converted into a colored final molecule by way of a differently colored intermediate molecule. This biochemical process is controlled by two genes. The dominant allele of the first gene, **A**, produces enzyme **A**, which controls the first step in the reaction. The dominant allele of the second gene, **B**, produces enzyme **B**, which controls the second step in the reaction. In each case the recessive homozygote produces a faulty enzyme which will not catalyze this reaction.

For a flower to be purple, both alleles **A** and **B** must be present. A red flower is produced when allele **A** is present but allele **B** is absent. If allele **A** is absent, then the flower will be white. This is summarized in the table.

Allele A	Allele B	Precursor molecule	Intermediate molecule	Final molecule	Color of flower
✗	✗	✓	✗	✗	white
✗	✓	✓	✗	✗	white
✓	✗	✓	✓	✗	red
✓	✓	✓	✓	✓	purple

key
✓ = present   ✗ = absent

epithelium: the cells which form the *tissue* covering an organ, either on its inside or on its outside. Substances which pass into and out of the body, the products of digestion, oxygen, water and carbon dioxide must go through epithelial cells. Not surprisingly, these cells are often specialized for exchange. Those in the small intestine, for example, have their free *plasma membranes* folded to form *microvilli*. This provides a very large surface area for the absorption of the products of digestion. Since they act as lining cells, epithelial tissues are very prone to damage and need constant replacement. Mitosis can often be seen.

Epithelial tissues are classified according to shape. The three basic types are shown in the diagram.

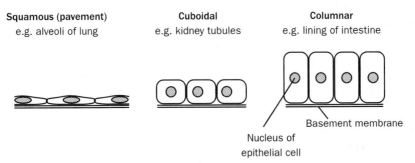

In addition to covering organs, epithelial tissue also forms *exocrine glands*.

erythrocyte: see *red blood cell*

esophagus: the part of the alimentary canal of a mammal which takes food from the mouth to the *stomach*. Food is usually chewed in the mouth and mixed with saliva. It is then swallowed and a wave of muscular contraction or *peristalsis* pushes it down into the stomach. When seen under the microscope, the esophagus has an *epithelium* consisting of several layers of cells in which there are mucous glands. Well developed layers of circular and longitudinal muscle in the wall assist with peristalsis.

essential amino acid: an *amino acid* which must be supplied in food as it cannot be made in the body. There are ten amino acids which are essential for children but only eight for adults. Proteins which contain essential amino acids in good proportions are called first-class proteins. Most foods derived from animals contain first-class proteins. Plant proteins, on the other hand, are second-class as they are often deficient in one or more of these essential amino acids. Because of this, people who are on strict vegetarian diets need to take care that they eat proteins from a variety of sources. In this way, they will obtain all the essential amino acids that they require.

> **hint** Essential amino acids cannot be made in the body. They are called essential because it is essential that they are present in the diet. Nonessential amino acids are just as important in the production of proteins but it is not essential that they are present in the diet because they can be synthesized in the body.

ester bond: a chemical bond formed as a result of condensation between glycerol and a fatty acid. There are three of these bonds in a *triglyceride* molecule.

estrogen: a female sex *hormone* produced by cells in the ovary. It is responsible for the changes which take place in the female body during puberty, such as the development of pubic hair, the growth of breast tissue and the broadening of the pelvic girdle.

estrous cycle

During the first half of the *menstrual* cycle, estrogen is secreted by the cells of the developing follicle. It has several effects on the reproductive system but, in particular, brings about growth of the lining of the uterus which becomes thicker and more glandular. Estrogen has a *negative feedback* effect on the level of another hormone, *FSH*. At the start of the cycle, there is a rise in the level of FSH secreted by the anterior lobe of the pituitary gland. FSH stimulates follicle development and estrogen is secreted by the follicle cells. There is, therefore, an increase in the amount of estrogen in the blood and this inhibits the release of more FSH. This is summarized in the diagram.

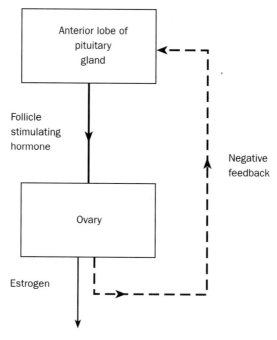

This effect enables estrogen to be used as an *oral contraceptive*, either on its own or with *progesterone*, in a combined pill.

The females of many mammals show a distinctive pattern of behavior around the time of ovulation. This is known as *estrus* and is triggered by the peak in estrogen secretion which occurs just before ovulation.

estrous cycle: the reproductive cycle in a female *mammal*. Different mammals have reproductive cycles of different patterns and lengths. Some, such as cattle, rats, mice and humans have cycles which follow, one after another, for the entire reproductive part of the animal's life. Others, such as dogs and sheep are only reproductively active at a particular time or times during the year. Whatever the pattern though, the estrous cycle is controlled by *FSH* and *LH*, hormones produced by the anterior lobe of the *pituitary gland*. Under their control the *ovaries* and other reproductive organs undergo a series of changes. These lead to ovulation and preparation of the body for possible pregnancy. If pregnancy does not occur, there is a short recovery period and then the cycle starts again. These events are summarized in the diagram.

100

The regulation of the estrous cycle is complex and the concentration of different hormones is controlled by *negative feedback*. A knowledge of this pattern of negative feedback has allowed us to alter the amounts of the hormones that control the cycle. Examples include the use of *oral contraceptives*, the treatment of infertility in humans, and the synchronization of reproductive cycles in farm animals.

estrus: the distinctive pattern of behaviour displayed by some mammals around the time of ovulation. Many female mammals live solitary lives and are only fertile for a short period of time. If they are to conceive, it is essential that they advertise their reproductive condition to potential mates. This is the purpose of estrus.

Estrus may involve:

* changes in the normal behavior pattern. Cattle, for example, are generally restless, walk about more and feed less

* the intense coloration and swelling of patches of skin, such as those on the chest and around the reproductive area in some monkeys

* the production of chemical signals or *pheromones*, such as those produced by a female dog.

ethene: a *plant growth substance* produced by most plant organs. Its best known effect is its involvement in the ripening of fruits such as bananas and tomatoes but it is also important in stimulating flowering in pineapples.

eukaryote: an organism with *eukaryotic* cells. Members of the four *kingdoms*, *Protoctista*, *Fungi*, *Animalia* and *Plantae* are eukaryotes.

eukaryotic: a term which refers to cells with a number of distinctive features separating them from prokaryotic cells. These are summarized in the table.

Eukaryotic cells found in members of the kingdoms *Protoctista*, *Fungi*, *Animalia* and *Plantae*	Prokaryotic cells found in members of the kingdom *Prokaryotae*
Large cells up to 50 µm in diameter	Cells small with a mean diameter under 5 µm
DNA linear and associated with proteins to form a true *chromosome*. Chromosomes found within a *nucleus*	Circular strands of DNA not associated with proteins and found in the cytoplasm. No nucleus present
Many membrane-surrounded organelles such as *mitochondria* present	Few organelles present and none are surrounded by a *plasma membrane*
Flagella (correctly known as undulipodia) have an internal arrangement of microtubules	Flagella lack system of microtubules

eutrophication: an increase in the quantity of plant nutrients. The term is generally used when freshwater lakes or rivers are enriched with nitrates and phosphates either as a result of the leaching of *fertilizer* from agricultural land or from sewage effluent. The flowchart shows how eutrophication can cause harm to an ecosystem by causing the death of many of its organisms.

> Increase in nitrate and phosphate levels

↓

> Increase in the growth of algae and other photosynthetic organisms. These die faster than they can be eaten

↓

> Increase in activity of *decomposers*. Their respiratory activity depletes the water of its oxygen supply

↓

> Other organisms such as fish and the invertebrate animals on which they feed die due to a lack of oxygen

 hint Bacterial decomposers respire. Eutrophication leads to increased bacterial respiration and it is this that lowers the oxygen concentration of the water.

excretion: the removal of waste products which have been produced as the result of *metabolic* processes in an organism. Although waste products such as *urea* and *uric acid* contain the element nitrogen, excretion also involves the removal of carbon dioxide produced as a waste product of respiration. The table summarizes some of the more important excretory substances in a mammal as well as the organs mainly responsible for their excretion.

Substance	Formed from	Organ mainly responsible for excretion
Urea	Breakdown of excess amino acids	*Kidney*
Carbon dioxide	Respiration	*Lungs*
Bile pigments	Breakdown of *hemoglobin*	*Liver.* Removed with the feces

exocrine gland: a *gland* which produces a secretion which leaves the organ through a duct. Glands, such as the salivary glands, the pancreas and the liver, which produce secretions involved in digestion, are exocrine glands. *Bile*, for example, is secreted by the *liver* and drains through the bile duct into the intestine.

exocytosis: a process which involves the transport of materials out of cells. Exocytosis takes place in many cells in the body. One example is provided by the cells in the mammary glands which are responsible for secreting milk. Proteins, lactose and mineral ions such as calcium are transported to the *Golgi apparatus*. This is an organelle which is responsible for the processing and packaging of substances produced by a cell. *Vesicles* are formed. They pinch off and move towards the surface of the cell. Here the membrane which surrounds them fuses with the *plasma membrane* and the contents of the vesicle are released outside the cell.

exogenous: coming from outside. Many organisms have a natural 24-hour *circadian rhythm* which governs their behavior. This has a considerable effect on animals. In hot, dry climates, for example, it may be an advantage to be nocturnal, hunting for food during the cooler hours of the night and resting in the shade during the day. Organisms have a natural biological clock that enables them to adapt to conditions like this. Experimental evidence suggests that in some organisms this mechanism can be influenced by exogenous factors, that is, changes in environmental conditions.

Plant growth substances are chemicals produced by plants which are involved in the control of various aspects of their growth and development. Much early experimental work was done by treating plants with these substances and observing the consequences. However, there was one major difficulty with this approach. It involved making the assumption that the response of a plant to exogenous plant growth substances would be the same as that to the smaller concentration produced by the plant itself.

exopeptidase: a protein-digesting enzyme which breaks the peptide bonds at the end of a polypeptide chain releasing either single amino acids or dipeptides. In mammals, these enzymes are associated with the final stages in protein digestion and are located on the *plasma membranes* of the cells which line the small intestine.

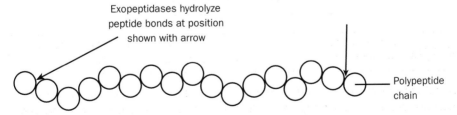

The action of exopeptidases

hint A simple diagram will help you to explain the difference between an endopeptidase and an exopeptidase. You can show clearly the positions of the bonds that are broken by these different enzymes.

103

exon: a piece of the *DNA* in a *gene* which codes for an *amino acid* sequence. A gene in a eukaryotic cell, such as one from an animal or a plant, consists of exons and pieces of DNA which do not give rise to amino acid sequences. The latter pieces of DNA are known as introns. During the process of transcription, the base sequence in the gene forms a template for producing an mRNA molecule. Before the mRNA leaves the nucleus however, it is edited, and the base sequences forming the introns are cut out. The mRNA base sequence therefore corresponds only to exons. The genes in many prokaryotic cells do not contain introns.

exoskeleton: a hard covering on the outside of an animal which acts as a *skeleton*. The body surface of an arthropod, for example, is covered in a cuticle which is composed mainly of *chitin*, a nitrogen-containing polysaccharide. (See *Arthropoda*.) In insects, this tough layer is further strengthened by the addition of a very hard type of protein. This effectively means that the exoskeleton of an insect is like a system of hollow tubes. Where they bend, the skeleton is thinner and forms a joint which acts as a hinge. The animal's muscles are attached to processes on the inside of this skeleton. One of the disadvantages of having a rigid covering on the outside is that it creates problems with growth. When arthropods grow, therefore, they must shed this outer skeleton. Only while the new skeleton is soft can an increase in size take place.

exotoxin: a poisonous substance or toxin produced and secreted by certain bacteria as they grow. Exotoxins are produced by a number of bacteria including those responsible for botulism, tetanus and diphtheria. Diphtheria has a short incubation period before a sore throat and fever develop. The multiplying bacteria produce an exotoxin which is secreted into their surroundings. This exotoxin inhibits protein synthesis and may result in damage to the heart and nerves. Successful treatment of diphtheria involves the use of an antibiotic to kill the bacteria together with an antitoxin.

expiratory reserve volume: the additional amount of air that can be breathed out after a person has breathed out normally. For an average adult male, this is about 1500 cm^3. See *lung capacities*.

exponential phase: see *log phase*

extensive food production: methods of food production which require relatively large areas of land. Nomadic herding and ranching of cattle and sheep are examples of extensive methods of production. This method contrasts with intensive farming where the animals are confined in a relatively small area. Extensive farming has a number of advantages over intensive farming:

- it is considered to be more ethically acceptable as animals are kept under conditions which are thought of as being closer to those in their natural environment

- relatively low stocking densities minimize the spread of disease. It also allows land which is considered unsuitable for other forms of agriculture to be used

- it does not produce large concentrations of waste material, such as urine and feces, which must be disposed of. Waste material is naturally recycled on the land itself

- there is less need to have a system which can transport large quantities of such materials as food concentrates, silage and straw to the site of production. Much more food can be supplied naturally.

external fertilization: *fertilization* which takes place outside the body. In many aquatic animals, both male and female *gametes* are released into the surrounding water. Because the chance of fertilization is not very high, animals which have external fertilization generally produce far more egg cells than those which have *internal fertilization*.

extracellular digestion: the chemical breakdown of food molecules which takes place outside cells. Many bacteria and fungi are *saprophytes* and feed by extracellular digestion. They secrete digestive *enzymes* onto the surface of their food. These enzymes break down complex molecules, such as those of starch and protein, into smaller, soluble ones which are then absorbed. In addition to this, much of the digestion which takes place in the gut of a mammal is also extracellular. Enzymes are secreted by glands such as the salivary glands and the pancreas. They break down the individual food molecules. This process occurs in the gut lumen, so is obviously extracellular.

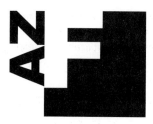

F_1: the offspring from parents which are *homozygous* for a particular characteristic. The term should only be used where it relates to a cross in which both parents are homozygous.

F_1 **hybrid:** offspring produced by crossing parents from strains which have been inbred. They are characterized by having a much greater productivity than either of their parents. Much of the early work on producing F_1 hybrids was carried out on maize. It had been known for some time that repeated self-fertilization of maize plants resulted in lower and lower productivity. Part of the reason for this was that a greater proportion of *homozygous* plants resulted and there was an accumulation of unfavorable *recessive* alleles. If two such lines of repeatedly self-fertilized plants are developed, it is likely that the unfavorable alleles that accumulate will be different in each case. Neither strain will give a very high yield. If they are now crossed, the offspring will be heterozygous for many of these alleles. Since the alleles concerned are recessive, yield will not be depressed. Look at the example below. Assume that each of the recessive alleles shown lowers the yield in the homozygous condition.

Inbred line A $\times$ Inbred line B

Genotype: **aabbCCddEEFF** Genotype: **AABBccDDeeff**

$\downarrow$

F_1 hybrid offspring

Genotype: **AaBbCcDdEeFf**

Lines A and B are both homozygous for alleles which lead to a low yield. The F_1 hybrids produced from this cross are heterozygous, however, so these recessive alleles will have no lowering effect on the yield. The yields of these plants will therefore be much higher than those of either of their parents.

There is one major disadvantage with the use of F_1 hybrid varieties. The seeds produced by the F_1 plants do not breed true. They cannot be kept and used the following year, since a large number of different *genotypes* will be produced. Many of these will result in plants having very low yields.

The technique of producing F_1 hybrids is widespread. Apart from maize, it has been used extensively to produce larger and more brightly colored garden flowers and to increase the numbers of young reared by female pigs.

facilitated diffusion: is a special case of *diffusion* involving a protein carrier molecule. Molecules such as glucose do not diffuse through the phospholipid layer in the *plasma membrane*. However, the membranes of cells contain carrier proteins. These proteins have two particularly important properties. They have binding sites for specific molecules. A glucose carrier protein, for example, will only bind to glucose molecules. Other molecules will not fit into the binding site. In addition, the carrier molecule can change shape many

106

times each second. In one form, the binding site will be exposed to the outside of the cell; in the other, it will be exposed to the inside. What happens is that the glucose molecules outside the cell are continually moving about, and one will eventually collide with the binding site of an exposed carrier protein. When the protein changes shape, the glucose molecule will be on the inside of the membrane where it will be released.

hint Errors are often made when listing the differences between facilitated diffusion and active transport. Both of these processes rely on carrier proteins. Active transport, however, takes place from a low to a high concentration and requires ATP. Facilitated diffusion occurs from a high to a low concentration and does not require ATP.

factor VIII: one of the factors or plasma *enzymes* involved in the process of *blood clotting*. Absence of factor VIII brings about an inherited condition known as hemophilia in which the blood takes a much longer time to clot. Hemophiliacs are usually male because the condition is *sex-linked* with the *recessive* allele concerned being carried on the X chromosome.

family: a level of *classification*.

fast muscle fiber: see *skeletal muscle fiber*

fat: a name usually given to *triglycerides* which are solid at temperatures below 20 °C. This is because they contain a high proportion of saturated *fatty acids*. Generally speaking, the triglycerides found in animals are fats. Those found in plants are oils and are liquid at temperatures of 20 °C.

fatty acid: molecules with a COOH group at one end and a hydrocarbon tail at the other. Fatty acids with small numbers of carbon atoms in this tail are produced when *cellulose* is digested by microorganisms living in the guts of *ruminants* and other herbivorous mammals. They can be absorbed into the blood and are important *respiratory substrates* in these animals. Fatty acids with long hydrocarbon tails are important constituents of fats and oils (*triglycerides*). The diagram below shows two fatty acids, stearic acid and oleic acid, which are commonly found in triglycerides.

Both have hydrocarbon chains with seventeen carbon atoms in them. Oleic acid, however, has a double bond present between two of the carbon atoms. Fatty acids with double bonds are said to be unsaturated. They have much lower melting points than fatty acids in which there are no double bonds. Triglycerides which contain unsaturated fatty acids therefore tend to be liquid at temperatures of around 20 °C. They are called oils and are found mainly in plants. Those which contain saturated fatty acids are solid at this temperature. They are called fats and are mainly found in animal cells.

Stearic acid

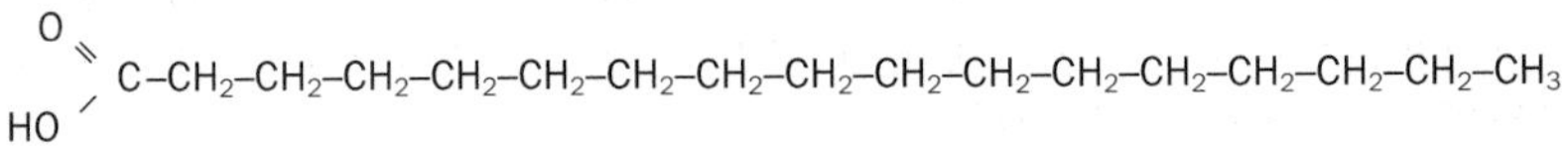

$$\underset{HO}{\overset{O}{\diagdown}} C-CH_2-CH_2-CH_2-CH_2-CH_2-CH_2-CH_2-CH_2-CH_2-CH_2-CH_2-CH_2-CH_2-CH_2-CH_2-CH_2-CH_3$$

Oleic acid

$$\underset{HO}{\overset{O}{\diagdown}} C-CH_2-CH_2-CH_2-CH_2-CH_2-CH_2-CH_2-CH=CH-CH_2-CH_2-CH_2-CH_2-CH_2-CH_2-CH_2-CH_3$$

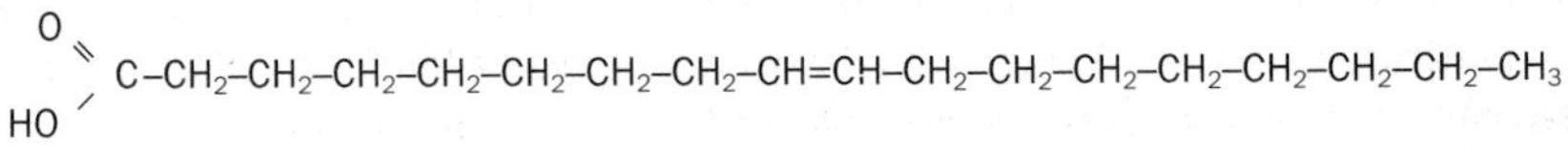

Fatty acids

feces: the waste products removed from the body during *egestion*. The main function of the last part of the gut or colon of a mammal is to remove water from the food that remains in the gut. This undigested material, together with dead cells which have been scraped off the gut wall and bacteria, form the bulk of the feces. The distinctive color is due to the presence of bile pigments.

female: the organism, or part of the organism, which is responsible for producing female sex cells or *gametes*. From whatever organism they come, female gametes have three characteristic features:

* they are large in size when compared to male gametes

* they are produced in much smaller numbers than the male gametes

* they do not move. The male gametes either swim or are moved in some way to the female gametes.

fermentation: a process which occurs in the absence of oxygen in which small organic molecules are produced from larger ones. In animals, in the absence of oxygen, glucose is converted to lactate. This is sometimes called *anaerobic respiration*, but it is an example of a fermentation. Fermentation by *microorganisms* has been exploited by humans for many thousands of years. The ancient Egyptians, for example, used yeast to make beer. With modern techniques of *biotechnology*, fermentations by microorganisms have been used to produce a variety of different products, ranging from foodstuffs, such as yogurt and cheese, to drugs such as the *antibiotic*, penicillin.

fertilization: the process which involves the fusion of male and female *gametes* to produce a *zygote*. In flowering plants, care must be taken to distinguish this process from *pollination* which is simply the transfer of pollen from the stamen to the stigma. Fertilization may take place outside the body, when it is known as *external fertilization*, or inside the body of the female, when it is known as *internal fertilization*.

fertilizer: a substance which adds mineral ions to the soil. There are two types of fertilizer. Chemical fertilizers are made either from naturally occurring rocks or by industrial processes. The main soil nutrients found in these fertilizers are nitrogen, phosphorus and potassium (NPK) which are usually present in fairly concentrated form. Organic fertilizers include farmyard manure and compost. They are usually slower acting as the nutrients they contain must be released by the action of soil fungi and bacteria. In both cases, however, the ions taken up by the plants are exactly the same.

When a crop is harvested, it is taken from the soil and any nutrients that the plants contain are removed as well. Because of this, fertilizers are essential to restore the chemical balance of the soil. Too much can be added, however. A *law of diminishing returns* operates and excess fertilizer does not always lead to extra productivity. In addition, rainfall may cause leaching of excess nutrients into nearby streams and lakes. This may result in *eutrophication*.

hint There is a surprising lack of knowledge of the role of fertilizers and pesticides in agriculture. *Fertilizers* are substances which are applied to the soil to provide crops with the required nutrients. *Pesticides* include herbicides, fungicides and insecticides. They are used to kill pests. Treat the two as entirely separate and don't begin answers with statements such as "Pesticides and fertilizers ...."

fetus: a young mammal in the later stages of development inside the *uterus* of its mother. In humans, the term "embryo" is usually used to refer to the two months following conception. After two months, once it is becoming recognizable as a human, it is called a fetus.

fetal hemoglobin: the type of *hemoglobin* found in the blood of a *fetus*. There is an important difference between the properties of fetal hemoglobin and those of adult hemoglobin. The *oxygen-hemoglobin dissociation* curve for fetal hemoglobin is further towards the left. This means that it can load up with oxygen at lower concentrations. This is important as the mother's blood which arrives at the placenta is not completely saturated. The blood of the fetus, because of its special sort of hemoglobin, however, is able to pick up oxygen and become almost saturated despite this. Shortly after birth, fetal hemoglobin is replaced by the slightly different sort of hemoglobin which is normally found in the blood of adults.

fibrin: a protein which is involved in the process of blood clotting. A protein called *fibrinogen* is present in blood plasma. When a wound occurs, this soluble protein is converted to insoluble fibrin. Fibrinogen can only be converted into fibrin in the presence of the enzyme, *thrombin*. Active thrombin is produced in the presence of a number of factors, which are released at the site of tissue damage. For further details of the process, see *blood clotting*.

fibrinogen: a protein, found in the blood, which is converted to fibrin when blood clots. Fibrinogen is a soluble protein which is made in the liver. When a wound occurs, the enzyme *thrombin* converts fibrinogen to fibrin. It does this by breaking off part of the fibrin molecule. The remainder of the molecule polymerizes to form long insoluble threads of fibrin. These form a mesh over the surface of the wound which traps *red blood cells* and forms a clot.

fibrous protein: a protein whose molecules consist of long chains of *amino acids*. The secondary structure of these proteins is important in determining their shape and they have little or no tertiary structure. Unlike globular proteins, fibrous proteins are insoluble. Fibrous proteins such as *collagen*, *elastin* and *myosin* have important structural functions in living organisms.

Fick's law: states that the rate of diffusion of a substance across a membrane is directly proportional to the surface area of the membrane and to the difference in concentration on either side, and inversely proportional to the thickness of the exchange surface. This can be written in simple mathematical terms as:

$$\text{rate of diffusion} \propto \frac{\text{surface area} \times \text{difference in concentration}}{\text{thickness of exchange surface}}$$

To a biologist, Fick's law provides a very useful framework against which to look at adaptations for diffusion. With gas exchange surfaces, such as the gills of a fish, the rate of diffusion needs to be as great as possible. In order to achieve this, the values on the top line of the expression need to be large, while those on the bottom line need to be small. A careful look at the structure of a gill will show adaptations which produce a large gas exchange surface and maintain as big a difference in concentration as possible. The gas exchange surface, on the other hand, is extremely thin. Considerations like these can be applied to all exchange surfaces. However, the converse also applies. A low rate of diffusion is required, for example, in the case of water loss from plants living in dry areas. Here various adaptations ensure that the figures on the top line will be low while those on the bottom line will be large.

109

hint Whenever you come across diffusion, think about Fick's law, even if it is not on your specification. What are the adaptations that ensure a large surface area, a large difference in concentration and a thin exchange surface?

Filicinophyta: the plant phylum that contains the ferns. Members of the Filicinophyta share the following features:

* like all plants, members of this phylum show an *alternation of generations*. In the Filicinophyta, although the dominant stage in the life cycle is a *sporophyte*, there is also a free-living *gametophyte* called a prothallus

* the young leaves of ferns are tightly coiled. As they grow, they uncoil

* mature leaves usually have spore-bearing structures called sporangia on the undersurface.

filter feeding: a method of feeding on tiny organisms or particles of food suspended in the water. Filter feeders are all aquatic animals. They have a method of producing a current of water which flows over the feeding surface. In molluscs and marine worms, the food is trapped in mucus. *Cilia*, tiny hair-like structures on the *plasma membranes*, then waft this mucus towards the mouth. Larger organisms, for example, some sharks and whales also have filter-feeding mechanisms, although the mechanism by which they trap food particles is somewhat different.

first convoluted tubule: the part of the *nephron* between the renal capsule and the *loop of Henle*. The fluid which has been forced by *ultrafiltration* into the renal capsule contains substances which are useful to the body as well as toxic waste. In the first convoluted tubule, active transport results in useful substances such as glucose, amino acids and mineral ions being reabsorbed into the blood, while most of the urea remains to be excreted in the urine. This is, therefore, a process of selective reabsorption.

hint "First and second convoluted tubule" are easier terms to learn than "proximal and distal convoluted tubule." You will be also less likely to confuse them.

flagellum (bacterial): a long, thin hair-like process on the surface of a bacterium which enables it to move. It is made of molecules of a *protein* called flagellin that are spirally arranged to form a tube. The entire flagellum has a wave-like appearance but it is rigid and does not beat. It is its rotating action that moves the cell along.

flagellum (eukaryotic): a long, thin hair-like process on the surface of a eukaryotic cell. If a cross-section is cut through one of these structures it will be seen to consist of a ring of nine filaments surrounding two more filaments located in the center, the so-called 9 + 2 arrangement. The filaments are made of proteins and enable the flagellum to beat and bring about movement in some single-celled organisms and *sperm* cells. Eukaryotic flagella are sometimes called undulipodia.

florigen: the name given to a plant growth substance which is thought to be responsible for flowering in plants. There are a number of pieces of evidence which support the idea of the involvement of a chemical in this process, such as the fact that changes in day length are detected by the presence of phytochrome in the leaves, while flowers form at the shoot tips. However, as yet, no one has actually been able to find this hypothetical substance.

110

flower: the structure on a flowering plant which contains the parts associated with sexual reproduction. For reproduction to be successful, pollen has to be transferred from the male parts of the same or a different flower to the female parts, a process known as *pollination*. There are a number of ways of doing this and although flowers are often divided into those that are wind-pollinated and those that are insect-pollinated, birds, bats and even water currents may be involved in the process. Flowers are adapted to the way in which they are pollinated. Because of this, there is a lot of variation in structure, but the diagram shows the main parts of a typical flower.

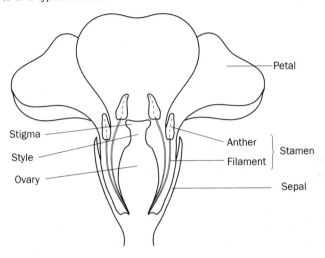

flowering: the process by which a plant forms flowers. The stimulus which normally triggers this is the relative proportion of daylight and darkness occurring over the course of a day. Three types of plants are recognized:

● long-day plants develop flowers when the day length is longer than a certain critical value. They include species which flower in summer, such as poppies and grasses

● short-day plants flower when day length is less than a certain amount. They include plants which normally flower in the spring, like primroses, as well as those like chrysanthemums whose normal flowering time is during the autumn. Tropical plants such as poinsettias are also short-day plants

● day neutral plants such as tomatoes. The length of daylight has no effect on these species.

The flowchart below and on page 112 summarizes the process by which flowering is triggered.

By changing the hours of daylight that a plant receives, it can be made to flower at a different time of the year. In this way, chrysanthemums, which normally flower in the autumn, can be made to flower at any time of the year.

> Light is absorbed by *phytochrome*, a pigment
> found in plant leaves

↓

111

> Phytochrome exists in two different forms, P_{FR} and P_R. After a period in the daylight, most of the phytochrome will have been converted into the form P_{FR}

↓

> If the amount of P_{FR} produced builds up above a certain level, it will either stimulate or inhibit flowering. In short-day plants, it inhibits flowering; in long-day plants, it stimulates it

↓

> It is suggested that a plant hormone passes from the leaves to the shoot tip to bring about flowering. Although this hormone has been called *florigen*, no one has been able to confirm its existence

fluid mosaic model: a model showing the way that molecules are thought to be arranged in a *cell membrane*. Even with an electron microscope it is not possible to see the molecular structure of a cell membrane, so it was necessary to construct a model in order to explain its various properties. Today, the most widely accepted model is the fluid mosaic model, originally suggested by Singer and Nicolson in 1972. The diagram shows the main features of this model.

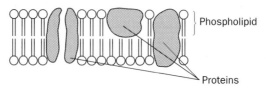

The fluid mosaic model of membrane structure. This is a simplified diagram. There are other molecules present such as cholesterol and glycoproteins, which have been omitted

The diagram is based on a double layer of *phospholipid* molecules within which protein molecules are scattered. Some of these proteins move freely within the phospholipid layers, while others are anchored rather more firmly to structures within the cell. The term "fluid mosaic" refers to the membrane's structure as a patchwork of different molecules which are free to change in position. The protein molecules in the membranes have many important properties among which are the following:

● they may act as *enzymes*. For example, there are important enzymes involved in carbohydrate and protein digestion on the *plasma membranes* of the epithelial cells in the *small intestine*

● they act as carrier proteins which are involved in the processes of *active transport* and *facilitated diffusion*

● they may act as specific receptors for *hormones*.

112

hint If asked to explain why a plasma membrane is described as a fluid mosaic, make sure you explain both why it is fluid and why it is a mosaic.

food chain: a sequence which represents the way in which energy is transferred from one organism to another in a *community*. An example of a food chain would be:

nettle plant $\rightarrow$ large nettle aphid $\rightarrow$ two-spot ladybird

The nettle plant is the *producer*. It converts the energy in sunlight into energy in the organic molecules in the plant. The large nettle aphid is the primary *consumer* that feeds on the nettle, while the two-spot ladybird is the secondary consumer. The arrows in a food chain represent the direction of energy flow. Because energy is lost at each stage by processes such as respiration, it is rare for any food chain to have more than five links.

Food chains are simplifications of what really occurs in a community. There are many different species of animal, for example, that feed on nettle plants, while two-spot ladybirds feed on a number of different species of aphid. Food chains therefore interlink with each other to form food webs.

hint Examiners often see food chains explained like this:

seeds $\rightarrow$ insect $\rightarrow$ bird $\rightarrow$ cat

They do not give much credit for this. There are thousands of different species of insect and most of them do not eat seeds at all. Similarly there are a lot of birds which do not consume insects. If you want to illustrate an answer with an example of a food chain, use a specific one.

food vacuole: a small vesicle found in the cytoplasm of the cells of some organisms which contains food particles. Food vacuoles are found in the cells of organisms, such as amoeba, which feed by *phagocytosis*. The *plasma membrane* engulfs particles of food and these are taken into the cytoplasm. *Lysosomes*, cell organelles which contain digestive enzymes, then fuse with the food vacuole and digest its contents. The products of digestion diffuse from the food vacuole into the cytoplasm.

food web: a diagram which shows the way in which all the different species of organism in a *community* depend on each other for food. A food web is made up of many interconnected *food chains*.

follicle stimulating hormone: see *FSH*

foramen magnum: the opening in the skull through which the spinal cord leaves the brain. It is where the vertebral column meets the skull. The position of the foramen magnum is important in interpreting evidence concerned with human evolution. It can be used to determine whether or not a fossil skull comes from an animal that was *bipedal*, walking on two legs, or quadrupedal and walked on all fours. In bipedal animals, the foramen magnum lies in a position underneath the skull; in quadrupeds, it is farther towards the back.

foramen ovale: a hole situated in the wall between the right *atrium* and the left atrium in the heart of a fetus. It is covered by a membranous valve. While a fetus is developing inside the mother's uterus, its oxygen supply comes from its mother via the placenta. Its lungs do not function. Blood entering the right atrium of the heart opens the valve and some passes through the foramen ovale into the left atrium and on into the aorta. The remainder goes into the pulmonary artery but is not transported to the lungs. Instead it goes through the *ductus arteriosus* into the aorta. At birth, the foramen ovale and the ductus arteriosus both close so all the blood then flows from the right side of the heart into the pulmonary artery and on to the lungs.

fovea: part of the *retina* of the eye which has no rod cells but very large numbers of cones. When attention is attracted to a particular object, the eye normally moves so that the image falls on the fovea. Because it has a large number of cone cells, it is the region of highest visual acuity. In other words, it is the part of the retina which enables the greatest degree of detail to be seen.

frame quadrat: see *quadrat*

fructose: a hexose sugar – that is a monosaccharide with six carbon atoms in each of its molecules. Fructose has the same molecular formula as *glucose*, $C_6H_{12}O_6$, but the atoms that make up the molecule are arranged in a different way. It is found naturally in many fruits and, together with glucose, forms the disaccharide, *sucrose*. Fructose is an important constituent of *diabetic* diets as it tastes sweet but its metabolism does not depend on *insulin*.

fruit: the structure which develops from an ovary after fertilization in a flowering plant. It encloses the *seeds*. Fruits are mainly concerned with the dispersal of seeds, and many have adaptations which ensure that the seeds are dispersed at a considerable distance from the parent plant. Two important methods of dispersal in plants are:

* animal dispersal. Many ripe fruits have succulent flesh and are brightly colored. Animals eat them and while the soft tissue of the fruit is digested, the harder seeds resist the action of the enzymes in the gut. Consequently, they are deposited back on the soil in the animal's feces. Other fruits have hooks or spines which result in them becoming temporarily attached to the fur or feathers of a passing mammal or bird.

* wind dispersal. Fruits often have devices such as tufts of hair or wings that increase their surface area and enable them to be blown by the wind. The fruits of other plants are capsules which move in the wind, shaking out large numbers of tiny seeds.

The development of fruits is controlled by *plant growth substances*.

FSH (follicle stimulating hormone): a hormone which is produced by the anterior lobe of the pituitary gland. In females, it is produced during the first part of the reproductive cycle. It travels in the blood to its target organ, the *ovary*, where it stimulates one or more of the immature follicles to develop. FSH is also produced in males, where it is necessary for the development of sperms.

fungi: the kingdom containing fungi (see *classification*). As well as mushrooms and toadstools, this kingdom contains a variety of smaller moulds and yeasts. Fungi are *eukaryotic* organisms which share the following features:

* as they do not possess chlorophyll, they are all *heterotrophic*. Many are *saprophytes* but a considerable number are *parasites*

* a mature fungus is made up of a number of thread-like structures called *hyphae*. These hyphae form a web known as a mycelium

* they have cell walls but these are made of *chitin*, not *cellulose*, as in plants. The main storage carbohydrate is the *polysaccharide*, *glycogen*

* they reproduce by means of spores.

galactose: a hexose sugar – that is a *monosaccharide* with six carbon atoms in each of its molecules. Galactose has the same molecular formula as *glucose*, $C_6H_{12}O_6$, but the atoms that make up the molecule are arranged in a different way. *Lactose*, the sugar found in milk, is a *disaccharide* formed by condensation of a molecule of glucose and a molecule of galactose.

gall bladder: a small muscular sac or bag found just under the liver. It stores *bile* between meals. Bile is produced in the *liver* and passes into the gall bladder. It is released through the bile duct into the first part of the small intestine. Release of bile from the gall bladder is controlled by a hormone, *cholecystokinin*.

gamete: a general name for a sex cell. Gametes are *haploid* cells which are formed by *meiosis* although, in plants and certain other organisms, there are various mitotic divisions between meiosis and the mature gamete. In *sexual reproduction*, gametes fuse to produce a diploid *zygote*. In most higher organisms, gametes are usually regarded as being either male or female. Male gametes move to the female ones and are generally smaller and produced in larger numbers.

gametogenesis: the process by which *gametes* are formed. In a mammal, gametogenesis takes place in the ovaries or the testes. Gamete-forming cells first divide by means of *mitosis*. Some of the cells that are formed simply go to replace these gamete-forming cells. Others develop into gametes. In a process which involves growth, *meiosis* and maturation, they form sperm or egg cells. In a flowering plant, gamete formation is more complicated but still involves meiosis. Although the detailed processes are different in animals and plants, there is an important similarity. In males, each gamete-forming cell develops to produce four individual daughter cells. In females, however, not all the products of meiosis go on to give rise to gametes. Usually only a single mature gamete will develop from each meiotic division. This leads to the production of large numbers of small gametes in males and small numbers of large gametes in females.

gametophyte: the *haploid*, gamete-producing stage in the life cycle of a plant. Plant life cycles show an *alternation of generations* in which a gamete-producing stage or gametophyte alternates with a spore-producing stage or *sporophyte*. The gametophyte produces the gametes which, in turn, will fuse and produce a sporophyte. In primitive plants, such as mosses and liverworts, the gametophyte is the dominant stage. The familiar green moss plant is therefore a gametophyte and will eventually produce either male or female gametes. In flowering plants, however, the sporophyte is the dominant stage in the life cycle and the gametophyte is very much reduced. Both male and female gametophytes are represented by a few nuclei found within the flower.

ganglion (plural ganglia): nervous tissue consisting of a small group of nerve-cell cell bodies and *synapses*. The function of a ganglion is mainly one of integration. The nervous

115

system in many invertebrate animals is made up of a nerve cord and a number of ganglia. In an earthworm, for example, there is a ganglion in each of the body segments. The nervous system of a mammal has a totally different structure, but there are ganglia in the *autonomic nervous system*. Additionally, the cell bodies of the *neurones* which convey sensory information from receptors to the spinal cord are found in the dorsal root ganglia.

gastric: an adjective meaning "to do with the *stomach*." *Gastric juice*, for example, is the digestive secretion produced by the stomach, while gastric glands are the glands in the lining of the stomach which are responsible for the secretion of gastric juice.

gastric juice: the secretion produced by glands in the lining of the stomach. Its main function is digestion and it contains a number of digestive enzymes. The most important of these is *pepsin* which acts on proteins to break them down into smaller polypeptides. It also contains *mucus* and hydrochloric acid. The hydrochloric acid has a number of functions:

* it converts *pepsinogen*, the inactive form in which the enzyme pepsin is secreted, to pepsin

* pepsin works best in acid conditions and has an optimum pH of between 1.6 and 3.2. Hydrochloric acid ensures that this pH is maintained in the stomach

* it kills some of the bacteria which may be ingested with food.

gastrin: a *hormone* produced by certain cells in the wall of the stomach. The flowchart shows how gastrin secretion is stimulated and its main effects.

> * The presence of amino acids and other products of protein digestion in the stomach
> * Stimulation by the *vagus* nerve
> * Distension of the stomach following a meal

$$\downarrow$$

> Secretion of gastrin into the blood

$$\downarrow$$

> Stimulates the production of acid and *pepsin* by glands in the wall of the stomach

gender-appropriate behavior: the different male and female roles in society. Sex differences are the fundamental biological differences between individuals. The possession of a penis and testes in a male or of a vagina and ovaries in a female are sex differences. Gender-appropriate behavior is more complex. It arises as a result of what a society expects and requires of its males and females. In primitive hunter–gatherer communities, for example, the role of the male was mainly as a hunter, while the female looked after the children and gathered much of the food which was obtained from plant sources.

gene: a length of *DNA* which occupies a specific position on a chromosome called a *locus*. The sequence of nucleotide bases which makes up a gene gives it a particular function. This usually means that each gene will code for a specific protein or polypeptide. In humans, for example, there are genes which code for a variety of different proteins including all the various *enzymes* which control the biochemical processes in cells, *hormones* such as insulin and growth hormone, and other proteins such as channel proteins in *plasma membranes* and *hemoglobin*. An individual gene may have more than one form that differs slightly from the others in the sequence of nucleotide bases. These different forms of a gene are known as *alleles*.

gene bank: see *genetic conservation*

gene mutation (point mutation): a change to one or more bases in the DNA that forms an organism's genetic material. There is a change in the *genotype* which may be inherited. Mutations occur randomly and any part of the DNA present in an organism may mutate. The rate at which mutation takes place can be increased by exposure to various *mutagens*, such as ultraviolet radiation, X-rays and a wide range of organic chemicals.

There are various forms of gene mutation. The table shows two of these.

Normal base sequence on the coding strand of DNA		G C A T T C C A G
Type of mutation		
Substitution	Replacement of one base by another	A C A T T C C A G
		G C C T T C C A G
Deletion	Removal of a base	C A T T C C A G

The normal base sequence shown here consists of nine *bases* which code for three amino acids: GCA codes for arginine, TTC for leucine and CAG for valine. Look at the first example of a mutation where there has been a substitution. The first base in the original sequence, guanine, has been replaced by adenine. The first three bases now read ACA, not GCA. This codes for a different amino acid, cysteine. This mutation will therefore result in cysteine replacing arginine in the protein which is produced. The second example of a substitution is a little different. Although the third base in the original sequence, adenine, has been replaced by cytosine, the altered sequence, GCC, also codes for arginine. To summarize, a mutation involving the substitution of a single base may affect a single amino acid in the protein for which it codes. It may have no effect at all.

Where a deletion is involved, however, there may be a much greater effect. In the example above, the first base, guanine, has been lost. This results in a change to all the following codes. The first six bases on this length of DNA will now code for two different amino acids. CAT codes for valine and TCC for tryptophan. A mutation involving deletion, therefore, can affect the rest of the base sequence and, therefore, all of the following amino acids.

gene pool: the total of all the alleles present in a particular population at a given time. If no selection takes place, the proportions of the *alleles* of a particular gene in a population will remain constant from one generation to the next. In such a population, it is possible to use the *Hardy-Weinberg expression* to predict the proportions of individuals with particular genotypes. In most populations, however, selection does take place. Some of the organisms in the population will have alleles which mean that they are better adapted to particular conditions. These will survive and breed, passing these alleles on to the next generation. This will produce a change in the proportion of alleles in the gene pool and may result in evolution.

gene probe: see *DNA probe*

gene therapy: involves the treatment of inherited conditions by altering a person's genetic makeup. There are three basic ways in which this might be attempted. The faulty gene could be repaired, it could be replaced by a normal gene or it could be left in place and a normal gene added as well. At present the most likely approach is the third of these options, to leave the faulty gene in place and add a normal one. Some experimental work has been carried out on treating the condition of *cystic fibrosis* in this way. Cystic fibrosis patients lack the *plasma membrane* protein that transports chloride ions. As a result they produce thick mucus in their lungs which has to be dispelled by physiotherapy. By introducing the gene for normal chloride ion transport proteins into the patient, it is hoped to be able to treat the condition successfully. Treatment like this involves the normal body cells. Changes cannot be inherited by the offspring. Gene therapy aimed at egg or sperm cells is not regarded as being ethically acceptable, as unknown and possibly harmful effects might be passed on in this way.

generator potential: the change which occurs in the electrical charge across the *plasma membrane* of a receptor as the result of a stimulus. There are many different sorts of receptor but most of them work in a similar way:

* at rest, there is a difference in electrical charge across the plasma membrane of a receptor. A sodium pump helps to maintain this difference

* if an appropriate stimulus is given to the receptor, the sodium pump stops working. This is a localized effect and it allows ions to pass through the membrane, producing a change in the electrical charge. This change is called a generator potential

* the bigger the stimulus, the bigger the generator potential. If the generator potential is big enough it will produce an *action potential* in the sensory nerve cell which leads from it.

genetic code: the way in which information about the sequence of amino acids which make up a protein is coded for by the bases on a molecule of *mRNA*. The genetic code has a number of characteristics:

* it is a triplet code. Each amino acid is coded for by a sequence of three bases. The reason for this is not difficult to appreciate. If each of the four bases found in an mRNA molecule coded for an individual amino acid, then it would only be possible to code for a total of four amino acids. If a group of two bases coded for an amino acid, then there would be 4×4 or 16 possible combinations. Since there are twenty amino acids which are found in proteins, this would not be enough. A group of three bases allows $4 \times 4 \times 4$ or 64 combinations. This is enough to code for all twenty amino acids

* the code is universal. The base sequence CGU, for example, codes for the amino acid, arginine, in all organisms. *Genetic engineering* would be impossible without this property. It has allowed us to take a gene from one organism and to insert it into an organism of a completely different species where it will continue to function in exactly the same way

* it is a *degenerate code*. Since there are 64 different combinations of three bases and only 20 different amino acids, it follows that some combinations will code for more than one amino acid. For example, the combinations, CGA, CGC, CGG and CGU all code for the amino acid, arginine.

118

- it is a non-overlapping code. Each sequence of three bases codes for a separate amino acid. There is no overlap in the coding sequence. Until recently, this was thought to be true of all organisms. Recently, overlapping codes have been found in the genes of some *viruses*.

genetic conservation: maintaining a range of genetic variation. Modern agriculture requires varieties which will be productive and profitable. As a consequence there has been a tendency towards fewer, more specialized breeds which have little genetic variation. If circumstances change, it is possible that genes which would normally allow the evolution of a better adapted plant or animal will no longer be present. Genetic conservation is the use of suitable techniques to maintain stocks of wild organisms or older breeds, so that their genetic potential is not lost. Among plants, this is relatively easily done by establishing *seed banks*. With animals, the task is rather more difficult. It is possible, however, to maintain small stocks of older breeds and some success has also been achieved with freezing gametes and embryos. The concept of maintaining genetic variation is also important in conserving wild animals. In the long term, the breeding success of small isolated populations is unlikely to be high enough to save vulnerable species, such as tigers, from extinction.

genetic counseling: informing parents who are at risk about the likelihood and consequences of a child inheriting genetic disease and discussing the various options that are open to them. In this way, they can arrive at an informed decision about a course of action. This will take into account their background and personal beliefs.

genetic engineering: an aspect of biotechnology which involves altering the genetic make-up of an organism. Genetic engineering is a rapidly expanding branch of biology and is now used to make a considerable range of useful substances. In producing a particular protein from genetically engineered microorganisms, a wide variety of different techniques are used, but they generally involve some or all of the stages shown in the flowchart.

Use of enzymes, such as *restriction endonucleases* and *reverse transcriptase*, and techniques, such as *electrophoresis*, to isolate the required gene. Alternatively an artificial gene may be made

↓

Insertion of the gene into a *vector* such as a *plasmid* or a *virus*. This process also makes use of enzymes including restriction endonucleases and *ligases*

↓

Vector used to insert gene into the DNA of the microorganism forming *recombinant DNA*

↓

cont'd

119

> Selecting microorganisms containing this gene
> and letting them reproduce to form *clones*
> all genetically identical and all
> containing the required gene

↓

> Separation of pure product by *downstream processing*

Using similar techniques, it is possible to insert genes into the DNA of plant and animal cells as well as into microorganisms.

genetic fingerprinting: a technique which can be used to distinguish between individuals by looking at similarities and differences in part of their DNA. Not all of the *DNA* present in the nucleus of a cell codes for a particular protein. Some of this "noncoding" DNA consists of short sequences of bases which may be repeated. The actual number of times these sequences are repeated varies from individual to individual. Genetic fingerprinting compares these sequences. The flowchart summarizes the main steps involved in the procedure.

> DNA is extracted from a suitable sample. In
> forensic work, this may be from blood, semen
> or even from the cells left on a cigarette end

↓

> Enzymes called *restriction endonucleases* are
> used to cut the DNA in the sample into smaller
> bits. Some of these will contain the repeated
> base sequences. The actual length of these
> pieces will be determined by the number of
> times that the particular base sequence is
> repeated in it. Few repeats will give rise to a
> small piece of DNA. A large number of repeats
> will produce a longer piece

↓

> The pieces of DNA are separated by
> *electrophoresis*. This produces a sheet of gel
> with bands of DNA arranged on it rather like
> the rungs on a ladder. The distance traveled
> by each piece of DNA will depend on its
> length. The smaller pieces will travel farther
> than the larger ones

↓

> Particular bands of DNA are now located by
> using a *DNA probe*. The position of these
> bands on the gel sheet can be used to
> compare individuals

genetic screening: testing for the allele that causes a particular inherited condition. The analysis of family trees can only provide information about the probability of inheriting a particular condition. It cannot tell whether a particular individual has the condition or not. To do this, genetic screening is necessary. Samples of cells are obtained. These may be gained from an adult by simply washing out the mouth and collecting the cheek cells. It is possible to get fetal cells either by *amniocentesis* or by *chorionic villus sampling*. These cells are then used to make a chromosome preparation and examine the DNA. A *DNA probe* can be used to identify the presence of the mutant allele responsible for the condition.

genetic marker: a gene whose effects are easily recognisable and is therefore used in genetic engineering as a sort of label. Genetic engineering involves isolating a particular gene or synthezising it artificially and putting it into the DNA of another organism. The processes involved are difficult and not very reliable. Furthermore, it may also be difficult to tell if the introduced gene that we are interested in has been successfully taken up by the host organism. Because of this a genetic marker is often used. In bacteria, it is often a gene for antibiotic resistance, whose effects are easily recognizable. If the marker gene is present there is a very high probability the gene we are trying to insert has also been incorporated.

genetically modified (GM): a term used to describe organisms whose *DNA* has been altered by the process of *genetic engineering*. There is considerable public concern over the use of organisms that have been genetically modified, particularly when used in human food. Some of the biological concerns include:

❋ Genes coding for antibiotic resistance are frequently used as *genetic markers* to show that another gene has been successfully incorporated into the DNA of the host organism. These resistance genes could spread to pathogenic bacteria, making them very difficult to control.

❋ Genes for *herbicide* resistance have been introduced into crop plants. Powerful herbicides can then be used which kill all species of weed but leave the crop unharmed. These genes might spread to closely related species of weed making them very difficult to control. The killing of all other species of plants will also have serious ecological consequences.

❋ Foreign proteins resulting from the use of genetic engineering techniques may act as *antigens* and increase the likelihood of *allergy*.

genome: all the genetic material in an organism. The term is usually used to describe the *genes* contained in its *chromosomes* but it can also be applied to the genes on the nucleic acid of a *virus* or to the genes in the DNA of *mitochondria* and *chloroplasts*.

genotype: the genetic makeup of an organism. The genotype describes the organism in terms of the *alleles* that it contains. There are some simple rules that should be followed in choosing the letters that will represent these alleles. They may seem complicated, but these conventions are followed by all examiners.

❋ A single letter should be chosen to indicate the gene, with the capital representing the dominant allele and the small letter, the recessive allele.

❋ If possible, the letter chosen should relate to one of the phenotypes. This makes it easier when it comes to translating genotypes into phenotypes when solving genetics problems.

* The letter should be chosen carefully so that there will not be any confusion between capital and small. It is best to avoid letters, such as c, o and s, where the only difference between the capital and the small is one of size.

* Some genes have more than two alleles (see *multiple alleles*). Where this is the case, the gene should be given a capital letter and each allele should be represented by an appropriate letter in superscript. So with human blood groups, I represents the gene and the alleles are I^o, I^A and I^B.

* Where there is *codominance*, the gene is represented by a single letter and the alleles as capitals written as superscripts. For example, some flowers are either white or red, if they are homozygous for a particular gene. If they are heterozygous, they are pink. In this case, the two alleles concerned are written as C^R and C^W.

genus: a level of classification. One or more different *species* belong to a genus. The scientific name given to a species contains two words. The first word gives the name of the genus and the second, the name of the species. *Rattus norvegicus* and *Rattus rattus* are both rats. Although they have been classified as different species, they are similar in many ways. They are both members of the genus *Rattus*.

geotropism: a growth movement of a whole plant organ in response to gravity. Roots are usually positively geotropic and, as a result, grow down into the soil, growing away from the light. Stems are generally negatively geotrophic and grow upwards. For further detail, see *tropism*.

germination: the process during which the food reserves present in a seed are broken down and the embryo starts to grow. Germination begins when a *seed* starts to absorb water. The carbohydrates, proteins and lipids which form the food store are broken down by enzymes. The root of the embryo or radicle emerges and starts to grow down into the soil, while the shoot or plumule grows upwards towards the light. Once the seedling starts to photosynthesize, germination is over.

Germination is controlled by two *plant growth substances*, *abscisic acid* and *gibberellin*. As a seed ripens, it loses water. This process is controlled by abscisic acid. Many seeds do not germinate as soon as they are produced. They undergo a period of dormancy during which they are prevented from germination by the abscisic acid still present in the seed. If they are exposed to particular conditions, such as a period of heavy rain in desert plants or the cold weather of winter in temperate species such as apples, the level of abscisic acid falls and germination can begin. At the same time, there is an increase in gibberellin. This switches on the genes which are involved in the production of the enzymes that bring about the breakdown of the food reserves of the seed.

gibberellin: a *plant growth substance* which has a number of different functions, one of the most important of which is the promotion of stem elongation. Gibberellins are also involved in *germination*. The flowchart opposite explains the part they play in this process.

Gibberellins and a knowledge of the way in which they work are important commercially:

* gibberellins may be used to produce seedless fruit such as grapes

* they can be used to alter the amount of space between individual grapes in a bunch. Since this results in the fruit being farther apart, they are not as likely to be damaged by fungal infections

* some substances are known to prevent normal plants being able to produce gibberellins. One of these substances is known as CCC. This is widely used to produce dwarf plants.

> A germinating seed such as a barley grain absorbs water

$\downarrow$

> Gibberellin diffuses from the embryo into the aleurone cells. These are the cells that surround the food reserves in the grain. The gibberellin switches on genes and the aleurone cells make large amounts of the enzyme amylase

$\downarrow$

> The amylase hydrolyzes the starch stored in the grain to produce sugars. These are used by the growing embryo

gill: an organ concerned with the exchange of respiratory gases. Gills are found in a variety of animals such as fish and certain insects which live in water. In fish, water is taken in through the mouth and pumped over the gill surface. Oxygen diffuses from the water into the blood, while carbon dioxide diffuses in the opposite direction. The gills show a number of adaptations which make them very efficient gas-exchange surfaces:

* there is a large surface area over which *diffusion* can take place. This is provided by the huge number of thin gill plates. Each gill plate has large numbers of *capillaries* which supply it with blood. These blood vessels provide a very large combined surface area

* the difference in the concentration of respiratory gases is maintained as high as possible. A *ventilation* mechanism ensures that a continual flow of oxygen-containing water is brought to the respiratory surface. The circulating blood, on the other hand, is constantly bringing to the gills blood which is low in oxygen and removing that which is rich in oxygen. In addition, a *counter-current system* helps to maintain an oxygen gradient between the water and the blood

* the total distance from the water surrounding the gills to the blood in the capillaries is very small. As a consequence there is a very short diffusion pathway.

> **hint** When describing the adaptations of gills for efficient gas exchange, use *Fick's law* to provide you with a framework for your answer. What are the adaptations that ensure a large surface area, a large difference in concentration and a thin exchange surface?

gland: a group of cells responsible for the secretion and release of a particular substance. *Endocrine* glands secrete *hormones* which always go directly from the cells into the blood. Other glands, however, pass their secretions into a duct. These glands are called *exocrine* glands. Salivary glands are exocrine glands. The *pancreas* is somewhat unusual, as it is

123

both an endocrine gland and an exocrine gland. It secretes the hormones *insulin* and *glucagon* which it releases directly into the blood. It also secretes pancreatic juice. This is released into the pancreatic duct from where it is carried to the small intestine. Like muscles, glands are *effectors*.

global warming: the rise in global temperature which is thought to have been brought about by an increase in the concentration of gases, such as carbon dioxide and methane, in the atmosphere. Processes such as the burning of fossil fuels, the clearing of forests and the weathering of limestone have led to an increase in carbon dioxide from about 270 parts per million two hundred years ago to approximately 350 parts per million today. Methane levels have also risen considerably. Radiation reflected from the earth's surface is trapped by these and other gases present in the atmosphere preventing the escape of heat energy. Carbon dioxide and methane are particularly effective "greenhouse gases" and an increase in their concentrations is thought to have resulted in an increase in the mean global temperature. It is difficult to predict exactly how global warming will affect biological processes such as crop production, but there is evidence to support the following suggestions:

* increased carbon dioxide and temperatures could lead to a higher rate of photosynthesis in some plants; it may have no effect or even lead to a lower rate in others. Rate of photosynthesis affects crop yield

* the life cycle of many insect pests is shorter at higher temperatures. This could lead to significantly greater damage to crops

* transpiration rate increases with temperature. This combined with major shifts in world climate could have a significant effect on crop distribution.

globular protein: a *protein* whose molecules consist of a long chain of *amino acids,* forming a polypeptide which is then tightly folded into a compact three-dimensional shape. It is the tertiary structure of these proteins that is most important in determining their shape. Unlike *fibrous proteins,* globular proteins are generally soluble. *Enzymes* and *hormones* such as *insulin* and *glucagon* are globular proteins.

globulin: a particular type of protein in blood plasma. Some globulins are made in the *liver.* They have a number of functions, such as transporting iron, lipids and certain hormones around the body. Other globulins are made by *lymphocytes* and are involved in the body's immune response. They are *antibodies.*

glomerulus: a small ball of capillaries inside a *renal capsule* in a *kidney.* The glomerular *capillaries* differ slightly in structure from those which are found in other organs in the body. The endothelial cells which line them are perforated by small pores. This makes them more permeable than capillaries found elsewhere. The pressure of the blood in the glomerular capillaries is also quite high. There are two reasons for this. First, the afferent arteriole which brings the blood from a branch of the renal artery is short and straight, so there is only a small drop in pressure. As well as this, the efferent arteriole through which the blood leaves the glomerulus is slightly smaller in diameter than the afferent arteriole which brings blood in. The pressure of the blood in the capillaries that make up the *glomerulus* forces fluid through the pores in the walls and the basement membrane into the space in the *renal capsule.* The pores are mainly responsible for determining what can be passed out of the blood to form the *glomerular filtrate* during this *ultrafiltration* process. Blood cells and protein molecules are too large to pass through.

glomerular filtrate: the liquid formed from filtering the blood into the kidney tubule. The pressure of the blood in the capillaries that make up the *glomerulus* forces fluid through the basement membrane into the space in the *renal capsule*. This fluid has the same composition as blood *plasma* except that it does not contain protein. Protein molecules are too large to go through the basement membrane.

glucagon: a hormone produced by the *islets of Langerhans*, a group of *endocrine* cells in the pancreas. The secretion of glucagon is stimulated by a fall in blood glucose concentration. The main effect of the hormone on the body is to activate enzymes in the liver which are responsible for the conversion of glycogen to glucose. It also stimulates the formation of glucose from other molecules such as amino acids.

hint Spell "glucagon" and "glycogen" correctly. Glucagon is a hormone and glycogen is a polysaccharide. There are no such substances as glucagen or glycogon!

gluconeogenesis: a biochemical process in which glucose is made from molecules which are not carbohydrates. The concentration of *glucose* in the blood is normally about 90 mg 100 cm^{-3}. Some of the organs in the body, such as the brain, can only use glucose as a respiratory substrate, so it is very important that the blood glucose concentration is maintained at a constant level. However, it will fall following a period of exercise or if a person has gone without food for any length of time. Under these conditions, the pancreas secretes *glucagon*. This hormone not only stimulates the liver to convert stored *glycogen* to glucose, but it brings about an increase in gluconeogenesis. Glucose is formed from molecules such as the *lactic acid* produced during *anaerobic respiration* and from amino acids.

glucose: a hexose sugar – that is, a *monosaccharide* with six carbon atoms in each of its molecules. Carbohydrates like this have the molecular formula $C_6H_{12}O_6$. The carbon atoms in this molecule are able to join up to form a ring-shaped structure. There are a number of different ways of arranging the other atoms and this results in different *isomers* being formed. Two of these isomers are α-glucose and β-glucose. They are shown in full and in simplified form in the diagram below.

α-glucose

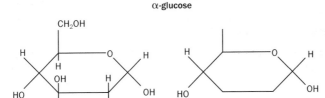

β-glucose

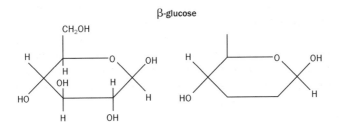

125

α-glucose is the substance commonly referred to as glucose. It is important as the molecule which is broken down to release energy in *respiration* and as the building block from which *starch* and *glycogen* are made. β-glucose is the sugar unit found in *cellulose*.

hint All high level students know that a glucose molecule contains six carbon atoms ... until they get into the examination.

glucose oxidase: an *enzyme* which catalyzes a reaction involving the oxidation of *glucose*. The reaction involved can be represented by the equation:

glucose + oxygen $\rightarrow$ gluconic acid + hydrogen peroxide

This reaction is used in biosensors which can be used to detect and measure the concentration of glucose in a solution.

glycerate 3-phosphate, GP: the three-carbon molecule which is formed when ribulose bisphosphate, RuBP, combines with carbon dioxide in the *light-independent reaction* of photosynthesis. Ribulose bisphosphate is a five-carbon molecule. It is able to combine with carbon dioxide to produce two molecules of glycerate 3-phosphate. Glycerate 3-phosphate is then reduced to triose phosphate. This requires *ATP* and reduced *NADP*, two substances which are produced in the *light-dependent reaction*.

Glycerate 3-phosphate is also formed during the process of respiration. It is an intermediate substance in the biochemical pathway of *glycolysis*.

glycogen: a storage carbohydrate found in animals. Glycogen is a *polysaccharide* formed by linking large numbers of *α-glucose* molecules into branched chains which are a characteristic feature of the structure of a glycogen molecule. In a mammal, large amounts of glycogen may be found in the liver. Some is also stored in muscles.

hint Spell "glycogen" and "glucagon" correctly. Glycogen is a polysaccharide and glucagon is a hormone. There are no such substances as glucagen or glycogon!

glycogen loading: increasing the amount of glycogen present in muscles before an athletic event by eating carbohydrate-rich meals. Exercise involves muscle contraction and this requires large amounts of energy in the form of *ATP*. *Respiration* produces ATP and to maintain a high rate of respiration in muscles for a considerable length of time, a good store of a suitable *respiratory substrate* is necessary. By eating carbohydrate-rich foods such as pasta before the event, more glycogen is stored in the muscles. This reserve can then be respired during the event and may lead to an improved performance.

glycogenesis: the conversion of glucose to glycogen. When the blood glucose concentration rises following a meal, *insulin* is produced by cells in the *pancreas*. This hormone has a number of effects on the body, all of which tend to reduce the blood glucose concentration. One of these is to activate enzymes which are responsible for the conversion of *glucose* to *glycogen*. Most tissues in the body contain some glycogen, but the main stores are in the muscles and the liver.

glycogenolysis: the conversion of glycogen to glucose. The blood glucose concentration will fall following a period of exercise or if a person has gone without food for any length of time. Under these conditions, the pancreas secretes *glucagon*. This hormone activates enzymes which are responsible for the conversion of *glycogen* to *glucose*. This process is also stimulated by *adrenaline*.

glycolipid: a substance whose molecules consist of a *lipid* and a *carbohydrate*. They form part of the *plasma membrane* surrounding a cell. Glycolipids are located on the outer surface of this membrane with the lipid parts of the molecules forming part of the lipid bilayer and the branched carbohydrate portions sticking out like antennae. Glycolipids have similar functions in membranes to *glycoproteins*. They help cells bind to each other to form tissues and they also allow cells to recognize one another by acting as *antigens*.

glycolysis: the part of the biochemical pathway of respiration in which a molecule of glucose is broken into two three-carbon pyruvate groups. Glycolysis is the first part of the respiratory pathway. It takes place in the cytoplasm of the cell and is common to both *aerobic respiration* and *anaerobic respiration*.

In glycolysis, phosphate groups are first added to the glucose molecule. In other words, the glucose molecule is phosphorylated. This phosphate is supplied by *ATP*. The phosphory-lated sugar is then broken down to give two pyruvate groups. The breakdown process produces a total of four ATP molecules so, altogether, there is a net gain of two ATP molecules. The complete reaction is an oxidation reaction and releases hydrogen. This is removed and used to reduce a *coenzyme* known as NAD. The complete process of glycoly-sis is summarized in the diagram.

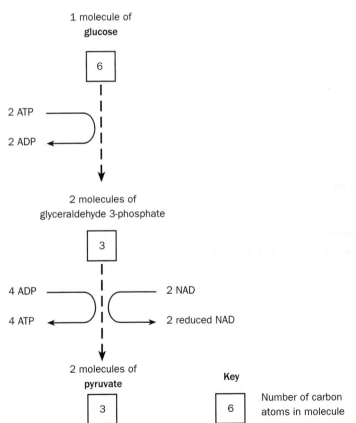

glycoprotein: a substance whose molecules consist of a *protein* and a *carbohydrate*. They form part of the *plasma membrane* surrounding a cell. The protein parts of the molecules penetrate part or all the way through the membrane while the branched carbohydrate portions stick out like antennae. Glycoproteins have similar functions in membranes to *glycolipids*. They help cells bind to each other to form tissues and they also allow cells to recognize one another by acting as *antigens*. Mucin is an important glycoprotein. It is the main component of mucus.

glycosidic bond: a chemical bond formed as a result of condensation between two *monosaccharides*. When two *glucose* molecules are joined in this way, a glycosidic bond is formed between carbon atom 1 of one molecule and carbon atom 4 of the other. This produces what is called a 1,4 glycosidic bond. *Polysaccharides* such as *amylose* are formed in this way. Their molecules consist of long, straight chains. Glycosidic bonds can also form between carbon atoms 1 and 6. Where this happens, such as in *amylopectin* and *glycogen*, molecules with branched chains are formed.

goblet cell: a type of cell which produces *mucus*. Goblet cells are pear-shaped cells which are found in the lining *epithelium* of, for example, the alimentary canal, the trachea and bronchi and parts of the reproductive system.

Golgi apparatus: an organelle which is responsible for the processing and packaging of substances produced by a cell. When seen with an *electron microscope*, the Golgi apparatus consists of a series of flattened sacs, each one enclosed by a membrane. These sacs are continually being formed on one side and pinched off to form small *vesicles* at the other. Cells which are involved in the secretion of substances contain particularly large amounts of Golgi apparatus. A variety of different processes are associated with this organelle:

* formation of *glycoproteins*. These are molecules made by combining a protein and a carbohydrate. An example of a glycoprotein is mucin. This is an important component of the mucus produced by the cells which line the gut, reproductive and respiratory systems

* packaging and secretion of particular proteins, such as the digestive enzymes produced by cells in the *pancreas*

* secretion of carbohydrates which make *cell walls* in plant cells

* formation of *lysosomes*.

gonad: a general name given to a gamete-forming organ in an animal. The gonads are the *ovaries* and the *testes*.

Graafian follicle: see *ovarian follicle*

Gram staining: a method of staining bacterial cells which enables two distinct types to be distinguished. The flowchart below summarizes the main steps in the process.

Bacteria smeared on a clean microscope slide and stained purple with crystal violet solution

↓

Slide washed with Gram's iodine solution then with ethanol

↓

Slide stained with the red stain, safranin

Gram-positive bacteria retain the purple dye when they are washed with ethanol. This hides the effect of the safranin, so they appear purple in color. Gram-negative bacteria lose the purple dye but take up the red one. They appear red. Whether or not the bacteria retain the purple dye depends on the structure of their cell walls. Gram-positive bacteria have thick cell walls; the cell walls of Gram-negative bacteria are much thinner.

grana: the stacks of membranes, looking rather like piles of coins, found inside *chloroplasts*. These membranes are associated with the light-dependent reaction of *photosynthesis* and contain the *chlorophyll*, *enzymes* and *carrier molecules* associated with this process.

granulocyte: a type of white blood cell or leukocyte which has a lobed *nucleus* and granular *cytoplasm*. The table summarizes the key features of the three main types of granulocyte.

Type of cell	Main features	Function
Neutrophil	The commonest type of granulocyte with fine granules in its cytoplasm	Engulf bacteria by phagocytosis
Eosinophil	Large granules stain pink with a stain called eosin	Involved in allergic responses
Basophil	Large granules stain with basic stains	Produce substances which cause inflammation, such as histamine

greenhouse effect: the way in which heat energy is trapped in the atmosphere leading to a warming of the environment. Some of the radiation reaching the earth from the sun has short wavelengths. It is reflected by the surface of the earth as radiation with longer wavelengths. This is absorbed by various gases in the atmosphere and warms them up. A number of the gases which are normally present in the atmosphere absorb radiation with longer wavelengths. These include water vapor, ozone, carbon dioxide and methane, but some of them are much more effective than others. The rise in carbon dioxide and methane levels brought about by human activities cause particular concern, because they readily absorb the reflected radiation. It is thought that this might lead to *global warming*.

grey matter: the inner layer of the *spinal cord*, shaped rather like a letter H, which is surrounded by the *white matter*. Grey matter consists of the cell bodies of nerve cells and unmyelinated axons.

gross primary production: the total amount of *energy* in the *biomass* that the *producers* in a *community* form as a result of *photosynthesis*. Gross primary production is usually described as the amount of energy in the plant biomass in a given area in a given time, and is expressed in units such as kJ m^{-2} year^{-1}. This allows a comparison of the production in different ecosystems. The gross primary production is the maximum amount of biomass produced. Some of this biomass goes into growth, the net primary production, but some is used in processes such as respiration which are needed to keep the producers alive.

gross primary production = net primary production + energy lost in respiration

growth of an organism involves the addition of more body substance. This usually means that the organism gets larger. Growth excludes, however, temporary changes such as those which result from *osmosis*. The pattern of growth differs between organisms. In animals it is described as being diffuse, with the process taking part in almost all parts of the body. A human child, for example, grows because most of its organs grow. Not only is a ten-year-old taller than a three-year-old but it has a larger liver, heart, kidneys and so on. Plants, however, tend to have particular regions of growth. These are called *meristems*. In mammals, a number of *hormones* are involved in the process of growth. These include *growth hormone*, *thyroxine* and the sex hormones, estrogen and testosterone. In plants growth is controlled by chemicals known as *plant growth substances*.

growth hormone: a *hormone* produced by the anterior lobe of the *pituitary gland*. Growth hormone controls the *growth* of children but we do not know if it has a function in adults. It acts on the *cartilage* cells at the ends of *bones* and in muscle cells where it stimulates growth.

growth rate: the amount of growth made by an organism in a given period of time. The root of a germinating seed may, for example, grow 9 millimeters in a 24-hour period. Its absolute growth rate would therefore be 9 mm per day. The graph shows the total amount of growth and the absolute growth rate of a human male over the first 18 years of his life.

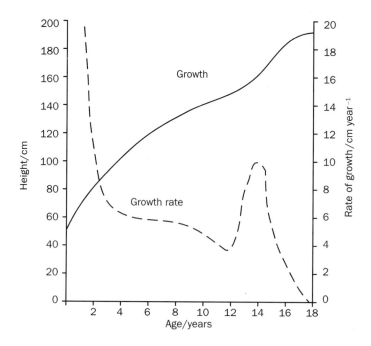

It is worth noting that the curve showing the total amount of growth increases steadily before reaching a maximum where it levels out. This fits in with our common appreciation of the pattern of human growth. We get bigger over the first 18 years of our lives; we don't decrease in size. Growth rate, however, is the amount of growth in a particular time. It can decrease as well as increase.

Growth rate gives a much better idea of what is happening at a particular time. A useful analogy might be a car journey. Growth is like the total distance traveled. It only gives information about how far you have got at a particular time. Growth rate, on the other hand, can be compared with speed. It provides much more information about progress at particular times on the journey.

Relative growth rate is determined in a different way. The table shows some figures relating to the growth of the individual from the graph shown on page 130.

Age/years	Height at beginning of year/cm
1	70
2	85
14	160
15	175

Between 1 and 2 years of age and between 14 and 15, the absolute growth rate is the same. It is 15 cm per year. However, the relative growth rate at the two ages is different. It may be calculated from the formula:

$$\text{Relative growth rate} = \frac{\text{Absolute growth rate}}{\text{Height at beginning of year}}$$

So the relative growth rate between 1 and 2 years of age will be:

$$\frac{15}{1} \text{ or } 15 \text{ cm year}^{-1} \text{ cm}^{-1}$$

while the relative growth rate between 14 and 15 years old will be:

$$\frac{15}{14} \text{ or } 1.07 \text{ cm year}^{-1} \text{ cm}^{-1}$$

guanine: one of the *nucleotide bases* found in nucleic acid molecules. Guanine is a purine which means that it has two rings of atoms in each of its molecules. When two polynucleotide chains come together, guanine always bonds with *cytosine*. The atoms of these two bases are arranged in such a way that three hydrogen bonds are able to form between them.

131

H-zone: a lighter area in the center of the *A-band* of a *myofibril* in a skeletal muscle. It is the area where the thick *myosin* filaments which make up the A-band do not overlap with the *actin* filaments.

habitat: the particular place where a *community* of organisms is found. Rocky seashores, the surface layer of the soil and human skin are all habitats for particular communities of living organisms.

habituation: a form of behavior in which an animal responds less and less to a repeated harmless stimulus. If, for example, a snail is allowed to crawl across a sheet of glass and the glass is tapped, the animal will withdraw into its shell. Eventually it will emerge again and carry on crawling. Tapping the glass again will cause it to withdraw once more into its shell. If this *stimulus* is repeated every time the snail emerges from its shell, eventually it will fail to respond. Other characteristics of this form of behavior include:

* habituation does not last very long if the stimulus is not repeated

* an animal will become habituated more quickly to a second stimulus if this is similar to the first.

Habituation has considerable survival value to the animal concerned. It allows it to adapt to neutral stimuli by not responding. Suppose the snail in the example above always reacted to the tapping stimulus by withdrawing into its shell. Despite the fact that the stimulus proved harmless to the animal, the snail would clearly have no time available for vital activities such as feeding.

haploid: a term which can refer to cells, organisms or stages in the life cycle in which the nuclei contain a single copy of each *chromosome*. The letter *n* is often used to represent the haploid number of chromosomes. In humans, for example, n = 23. In animals and plants, the sex cells or gametes are haploid. On fertilization, the nuclei of the male and female gametes fuse together and the *diploid* number of chromosomes found in the body cells is restored.

Hardy–Weinberg expression: a mathematical expression which can be used to predict the proportions of individuals with particular genotypes in a population. It relies on the basic idea that, in the absence of *selection*, the proportions of the *alleles* of a particular gene in a population will remain constant from one generation to the next. This involves making a number of assumptions. These are:

* it is a large population

* mating is random. For example, individuals with a particular genotype do not tend to mate only with other individuals of the same *genotype*

* there is no selection or *mutation*. This would affect the frequency of particular alleles

132

- there is no pattern of immigration or emigration which would again affect the frequency of particular alleles.

Let us consider a specific example. In humans, the shape of the earlobe is determined by a particular gene. It has two alleles. E is the allele for free earlobes while e is the allele for attached earlobes. If we consider all the earlobe alleles in a particular population, a certain proportion will be the dominant allele, E, and a certain proportion will be the recessive allele, e. We will represent the frequency or proportion of the dominant allele by the letter, p, and the frequency of the recessive allele by the letter, q. Since the earlobe alleles must be either E or e, we can say that:

$$p + q = 1$$

The Hardy–Weinberg expression gives the frequency of the genotypes of the individuals in the population. It is represented by the equation:

$$p^2 + 2pq + q^2 = 1$$

where p^2 is the frequency of the dominant homozygote (in the case of the example given here, individuals with the genotype, EE),

$2pq$ is the frequency of heterozygotes with the genotype Ee,

and q^2 is the frequency of the recessive homozygote, ee.

This formula can be used to calculate the expected frequency of individuals with a particular genotype in a population.

Worked example: In humans, earlobes may be attached to the side of the head or they may hang free. In a population, 51% of all individuals had free earlobes. What percentage of this population would you expect to be heterozygous for the earlobe gene?

Since the allele for free earlobes is dominant to that for attached earlobes:

- individuals with free earlobes will have either of the genotypes EE or Ee. The frequency of these individuals in the population will be 0.51.
- individuals with attached earlobes can only have the genotype, ee. The frequency of these individuals in the population will be $1 - 0.51$ or 0.49.

The Hardy–Weinberg expression is:

$$p^2 + 2pq + q^2 = 1$$

In this expression, the frequency of individuals with the genotype, ee is given by q^2, so, combining this information:

q^2 $= 0.49$

and q $= 0.7$

Since $p + q = 1$

$p + 0.7$ $= 1$

p $= 1 - 0.7$

p $= 0.3$

Now that p and q have been calculated, it is quite simple to find $2pq$, the expected frequency of the heterozygotes in the population:

$2pq = 2 \times 0.3 \times 0.7$

$2pq = 0.42$ or 42% of the population

hint The basic steps in calculations involving the Hardy–Weinberg expression are:
- Work out the frequency of the recessive homozygote. This is q^2.
- The square root of q^2 will give the value of q.
- Find p from the equation $p + q = 1$
- Now put the values of p and q into the Hardy–Weinberg expression to find the information required.

harvestable dry matter: the dry mass of the part of a crop that is normally eaten. This is a useful measure of crop yield for agricultural scientists, since it measures the productivity of the part of the plant that is economically important, such as wheat grain, or potato tubers. The water content of any crop is extremely variable. By measuring dry matter, problems associated with this variable water content are eliminated.

harvestable protein: a measure of the yield of protein that may be obtained from a particular crop. Cereal and root crops, such as rice and cassava, are grown mainly as carbohydrate sources. They have relatively low yields of harvestable protein. The figures for crops of leguminous plants, such as soya beans and groundnuts, are much higher.

hay fever: a type of *allergy* resulting from exposure to *pollen*. The pollen usually responsible comes from wind-pollinated plants such as grasses and trees. Since many of the plants responsible for producing this pollen flower in late spring and summer, hay fever is most frequent at this time of year. The symptoms include sneezing and a running nose produced by the release of histamine.

health: defined by the World Health Organisation as "a state of complete physical, mental and social well-being." Like many everyday words, it is hard to define in precise biological language. Even with this definition it is difficult to know exactly what is meant by a "state of complete well-being." The best approach is probably to list its characteristics. The term health:

- means more than being free of disease
- describes a condition where all of the body's organs and systems are working properly
- takes into account mental as well as physical health.

heart cycle: see *cardiac cycle*

heme: an iron-containing chemical group. A molecule of human hemoglobin is made up of four units. Each of these units contains a heme group and globin, a polypeptide. *Hemoglobin* is able to combine with oxygen to form oxyhemoglobin. In this reaction, the oxygen becomes attached to the iron in the heme group. Heme is also part of another respiratory pigment, *myoglobin*, as well as being found in *cytochromes*, important enzymes associated with respiration.

After about 120 days in the circulation, red blood cells die and the *hemoglobin* which they contain is broken down. The heme group is split off from the globin part of the molecule. The iron from the heme group is used to make more hemoglobin, while the rest of the group is converted into *bile pigments* which are excreted in the bile.

hemocoel: the blood-filled body cavity found in insects and other members of the phylum *Arthropoda*.

hemocytometer: a specially marked slide, originally designed for counting the number of cells present in a sample of blood. The marked surface of the slide is exactly 0.1 mm deep and it has, within this area, a number of very fine lines engraved on the surface. This means that the number of cells in a particular volume of blood can be counted when examined with a microscope. Multiplication by the appropriate factor will give the number of cells per cubic millimeter. Although hemocytometers are being replaced by electronic methods in hospitals, they are extremely useful in biology laboratories for counting *microorganisms* such as bacteria or yeast cells.

hemoglobin: a globular *protein* found in the blood of many animals where it is mainly responsible for transporting oxygen. Each hemoglobin molecule consists of four polypeptide chains and each chain is associated with an iron-containing heme group. It is to these heme groups that oxygen molecules bind. Since there are four heme groups, it follows that a single molecule of hemoglobin can transport four molecules of oxygen. A graph can be plotted of the amount of oxygen carried by the hemoglobin in the blood against the concentration of oxygen present. This graph is called an *oxyhemoglobin dissociation curve*.

There are many different types of hemoglobin which result from the sequence of amino acids which make up their polypeptide chains. They have slightly different oxygen-carrying properties, which can be related to the environment in which the animal lives. In addition to its oxygen-carrying function, hemoglobin is an important *buffer* and helps to keep the blood at a constant pH.

> **hint** When you are asked to explain how oxygen is transported by hemoglobin, make sure that you mention the oxyhemoglobin dissociation curve.

hepatic: an adjective meaning "to do with the *liver*." The hepatic artery, for example, is the blood vessel which supplies oxygenated blood to the liver.

hepatic portal vein: the blood vessel which goes from the small intestine to the *liver*. It plays a very important part in taking the products of *digestion* to the liver for further processing. Most *veins* in the body start from a *capillary* network. Blood drains into larger vessels known as venules and then into the veins. These return the blood to the heart. Portal veins are different, however. Not only do they start with a capillary network but they also end in one. Because of nutrients which have been absorbed from the small intestine, the concentration of dissolved substances, such as glucose in the hepatic portal vein, varies much more than in other blood vessels. After a meal, the amount present can rise to quite a high level, as it is absorbed from the small intestine and taken to the cells of the liver.

> **hint** Make sure you know about the three main blood vessels taking blood to and from the liver:
> * the hepatic artery takes oxygenated blood to the liver
> * the hepatic portal vein takes blood from the small intestine to the liver
> * the hepatic vein returns blood from the liver to the heart.

herbicide: a substance which is used to kill weeds. Weeds are economically very important, as they can severely reduce crop yields by competing for light, water and inorganic ions. From an agricultural point of view, it is therefore very important that they are controlled. There are many different types of herbicide available, but they can be classified by the way in which they act:

- contact herbicides only kill those parts of the plant with which they come into contact. They are not very effective in dealing with the underground roots and stems of weeds such as thistles and nettles

- *systemic herbicides* are absorbed by leaves and roots and transported to all parts of the plant where they cause death by interfering with the natural growth pattern. They have little effect on monocotyledonous plants, such as cereals and other grasses, but they rapidly bring about the death of dicotyledonous weeds. They are very suitable, therefore, for controlling weeds on lawns or in cereal crops.

- residual herbicides can be added to bare soil before the crop emerges or can be used in providing total weed removal from small areas of ground. They remain active in the soil and kill the seedlings as they emerge.

herbivore: an animal that feeds on plants. In a *food web*, an herbivore is a primary consumer.

heterotrophic nutrition: a method of nutrition in which an organism gains its nutrients from complex organic molecules. These are broken down into simple soluble molecules by digestive enzymes and then built up again to form the organic molecules which the organism requires. As well as this being the means of nutrition shown by free-living animals, it is also shown by *saprophytes* and *parasites*. Heterotrophs are the *consumers* in food chains and, as such, they all ultimately depend on *autotrophs*.

heterozygote: an organism in which the *alleles* of a particular *gene* differ from each other. In peas, pod color is determined by a gene with two alleles. The allele for green pods, G, is dominant to the allele for yellow pods, g. The heterozygote will have the *genotype*, Gg. It will have green pods, since the allele for green pods is the dominant one.

hexose: a sugar that has six carbon atoms in each of its molecules. Hexoses are *monosaccharides*: each molecule consists of a single sugar unit. Biologically important hexose sugars include *glucose* and fructose.

hierarchical group: the social structure of a group of organisms in which high-ranking individuals control the behavior of lower ranking ones. The highest ranking animal, often called the α animal is dominant to all other animals in the group. The next ranking individual is the β animal. He or she is dominant to all the other members of the group with the exception of the α animal. Hierarchical groups are maintained by complex behavior patterns and help to produce a stable social structure. *Primates* such as chimpanzees and gorillas have this sort of social structure.

histocompatibility: tissues which can be transplanted from one person to another without being rejected by the immune system are said to show histocompatibility. In the same way that red blood cells have *antigens* on their surfaces (see *blood type*), other tissues have similar molecules called histocompatibility antigens on their *plasma membranes*. These antigens are determined genetically. There is a great range in the histocompatibility genes that two people may possess, even if they are closely related. However, related individuals are much more likely to be genetically similar and have compatible antigens. The closer the match that can be made between donor and recipient, the more likely it will be that a transplant will be successful. If these antigens are not compatible, there is a strong possibility that the transplanted organ or tissue will be rejected.

histone: a type of protein which is found with DNA in chromosomes. In eukaryotic cells, such as those of animals and plants, half of the total mass of each *chromosome* is due to the *DNA* it contains and half to protein. Histones are the commonest type of protein and are important in the packaging of DNA to form chromosomes. Prokaryotic cells such as *bacteria* do not contain histones.

HIV (human immunodeficiency virus): the virus which is responsible for AIDS (acquired immune deficiency syndrome). Each *virus* consists of an enzyme, called *reverse transcriptase*, and an *RNA* molecule which carries the viral genes. These components are surrounded by a protein coat. Finally, the whole virus particle is enclosed in a piece of *plasma membrane* which comes from its human host. The virus is transmitted in blood plasma and sexual fluids. When a person is infected by HIV, proteins in the surrounding cell membrane of the virus bind to specific proteins on the surface of a type of white blood cell known as a T-helper cell. The membrane surrounding the virus fuses with that of the white blood cell, and the RNA molecule and the enzyme enter the host cell. Reverse transcriptase is an enzyme which allows a *DNA* copy to be made from the RNA of the virus. This DNA copy is now to be inserted into the human DNA in the white blood cell where it can remain for a long period of time. Sooner or later, however, new viruses are made using the genetic information on this inserted piece of DNA. These viruses go on to infect other white blood cells.

holozoic nutrition: the type of nutrition found in animals. Holozoic nutrition involves:

* *ingestion* of solid food
* *digestion*, a process involving the chemical breakdown of large insoluble molecules into smaller soluble ones by enzymes in the gut or cells of the animal
* *absorption* of these small soluble molecules into the body
* *assimilation* in which the molecules which have been absorbed are built up into more complex molecules in the body of the animal concerned.

homeostasis: the maintenance of constant internal conditions. All living organisms regulate their internal environment and show homeostasis to some extent. Even a simple organism like an amoeba has a contractile vacuole which regulates the amount of water in the cytoplasm. Mammals, however, have a very extensive range of homeostatic mechanisms. These maintain the levels of a wide range of features such as temperature, pH, water potential and blood glucose concentration. Many of these mechanisms rely on *negative feedback*. This is the process in which a departure from the set level is detected by receptors. These convey information to effectors which bring about a return to the original value.

There are a number of reasons why homeostasis is particularly important to living organisms. These include:

* Biochemical reactions in living organisms are controlled by *enzymes*. Fluctuations in pH and temperature have a marked effect on enzyme-controlled reactions and, in extreme cases, can lead to denaturing of enzymes and other proteins.
* Most biochemical reactions are reversible and an equilibrium is maintained between the various substances in a cell. If there are major fluctuations in the concentrations of these substances, it is impossible to maintain equilibrium.
* The external environment in which many organisms live shows quite wide fluctuations in abiotic factors such as temperature. A constant internal environment allows a

considerable degree of independence from these and allows organisms such as mammals to live in areas ranging from the arctic to the tropics.

* Water can move into or out of cells by *osmosis*. By maintaining a constant water potential in the surrounding tissue fluids, osmotic problems are avoided.

Since mammals control such a range of conditions, homeostasis forms one of the important synoptic themes that links aspects of biology. The spider diagram shows some of these links. You can use the cross-referencing system in this book to add greater detail to the diagram.

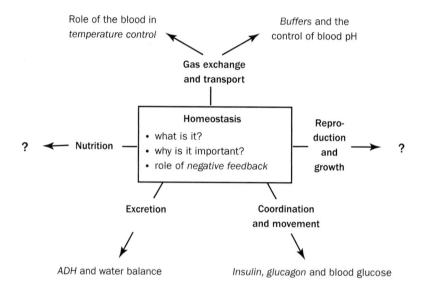

home base: the central core area of the *home range*. In *hunter-gatherer* societies the home base is the place to which the group returns at the end of the day.

home range: the area in which a group of animals normally lives and carries out its day-to-day activities, such as getting food and water. A part of the home range which is defended against other members of the same species is known as a *territory*.

hominid: modern humans and their closely related fossil relatives. Unfortunately, there is some confusion over the exact meaning of this term. Some biologists now think that, as humans are much more closely related to the gorilla and chimpanzee than was originally thought, these two animals ought to be classified as hominids as well. There is some uncertainty over which definition we ought to adopt. At this stage, however, it is probably best to regard hominids as the group containing modern humans and their direct relatives only. If this approach is adopted, then hominids can be characterized by two particular features. They are able to walk upright and they have relatively large brains. Associated with these features is another very important characteristic, the ability to use tools.

The first hominids appeared in the fossil record around four million years ago. From these ancestral species, two distinct genera appeared, *Australopithecus* and *Homo*. The diagram shows one way of interpreting the evidence to show how these two groups might have evolved, although it must be remembered that the fossil record is far from complete and this is only one of a number of theories.

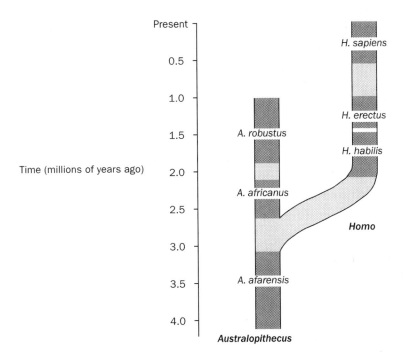

The evolution of hominids

Homo: the genus to which modern humans and their immediate ancestors belong. Their most obvious characteristic is that they have a large brain when compared to the size of the body. The large size of the brain is associated with changes in the shape of the skull. The cranium or brain case is large in relation to the face which is consequently much flatter than in early hominids. Many of the characteristics which help to define the genus, such as the use of language and the development of complex societies, are behavioral rather than structural and create difficulties because they do not leave any direct fossil evidence.

homologous: structures which have different functions but a common evolutionary origin. An example is provided by the wing of a bird and the front leg of a dog. At first sight they look very different and they have very different locomotory functions. However, they are both variations of the basic vertebrate limb. Careful examination of the skeleton of both of these limbs reveals this similarity.

homologous chromosomes: chromosomes that are capable of pairing during the first division of *meiosis*. A pair of homologous chromosomes will have the same sequence of gene loci (see *locus*), although the individual *alleles* are not always identical. For example, in humans, there is a protein which is responsible for the transport of chloride ions. The gene responsible for producing the protein concerned has been identified as being on chromosome number 7. Both copies of chromosome 7 present in a particular cell will have the locus for the gene for this protein. However, if a person is *heterozygous* for the gene, one chromosome will have a copy of the allele for making the chloride-ion transporting protein

139

while the other chromosome will have the allele for the defective protein which, in its homozygous state, will give rise to the condition known as *cystic fibrosis*.

homozygote: an organism in which the *alleles* of a particular *gene* are identical to each other. In peas, pod color is determined by a gene with two alleles. The allele for green pods, G, is *dominant* to the allele for yellow pods, g. There are two different homozygotes. The dominant homozygote has the genotype GG and green pods. The *recessive* homozygote has the genotype gg and yellow pods. Homozygotes may be described as pure breeding as, when crossed, they produce offspring identical to their parents.

hormone: a molecule which is transported in the blood and acts as a chemical messenger. Hormones are produced by the various cells and glands that make up the *endocrine system*. The blood transports them throughout the body, but they only produce responses in certain target cells. Although other sorts of molecules are involved, most of the hormones found in a mammal are either proteins, such as *insulin*, or *steroids*, such as *estrogen* and *proges-terone*. The type of chemical determines the way in which the hormone works. Protein hormones attach to receptor molecules on the *plasma membrane*. This triggers off a series of reactions leading to the activation of particular enzymes within the cell. Because there is a *cascade effect*, small amounts of hormone can lead to the activation of relatively large numbers of enzyme molecules and the production of even larger amounts of product inside these cells. Steroid hormones, on the other hand, work in a different way. They penetrate the plasma membrane and, once inside the cell, usually activate particular genes.

hormone replacement therapy (HRT): treatment with female sex hormones to overcome the unpleasant symptoms which occur after a woman passes through menopause or after her *ovaries* have been removed surgically. As a woman reaches the end of her reproductive life, her ovaries stop working. The lowered levels of *estrogen* and *progesterone* result in unpleasant symptoms which may include circulatory and psychological disorders as well as an increased tendency to develop *osteoporosis* and some types of heart disease. A combi-nation of estrogen and progesterone, or estrogen only, is prescribed in the form of tablets, implants or skin patches.

human chorionic gonadotrophin (hCG): a *hormone* produced during the early stages of pregnancy by the cells which will become the *placenta*. Following ovulation, the *corpus luteum* in the ovary produces the hormone *progesterone*, which is responsible for maintain-ing the lining of the uterus to provide a suitable environment for the developing fetus. If fertilization does not take place, the corpus luteum starts to break down, progesterone is no longer produced and menstruation takes place. If fertilization takes place, therefore, it is important that the corpus luteum continues to produce progesterone. Maintenance of the corpus luteum is the main function of human chorionic gonadotrophin. Pregnancy can be diagnosed very early by testing for the presence of this hormone. It usually appears in mea-surable quantities in the urine about two weeks into the pregnancy.

human genome project: an international research project involving the determination of the base sequence of the genes on all 23 pairs of human *chromosomes*. Since there are about a hundred thousand genes and 3 000 million base pairs in the human genome, this is a massive undertaking. The knowledge derived so far from this project has resulted in the identification and sequencing of many of the genes associated with human genetic disorders. In theory, it could lead to the development of successful treatments. There are fears, however, that the information revealed could be abused.

hunter-gatherer society: a human society in which the main means of getting food involves hunting animals and gathering wild plants. Until about 9 000 years ago, most human societies were hunter-gatherer societies. There are a few groups of people living today who still get a substantial proportion of their food in this way. Evidence from these modern people and from archaeological remains suggest that early hunter-gatherer societies had a number of features in common:

* they involved relatively small groups of 40 or so people living in a closely integrated community
* individual women tended to have relatively few children going four to six years between births
* the incidence of infectious disease, particularly in earlier societies, was extremely low
* there was a marked division of labor with the men hunting and the women and older children mainly responsible for gathering plant foods
* members exploited an area around a *home base* to which they returned in the evening. When this area had been exhausted, they moved on. Largely because of the need to move from place to place, they had little by way of personal belongings.

Huntington's disease (Huntington's chorea): an inherited condition in which an affected person shows symptoms which include involuntary jerky movements and progressive mental degeneration. These symptoms do not develop until the person reaches middle age. Huntington's disease is a dominant condition and the gene responsible has been identified as being located on chromosome 4.

hydrogen bond: a particular type of chemical bond. It is because of hydrogen bonding that water molecules stick to each other by *cohesion*. It is also an important bond helping to form and maintain the three-dimensional structure of a number of biological molecules, such as *proteins*, *polysaccharides* and *DNA*. One of the features of hydrogen bonds is that they require relatively little energy to break.

hydrolase: a type of enzyme which is responsible for catalyzing reactions which involve the breakdown of larger molecules into smaller ones by hydrolysis. Digestive enzymes are all hydrolases. A number of important hydrolases are mentioned elsewhere in this book. For further information, see *amylase*, *lipase* and *protease*.

hydrolysis: the breaking down of a large molecule into smaller molecules by the addition of water. The digestive enzymes found in the gut are all *hydrolases*, that is, they hydrolyze their substrates. In this way, starch is digested to maltose and glucose, fats to glycerol and fatty acids and proteins to amino acids. Chemically, hydrolysis is the opposite reaction to *condensation*.

> **hint** Sort out the meaning of the terms hydrolysis and condensation. Hydrolysis is the breakdown of larger molecules into smaller ones and involves the addition of water. *Condensation* is the joining of smaller molecules to form larger ones and involves the removal of a molecule of water.

hydrophyte: a plant which is adapted to living in water. General features shown by hydrophytes include:

* stems contain *aerenchyma*, a tissue that contains a mixture of cells and large, air-filled spaces. Aerenchyma provides buoyancy and its air-filled spaces form a pathway

allowing the diffusion of oxygen and carbon dioxide to parts of the plant that are under water

* the leaves of many hydrophytes float on the water surface; the stomata are generally on the upper surface of these leaves

* support is provided by the water. The stems of hydrophytes contain little if any of the woody tissue that provides support in larger land-living plants.

hydroponics: a system of growing plants without soil. Soil provides a useful medium for the roots to support and anchor the plant, as well as a supply of water and nutrients, but it is not actually needed for growth. Hydroponic culture has been used very successfully in growing greenhouse crops like tomatoes. The plants are grown in a series of polythene-lined channels along which flows a stream of nutrient solution which provides the necessary mineral ions. As the solution is very shallow, the roots are able to obtain the oxygen they need. In such systems, it is usual to monitor environmental conditions very carefully and, as a result of this, it is an economical way of growing greenhouse crops and produces vigorous, even growth of the plants.

hydrostatic skeleton: the sort of skeleton that is found in soft-bodied animals such as worms. A worm has a body cavity or *coelom* which contains a liquid known as coelomic fluid. Around the outside of this are two layers of muscle: longitudinal muscle which goes along the length of the animal and circular muscle which runs around the circumference. Squeezing a liquid will not change its volume. Therefore when the muscles contract, the animal will not get smaller, it will change its shape. Contraction of the longitudinal muscles will result in it becoming shorter and fatter, while contraction of the circular muscles results in it becoming longer and thinner. This change in shape allows the animal to move.

hypertension: said to occur when arterial blood pressure remains high even when a person is at rest. Hypertension can have serious health effects:

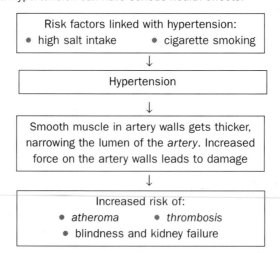

Risk factors linked with hypertension:
* high salt intake   * cigarette smoking

↓

Hypertension

↓

Smooth muscle in artery walls gets thicker, narrowing the lumen of the *artery*. Increased force on the artery walls leads to damage

↓

Increased risk of:
* *atheroma*   * *thrombosis*
* blindness and kidney failure

hypertonic: one solution is hypertonic to another if it has a higher solute concentration. If cells are placed in a hypertonic solution, the concentration of dissolved substances in the cytoplasm of the cell will be lower than that in the surrounding solution. The *water potential* of the cytoplasm will, therefore, be higher than the water potential of the surrounding solution. As a result, water will move out of the cell by *osmosis*. In plant cells, this will result in *plasmolysis*.

hint Hypo means below or less; hyper means above or higher. So the hypothalamus is the part of the brain below the thalamus; hypothermia is the condition brought about when a person's body temperature falls dangerously below its normal level; and a hypotonic solution is one with a lower concentration. Hypertension results from the blood pressure being higher than normal, and a hypertonic solution is one with a higher concentration. So, hyper means more, hypo means less . . . and hyper has more letters than hypo!

hyperventilation: involves breathing more deeply and faster than is necessary to meet the body's respiratory requirements. As a result, the concentration of oxygen in the blood goes up, while the concentration of carbon dioxide falls. If a person hyperventilates for two minutes or so, then allows breathing to continue normally, there is a period in which breathing stops altogether. This is called apnea. This is followed by a few shallow breaths and then another complete stoppage. Eventually the breathing pattern returns to normal. The reason for this sequence of events is that the low concentration of carbon dioxide in the blood (the result of hyperventilation) causes the apnea. When breathing stops, the oxygen level falls and this stimulates the respiratory centers. A few shallow breaths is then enough to provide sufficient oxygen to remove this effect, but there is still not enough carbon dioxide present to stimulate the normal breathing pattern.

hypothalamus: an area on the floor of the *forebrain* which has a number of important functions. These include:

* the control of many aspects of *homeostasis*. For example, the hypothalamus contains receptors which monitor the temperature of the blood. This information is used by the temperature control centers, which are also in the hypothalamus, to coordinate the body's response by either increasing the amount of heat that is lost or by conserving it within the body

* integration of the nervous and hormonal systems. The reproductive behavior of some mammals is initiated by a change in day length. Information concerned with this reaches the hypothalamus from sensory receptors. This information stimulates the hypothalamus to produce subsances known as release factors. These travel in the blood to the anterior lobe of the *pituitary* gland, where they result in the production of the hormones which control the sexual development and activity concerned with breeding

* coordination of the *autonomic nervous system*.

hypothermia: a condition in which the body temperature is lower than normal. In humans, prolonged exposure to the cold can produce hypothermia, particularly in the very young and the elderly. There is a slowing down of metabolic and physiological processes. Breathing and heart rate fall, blood pressure drops and finally the person may lose consciousness. Reducing the body temperature or induced hypothermia is sometimes used when surgical operations are carried out on the heart or brain. The advantages are that the heart can be stopped for relatively long periods, as the rate of respiration is much lower and the tissues require less oxygen. In addition, a low blood pressure reduces the dangers of extensive bleeding.

hypotonic: one solution is hypotonic to another if it has a lower solute concentration. If cells are placed in a hypotonic solution, the concentration of dissolved substances in the cytoplasm of the cell will be higher than that in the surrounding solution. The *water potential* of the cytoplasm will, therefore, be lower than the water potential of the surrounding solution. As a result, water will move into the cell by *osmosis*.

hypoxia: a deficiency of oxygen reaching the tissues. There are a number of reasons why the amount of oxygen reaching the tissues may be inadequate to meet the needs of respiration. These include:

- a low partial pressure of oxygen in the blood. This may happen at high altitudes, where there is not enough oxygen present to saturate the *hemoglobin*. Some diseases which affect the lungs can also result in hypoxia. These include *emphysema* and pneumonia

- the amount of hemoglobin in the blood is insufficient to transport the oxygen. This occurs when a person is anemic

- reduced blood flow to the tissues concerned.

hyphae: the thread-like structures which make up the body mass or mycelium of a fungus (see *Fungi*). Each hypha is surrounded by a cell wall in which the nitrogen-containing polysaccharide, *chitin*, forms an important part. In the cytoplasm are one or more nuclei and a number of vacuoles. The cell wall of the tip of a hypha is thin and this is where growth takes place. When living hyphae are observed, the cytoplasm is seen to be constantly moving, flowing towards the tip and away from it. This is known as cytoplasmic streaming and is the process which transports food materials to the growing tip. Further back along the hypha, the walls are much more rigid and growth is not possible here.

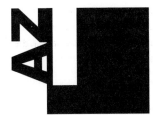

I-band: one of the light bands which can be seen running across a *myofibril* in a skeletal muscle. It corresponds to the location of filaments of the protein *actin*.

ileum: the last part of the small intestine of a mammal. In humans it is a long tube about five meters in length. Its inside surface is covered in folds. Each of these folds has many tiny finger-like projections known as villi (see *villus*) on its surface. In turn, the epithelial cells which line these villi possess large numbers of *microvilli*. The ileum, therefore, has an extremely large surface area both for the attachment of enzyme molecules and for absorption of the products of digestion. The cells that line the ileum contain the enzymes responsible for the final stages of protein and carbohydrate digestion. These enzymes are present in the *plasma membrane* and the cytoplasm of the epithelial cells. The villi contain large numbers of capillaries which take the amino acids and glucose produced by digestion to the *hepatic portal vein* and the *liver*. Lacteals are small lymph vessels. They absorb the products of fat digestion. Layers of circular and longitudinal muscle enable the digested food to be pushed along the ileum by waves of muscle contraction called *peristalsis*.

> **hint** When describing the adaptations of the ileum that ensure efficient absorption, use Fick's law to provide you with a framework for your answer. What are the adaptations that ensure a large surface area, a large difference in concentration and a thin exchange surface?

immobilization: a technique in which enzymes or microorganisms are bound to larger particles before being used in *biotechnological* processes. There are several advantages in using immobilization in enzyme-controlled reactions. If the enzyme is attached to a large insoluble particle instead of being free in solution, it is much easier to recover at the end of the reaction. This is not only cheaper, as the enzyme can be used again, but also helps to avoid contamination. In addition, immobilization can make enzymes tolerant of higher temperatures and a wider pH range.

> **hint** Immobilized enzymes are enzymes which are bound to larger particles before they are used in biotechnological processes. Immobilization is not about stopping enzymes from functioning by raising the temperature or adding an inhibitor.

immunization: an artificial method of producing immunity. Passive immunization involves the injection of *antibodies* made by another organism. It is useful in circumstances where the body would not have enough time to produce antibodies of its own. Passive immunization is used to treat snakebite and certain diseases such as tetanus. Antibodies produced by one organism are foreign proteins when they are introduced into the body of another. They act as *antigens* and are usually destroyed rapidly. Protection by passive immunization is therefore only short-lived.

145

Active immunization involves stimulating the body to produce its own antibodies. This can be achieved by injecting live microorganisms (which have been treated in such a way as to make them harmless), dead microorganisms or the toxins produced by microorganisms. Active immunization is often referred to as vaccination.

immunity: the way in which white blood cells and the *antibodies* that some of them produce enable the body to combat infection.

immunosuppression: lowering or suppression of the activity of the immune system which results in the body being more susceptible to certain diseases. *AIDS* results from the destruction of a particular group of white blood cells by the human immunodeficiency virus (*HIV*). As a result, patients are much more likely to become infected by pathogenic organisms which do not normally affect healthy people. In addition, conditions such as athlete's foot, a fungal infection of the skin between the toes, which are usually no more than mild infections in healthy individuals, may be extremely severe in AIDS patients.

If an organ such as a kidney or heart is transplanted from one individual to another, it is very likely that it will be rejected by the recipient's immune system. In order to reduce the risk of this happening, the patient is treated with immunosuppressive drugs such as cyclosporin. Unfortunately, this treatment increases the susceptibility of the patient to infections and to certain types of cancer.

Immunosuppressive drugs are also used in the treatment of *autoimmune diseases* like rheumatoid arthritis.

implantation: the process by which a developing mammalian embryo becomes attached to the wall of the uterus. The zygote which results from fertilization of the egg cell divides and develops into a hollow ball of cells called a *blastocyst*. Implantation involves the blastocyst burrowing into the wall of the *uterus*. In humans, implantation involves or leads to:

* hormonal changes in the mother which prevent the breakdown of the *corpus luteum* and prevent menstruation from taking place. A hormone known as *human chorionic gonadotrophin* (HCG) is produced. Its main function is to maintain the corpus luteum. Pregnancy can be diagnosed very early by testing for the presence of this hormone. It usually appears in measurable quantities in the urine about two weeks after fertilization

* the development of mechanisms which prevent the immune system of the mother from rejecting the developing embryo

* growth of the embryo and the development of the placenta.

imprinting: a form of learned behavior shown by the young of a number of species. If orphaned lambs are hand-reared, they learn to follow the person who looks after them. When they are returned to the flock, they will still try to find the person who reared them and stay close by. Similar behavior can be seen in birds, such as ducks and geese, where the young leave the nest almost immediately after hatching and, to have any real chance of survival, need to remain near the parent bird.

impulse: see *nerve impulse*

in vitro: refers to experiments carried out in test-tube conditions rather than in an intact living organism. Inside a living cell, there may be many hundreds of biochemical reactions taking place at the same time. It is obviously much simpler to look at individual reactions and study these on a test-tube scale. Care must be taken, however, in interpreting the results of such investigations. Just because something happens in a test tube, it does not mean that it will occur in exactly the same way in a living organism.

in vivo: refers to experiments carried out on living materials rather than in test-tube conditions (see *in vitro*).

index of diversity: a mathematical way of expressing diversity. There are many different ways of doing this, but one relatively simple one is shown in the example below.

Worked example: The table shows the number of birds of different species encountered on a walk across a park.

Species	Number of birds of this species encountered
Magpie	11
Black-headed gull	4
Carrion crow	4
Blackbird	1
Starling	37
House sparrow	7
All species	64

An index of diversity may be calculated from the formula:

$$d = \frac{N(N-1)}{\sum n(n-1)}$$

where N = total number of organisms of all species and n = total number of organisms of a particular species

$$d = \frac{(64 \times 63)}{(11 \times 10) + (4 \times 3) + (4 \times 3) + (1 \times 0) + (37 \times 36) + (7 \times 6)}$$

$$d = \frac{4032}{110 + 12 + 12 + 0 + 1332 + 42}$$

$$d = \frac{4032}{1508} = 2.7$$

On its own this figure does not mean very much, but it does allow the diversity to be compared with other areas. In this particular case, the diversity is fairly low, but it is higher than the value for the city center.

hint The key to calculating the index of diversity correctly is to remember that Σ means "the sum of."

indicator species: an organism which is very sensitive to a particular *abiotic* factor. The presence or absence of this organism can therefore be used as an indicator of environmental conditions. Indicator species can be used in monitoring pollution. For example, numerous species of lichens are found growing on tree trunks, rocks, stone walls and gravestones all over the country. They are very sensitive to the level of sulphur dioxide in the atmosphere. It is possible to produce a scale of increasing sulphur dioxide concentration and correlate this with the presence of particular lichens. By identifying which of these indicator species is present in a particular area, a good measure of the mean sulphur dioxide level can be obtained.

induced fit hypothesis: a model used to explain the way in which enzymes work. The *lock and key hypothesis* is a rather simple model which suggests that the active site of an enzyme has a rigid shape into which the enzyme molecule fits. In the induced fit model, the shape of the active site only corresponds generally to the shape of the substrate molecule. The substrate molecule then produces changes in the active site which results in the two fitting together exactly. In other words, the substrate induces the enzyme to fit. This is really no more than a modification of the lock and key hypothesis explaining how the enzyme substrate complex is formed.

innate behavior: an aspect of behavior that an animal shows from birth. In other words, it is genetically determined. In practice, it is very difficult to say how much of an animal's behavior is a direct result of its genes and how much is influenced by its environment or by learning from other members of its species. It is probably safe to say that simple responses, such as *taxis* and *kinesis*, are innate, as they are always shown by the organism concerned in appropriate circumstances. Other aspects of behavior may be partly innate but are modified by learning. Many birds, for example, can sing a recognizable song even when reared in isolation. This song, however, is modified once they hear that produced by other individuals. Such behavior is part innate and part learned.

innate releaser mechanism: the mechanism involved when a specific external stimulus triggers a particular pattern of behavior. As an example, the red belly of a male stickleback will stimulate a rival male to demonstrate aggressive behavior.

inspiratory reserve volume: the additional amount of air that can be breathed in after a person has breathed in normally. For an average adult male, this is about 1500 cm^3. See *lung capacities*.

insulin: a hormone produced by the *islets of Langerhans*, a group of *endocrine* cells in the pancreas. The secretion of insulin is stimulated by the rise in blood glucose concentration which follows a meal. The hormone has a number of effects on the body, all of which tend to reduce the blood glucose level. Two of these effects are:

* insulin speeds up the rate at which glucose is taken into cells from the blood. Glucose normally enters cells by *facilitated diffusion* through special protein carrier molecules in the *plasma membrane*. Cells have extra carrier molecules present in their cytoplasm. Insulin causes these carrier molecules to be sent to the membrane where they increase the rate of glucose uptake by the cell

* it activates enzymes which are responsible for the conversion of glucose to glycogen.

Insulin used to treat diabetics is now produced by *genetic engineering*.

integrated pest management: managing pests by combining the use of chemical pesticides and other methods of control. Sugar beet is harvested at the end of its first year of growth. Sugar beet seed, however, is only obtained from two-year-old plants. Aphids are an important pests of sugar beet because they spread the virus disease, beet yellows. These aphids come from sugar beet plants that have survived the winter.

148

Growing beet for sugar a long way away from beet that is grown for seed (and from closely related plants) reduces the number of aphids reaching the crop. However, when the number of aphids exceeds a certain number, insect sprays are used. The advantage of this method of pest management is that much less insecticide is needed, saving on cost and minimizing environmental damage.

intensive food production: the production of large amounts of food from relatively small areas of land. Much of the farming of pigs, cattle and sheep is intensive. This method contrasts with extensive farming, where the animals are more thinly spread over a wider area. Intensive farming has a number of advantages over extensive farming:

* it enables a number of measures to be adopted that result in greater profitability

* it allows control of the environment. By providing shelter or heating, for example, less food goes towards maintaining body temperature and more into production

* because it is possible to monitor individual animals, reproduction can be better controlled. It is much easier, for example, to identify estrous behavior in cattle and pigs and to make sure that artificial insemination takes place at the correct time.

interferon: a substance produced by the cells of a mammal when they are attacked by *viruses*. There are, in fact, a number of different interferons, each produced by a different sort of cell. They do not have an affect on the viruses themselves, but they appear to protect other cells from infection. Interferons have a range of other effects on the body and it is thought that they could have a number of possible uses in medicine, including the treatment of certain cancers.

internal fertilization: *fertilization* which takes place within the body of the female. In animals, *gametes* are small. This means that they have a large surface area to volume ratio and they dry out rapidly when exposed to air. This, and the fact that male gametes have flagella which enable them to swim, means that they need an aquatic environment if fertilization is to take place. For animals living on land, this is provided by the male of the organism concerned introducing sperm cells into the female reproductive tract. Fertilization will therefore take place internally.

interspecific competition: *competition* between organisms of different species. The main reason why *weeds* are removed from crops is that they compete with crop plants for resources such as light, water and soil nutrients. Crop yield falls as a result of competition.

intraspecific competition: *competition* between organisms of the same species. Farmers plant seeds of crop plants at particular population densities. If they plant them too far apart, they will not get the best yield. If, on the other hand, they are planted too close together, they will compete with each other for light and the substances they need in order to grow. Each will get less than the required amount and the yield from each individual plant will be less. The aim is to plant seeds at the optimum density, so that the best possible crop yield can be obtained. The same principle holds whether yields of crop plants are being discussed or the number of animals kept in a field.

hint Inter means between different things while intra means within something. These prefixes are often used when referring to competition. So interspecific competition is competition between different species. Intraspecific competition is competition within a species. Remember, Intercity trains go between different cities. Might help!

intron: a piece of the *DNA* in a *gene* which does not give rise to an amino acid sequence. A gene in a eukaryotic cell, such as one from an animal or a plant, consists of introns and the pieces of DNA which code for the amino acids which make up proteins. Pieces of DNA that code for proteins are known as *exons*. During the process of transcription, the base sequence of the gene forms a template for producing an mRNA molecule. Before the mRNA leaves the nucleus, however, it is edited, and the base sequences forming introns are cut out. The genes in many prokaryotic cells (see *Prokaryotae*) do not contain introns.

involuntary muscle: muscle which is not under conscious control. The muscle movements of most of the internal organs of the body are brought about by this type of muscle. Involuntary muscle forms the outer wall of the uterus or *myometrium*; it produces *peristalsis* in various parts of the gut and its presence in the walls of blood vessels allows them to change diameter and divert blood flow according to the needs of the body. Involuntary muscle contracts more slowly than *skeletal muscle* and has two particularly important properties. It is able to contract spontaneously if it is stretched and it also responds to substances released from nerve endings or to *hormones* brought to it in the circulation. Involuntary muscle is also known as visceral muscle or smooth muscle.

iodine: a micronutrient required by mammals. It is an important part of *thyroxine*, the hormone secreted by the thyroid gland which helps to control the *basal metabolic rate* of the body. If there is not enough iodine in the diet, the thyroid gland enlarges producing a condition known as goiter.

ion: a charged particle. *Atoms* or parts of *molecules* can lose one or more *electrons*. If this happens, they acquire a positive charge and are known as anions. Other atoms or parts of molecules can gain electrons. They have a negative charge and are known as cations. Many substances in the body exist as ions. One example is provided by what happens to carbonic acid formed in the reaction between carbon dioxide and water.

carbon dioxide + water $\rightarrow$ carbonic acid $\rightarrow$ hydrogen ions + hydrogencarbonate ions

$$CO_2 + H_2O \rightarrow H_2CO_3 \rightarrow H^+ + HCO_3^-$$

The important thing to note is that the positively charged hydrogen ion and the negatively charged hydrogencarbonate ion behave in different ways.

ion-channel protein: a type of *protein* found in the plasma membrane of a cell, which allows the passage of *ions* into or out of the cell. Some of these proteins are able to open and close and are thus described as being "gated." There are important gated ion-channel proteins in the *plasma membranes* of nerve cells. Changes in the electrical charge across the membrane bring about the opening and closing of these ion channels. This allows sodium ions to flow into the cell and potassium ions to flow out, giving rise to an *action potential*. Other gated channels are found in nerve *synapses*. They open in response to the presence of neurotransmitters.

ionizing radiation: radiation produced when radioactive elements decay. Some of this radiation consists of electromagnetic waves and some involves parts of atoms moving at high speed. Radiation of this type is called ionizing radiation because, when it strikes atoms in materials through which it passes, it causes the loss of electrons so that the atoms concerned are ionized. Cells are very vulnerable to ionizing radiation. High doses kill them, but even small amounts can cause damage to *DNA* molecules. This may result in increased *mutation* or, in some circumstances, may lead to *cancer*.

iron: a micronutrient required by animals and plants. Iron has a number of important functions. Iron is:

⊛ contained in cytochromes, molecules which form part of the *electron transport chain*. A cytochrome is a protein which is combined with another chemical group containing iron or copper

⊛ part of hemoglobin

⊛ necessary for enzymes like catalase to work

⊛ essential for the synthesis of *chlorophyll*.

islets of Langerhans: small groups of cells found in the *pancreas* which are responsible for the secretion of hormones involved in the control of blood glucose. The pancreas is an important organ found in the abdomen. Most of the cells it contains are responsible for the secretion of digestive enzymes, but scattered through the pancreatic tissue are the small groups of cells which make up the islets of Langerhans. The islets contain different types of cells. Small numbers of large α-cells secrete the hormone *glucagon* while the more numerous but smaller β-cells produce *insulin*.

isomers: molecules which contain the same types and numbers of atoms, but these are arranged in different ways. All *hexose* sugars have a molecular formula which may be written as $C_6H_{12}O_6$, so each molecule will contain 6 carbon atoms, 12 hydrogen atoms and 6 oxygen atoms. These atoms, however, can be arranged in different ways to give different sugars. The diagram shows two of them, α-glucose which is the basic building block for starch and β-glucose, the basic building block for cellulose.

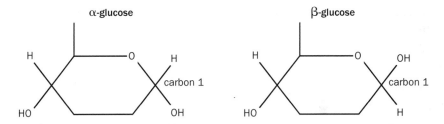

Isomers of glucose: both of these molecules have the formula $C_6H_{12}O_6$. Note the different positions of the −H and −OH groups on carbon 1

isotonic: solutions which have the same solute concentration as each other. If isotonic solutions are separated by a partially permeable membrane, the *water potential* will be the same on either side. There will be no net osmotic movement of water between the two solutions. The amount of water which moves in one direction will be exactly balanced by the amount which moves back in the other.

Mitochondria may be separated from the rest of the organelles present in a liver cell by using the technique of *cell fractionation*. The first stage in the procedure is to grind up or homogenize the liver tissue in an isotonic *buffer* solution. The buffer solution maintains a constant pH. Being isotonic prevents damage to the mitochondria through water entering or leaving by *osmosis*.

isotope: a form of an element which has a different number of neutrons in the nuclei of its atoms. Because of this, different isotopes differ in mass. This property can be useful in

151

isotope

marking or labeling atoms to see what happens to them during particular biological processes. Meselson and Stahl used two isotopes of nitrogen, ^{14}N and ^{15}N, to provide evidence for the semiconservative replication of *DNA*. Some isotopes emit particles from their nuclei as they break down. These are referred to as being radioactive and can be detected by using a Geiger counter or because they will blacken a photographic plate. Radioactive isotopes have proved to be very useful in investigating physiological processes, such as the transport of various substances in *phloem*, and in investigating biochemical pathways, such as the *light-independent reaction* in photosynthesis.

kidney: an organ responsible for the formation of urine in a mammal. A human kidney is about 10 cm in length and divided into an outer, dark-colored cortex and an inner, lighter medulla. Each kidney contains approximately a million kidney tubules or *nephrons* which produce the urine.

kidney tubule: see *nephron*

kinesis: a form of behavior in which the response of the animal is proportional to the intensity of the stimulus. In dry conditions with low humidity, a woodlouse moves faster and changes direction much less. When the humidity is higher, it moves slower and turns much more. The advantage of this pattern of behavior to the woodlouse is that it helps the animal to stay in favorable, humid conditions.

kingdom: the first stage in the biological system of *classification*. All living organisms can be classified into one of five kingdoms:

* *Prokaryotae*   Simple organisms without nuclei in their cells. Prokaryotae include *bacteria* and other related organisms

* *Animalia*   As the name suggests, this is the animal kingdom. Care should be taken that only members of this kingdom are referred to as animals

* *Plantae*   The plant kingdom

* *Fungi*   The group of organisms that includes molds and yeasts

* *Protoctista*   A number of different organisms are placed in this kingdom, mainly because they don't fit anywhere else! The kingdom Protoctista contains single-celled organisms such as amoeba and algae ranging from those with only one cell to the seaweeds.

Koch's postulates: a set of basic principles put forward by the nineteenth century biologist, Robert Koch, used for confirming that a particular microorganism is the pathogen that causes a disease. There are many different microorganisms found on and in the human body. Most of these do no harm, but some cause disease. Simply finding a bacterium or virus present in a person with a disease does not, therefore, prove that it causes the illness concerned. Koch suggested four principles that would identify the pathogen:

* The microorganism must always be present in a person with the disease. If a person does not have the disease, then the microorganism must be absent.

* It should be possible to isolate the microorganism from an infected person and culture it.

153

- The disease should develop when the cultured microorganism is introduced into a healthy person.

- The microorganism can be re-isolated from the new host.

Krebs cycle: a cycle of biochemical reactions which forms an important part of the pathway of *aerobic respiration*. Acetylcoenzyme A produced in the *link reaction* may be regarded as containing two carbon atoms which are involved in the process of respiration. This two-carbon fragment is fed into Krebs cycle, where it combines with a four-carbon compound to form a six-carbon compound. The six-carbon compound is broken down via a number of intermediate compounds to produce the four-carbon compound once again. This process is shown in the diagram.

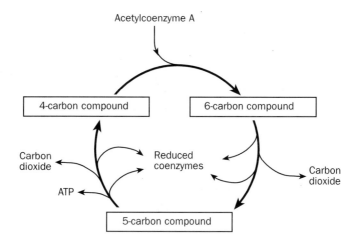

Krebs cycle involves both decarboxylation and oxidation. Decarboxylation is simply the removal of carbon atoms as carbon dioxide. Oxidation, as elsewhere in the process of aerobic respiration, involves the loss of hydrogen which is used to reduce *coenzymes*. The importance of Krebs cycle lies in its ability to produce reduced coenzymes from which ATP may be generated in the *electron transport system*.

---

hint Don't get bogged down with the biochemical details of the Krebs cycle. You don't need to learn the names of all the intermediate compounds and you certainly won't get credit for writing them down in an examination. Check the question and find out exactly what you need to know.

lactase: an *enzyme* that breaks down the milk sugar, *lactose*. It is a hydrolase and hydrolyzes lactose forming two *monosaccharides*, *glucose* and *galactose*. In most mammals, it is produced in the small intestine only when the animal is young. Most adult mammals do not produce lactase and cannot, therefore, digest lactose. If they are given milk, it is likely to make them ill (see *lactose intolerance*).

lactate: formed as a waste product in animal cells during *anaerobic respiration*. The first step in this process is *glycolysis*. In glycolysis, glucose is converted to pyruvate and a limited amount of energy is transferred to produce *ATP*. Oxidation is involved and this is associated with the reduction of a coenzyme called NAD. When oxygen is available, this reduced NAD is reconverted to NAD in the reactions of *aerobic respiration*. In anaerobic respiration, however, the reduced NAD is reconverted to NAD during the formation of lactate from pyruvate. This is shown in the diagram below.

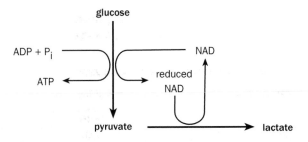

The formation of lactate in anaerobic respiration

The buildup of lactate in muscles is one of the factors which contributes to fatigue during vigorous exercise. Only so much can be tolerated before we have to take a rest. Once exercise stops and sufficient oxygen becomes available again, this lactate is removed. There are a number of metabolic pathways involved in lactate removal. These include conversion to *carbohydrate* in the liver and oxidation in the respiratory pathway.

 hint Remember, the end product of anaerobic respiration in humans and other animals is lactate, not ethanol. Athletes don't fall down drunk at the end of a race!

lactation: milk production by the mammary glands of a female mammal. Lactation involves three major processes. These are:

* the preparation of the tissues of the mammary gland which takes place during pregnancy. Under the control of the hormones produced by the *placenta* and the ovary,

the glands and ducts that make up the mammary glands develop and begin their secretory activity

* the secretion of milk which is controlled by the hormone, *prolactin*

* the release of this milk when the young animal suckles. Another hormone, *oxytocin*, plays an important part in this process.

lactic acid: a substance formed by *anaerobic respiration* in animal cells. In the conditions within the body, lactic acid is present as *lactate* ions. Because of this, the term lactate, is often used instead of lactic acid.

lactose: the main *carbohydrate* found in milk. Lactose is a *disaccharide* which is made in the mammary glands from two monosaccharides, glucose and galactose.

lactose intolerance: the inability to digest lactose, the sugar found in milk. The enzyme lactase is found in the small intestine of young mammals. It digests lactose, breaking it down to the monosaccharides, glucose and galactose. In many adults, however, this enzyme is not present. These people are said to be lactose intolerant and suffer considerable discomfort and diarrhea if they drink milk. The condition is inherited and is very common in many parts of the world. Approximately 70% of the black American population, for example, is lactose intolerant.

lag phase: the early part of a *population growth curve*. Organisms may increase in size but they do not increase in number. If yeast is introduced to a sucrose solution, the lag phase represents the time when the cells are adapting to their new medium. During this stage, various *genes* are switched on, mRNA is synthesized and the relevant *enzymes* are produced and secreted.

hint When you are asked to explain the slow initial growth of a population during the lag phase, bring in some relevant detail. An answer which merely states that the organisms are getting used to the area or the conditions is unlikely to gain credit.

larva: an immature stage in the life cycle of an animal. Many larvae are very different in appearance from the adults. This is particularly true of the larval stages of insects, such as the caterpillars of moths and butterflies. Larvae may develop gradually into adults or there may be a more rapid transition known as *metamorphosis*. The possession of a larval stage enables an animal to be adapted in different stages in its life cycle.

* The larval stage in the life cycle of insects is usually the main growth stage. As a consequence, many insect larvae are specialized for feeding on protein-rich foods. The caterpillars of most moths and butterflies, for example, eat the protein-rich young leaves of plants. Adult butterflies do not grow. They can survive on nectar which, although containing carbohydrates, has very little protein.

* Sessile animals do not move. They spend their adult lives in one place. Many sessile animals have larvae which normally help to ensure effective dispersal.

law of diminishing returns: a basic economic rule usually associated with the application of fertilizer to crop plants. The growth of many crop plants is limited by soil nutrients such as nitrates and phosphates. Therefore, if fertilizer is added, yield is increased. When a certain point is reached, however, the addition of even more fertilizer results in no further increase in crop yield. There are several reasons for this which include:

- other factors may be limiting plant growth. Addition of fertilizer during the winter, for example, would have little benefit. The low temperatures would limit crop growth and most of the fertilizer added would simply be leached from the soil when it rained

- if high levels of nitrate are added to cereal crops, the leafy parts of the plant grow as well as the developing grain. The plants are then much more likely to be blown over in summer storms. This will result in difficulties with harvesting and a lower yield

- high concentrations of fertilizer can reduce germination and can also result in reduced root growth.

leaf area index: is the total plant leaf area divided by the area of the ground under these leaves. It is a useful measure of the photosynthetic surface of a crop and it has been found that a leaf area index of between 4 and 7 would enable a crop to absorb most of the available light. Since the growth of any plant is affected by its ability to absorb light and photosynthesize, it is important that crops reach their maximum leaf area index as soon as possible after planting. Knowing the leaf area index enables agricultural scientists to calculate the productivity of a crop from the simple formula:

Productivity = leaf area index × *net photosynthesis* × time

lean body mass: a measure of the yield or productivity of domestic animals such as cattle and sheep. Measuring the body mass of the animal is not very precise: firstly, because the contents of the gut and bladder can make a lot of difference and secondly, because the proportion of fat may vary considerably over a relatively short period of time. The lean body mass is its empty body mass minus the mass of the body fat.

learning: a process which involves changes in behavior which result from experience of new situations. At one stage it was thought that behavior was either learned or it was *innate* and shown by the animal from birth. In practice, it is very difficult to say how much of an animal's behavior is innate and genetically determined and how much is influenced by its environment or by learning from other members of its species. Several types of learned behavior are recognized. These include *habituation, imprinting, classical conditioning* and *operant conditioning.*

leguminous plants: an agriculturally important group of plants which contain species such as peas, beans and clover. They have seeds which are contained in pods and are generally rich in protein. They also have nodules on their roots which are associated with *nitrogen fixation.* These nodules contain bacteria which belong to the genus *Rhizobium.* The bacteria can convert nitrogen in the air into ammonium compounds, some of which they use and some of which they pass on to the plant itself. In return, the bacteria gain carbohydrates which are made by photosynthesis. This relationship is a good example of *mutualism.* Leguminous plant crops are often plowed back into the soil as a sort of green manure, which will rot down to increase soil nitrate levels.

lens: a transparent structure at the front of the eye which helps to focus light rays on the *retina.* Because it is able to change shape, the lens is able to change the amount of refraction, allowing both close and distant objects to be brought into focus. The lens is composed of living cells and these require a supply of nutrients. The *aqueous humor* provides these nutrients.

lenticel: a small pore found on the surface of a plant stem. In older stems, a layer of cork develops around the outside. Mature cork cells are impermeable to water and respiratory gases. If this cork layer completely covered the stem, the cells inside would die as they would be unable to gain oxygen from or lose carbon dioxide to the atmosphere. Lenticels are small pores in the surface of the cork layer. They are very loosely packed with cells, so respiratory gases can diffuse through the large intercellular spaces from the atmosphere to the cells of the stem.

leukocyte: an alternative name for a white blood cell or white cell. There are several different types of leukocyte but all of them are involved in some way in helping the body to combat infection. They are found in smaller numbers than red blood cells and, because their *cytoplasm* is transparent, they need to be stained before they can be examined in detail. The term white blood cell is a little misleading because some types of leukocyte are able to leave the blood and are found in the lymphatic system and other parts of the body. The table shows the characteristics of some different types of leukocyte.

Type of leukocyte	Appearance	Function
Lymphocyte	Relatively large, round nucleus and small amount of cytoplasm	B-cells or B-lymphocytes secrete antibodies. T-cells or T-lymphocytes have a number of functions. These include killing infected cells and controlling different aspects of the immunological process.
Monocyte	Large, kidney-shaped nucleus	These cells are phagocytes and engulf bacteria
Granulocytes	Possess granular cytoplasm and a lobed nucleus	These have a number of different functions. Some engulf bacteria by phagocytosis. Others are involved in allergies and in inflammation.

LH (luteinizing hormone): a hormone which is produced by the anterior lobe of the pituitary gland. In females, it is produced during the first part of the reproductive cycle. It travels in the blood to its target organ, the *ovary*, where it stimulates ovulation and causes the follicle cells which remain in the ovary to develop into the *corpus luteum* which secretes *progesterone* during the second half of the reproductive cycle. LH is also produced in males, where it stimulates *interstitial cells* in the testes to produce *testosterone*.

ligament: a strip of *connective tissue* which attaches bones to each other. Some ligaments contain a lot of closely packed fibers of a protein called *collagen*. The properties of collagen make it ideally suited to its functions in a ligament. It is flexible but very resistant to stretching. Other ligaments, such as those between the vertebrae in the spine, contain elastin. This is another protein which, as its name suggests, can be stretched more readily.

ligase: a type of enzyme which is used to join lengths of DNA together. Ligases are used in *genetic engineering* to produce *recombinant DNA*.

light-reaction: see *light-dependent reaction*

light-dependent reaction: the process in which light energy is absorbed by chlorophyll and used to produce ATP and reduced NADP during photosynthesis. Light striking a *chlorophyll* molecule causes some of its electrons to become excited. These are lost from the chlorophyll molecule and pass through a series of molecules which act as electron acceptors. As a result, energy is progressively lost and some of this is used to make *ATP* in the process of *photophosphorylation*. The electrons are eventually accepted by a coenzyme known as *NADP* and convert this to reduced NADP.

Another reaction is also involved. This is *photolysis*. Water molecules break down to produce electrons, hydrogen ions and a molecule of oxygen. It can be summarized by the equation:

$$2H_2O \rightarrow 4H^+ + 4e^- + O_2$$

The electrons produced as a result of photolysis replace those which were lost from the chlorophyll molecule as a result of it absorbing the light energy. The hydrogen ions or protons help in the formation of reduced NADP and the oxygen is given off as a waste product. The ATP and reduced NADP produced as a result of these light-dependent processes play an important part in the conversion of carbon dioxide to carbohydrate in the *light-independent reaction*.

light-independent reaction: the process by which carbon dioxide is reduced to form carbohydrate during photosynthesis. The biochemical reactions that form the light-independent reaction of photosynthesis are summarized in the diagram.

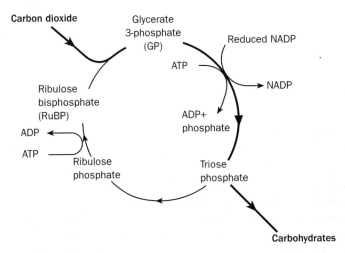

Ribulose bisphosphate, RuBP, is a five-carbon compound. It combines with a molecule of carbon dioxide to produce two molecules of the three carbon-compound, glycerate 3-phosphate or GP. In turn, GP can be converted into the sugar, triose phosphate. This requires the reduced NADP and ATP produced in the *light-dependent reaction*. Some of this triose phosphate is then converted into other carbohydrates, such as glucose and starch, while the rest is used to form more ribulose bisphosphate. This cycle of reactions involving ribulose bisphosphate is known as the *Calvin cycle*.

hint Light-independent reactions are so called because they are independent of light. However, they cannot take place without the ATP and reduced NADP produced in the light. Because of this, light-independent reactions cannot continue for long in the dark.

159

lignin: a complex biological polymer found in some plant *cell walls*. A newly formed plant cell has a cell wall containing *cellulose* microfibrils cemented together with a variety of other molecules. This is called the primary cell wall. In some cells, such as those in the *sclerenchyma* and *xylem*, lignin is then laid down. It further helps to cement the cellulose microfibrils together and makes the wall even stronger and more resistant to the forces on it. However, it also makes the wall impermeable, so important substances are unable to pass through. As a consequence, cells which are heavily lignified do not have living contents.

limiting factor: a variable which limits the rate of a particular process. If it is increased, then the process will take place at a faster rate. The graph shows how *substrate* concentration affects the rate of an enzyme-controlled reaction.

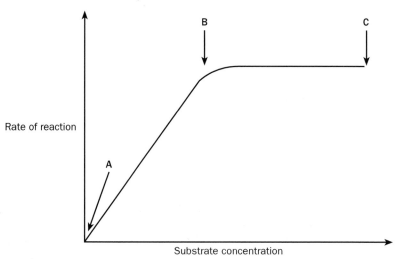

At low substrate concentrations, between points **A** and **B** on the graph, the rate of the reaction is limited by substrate concentration. An increase in substrate concentration will produce an increase in the rate of the reaction. Between points **B** and **C**, however, increasing the substrate concentration has no effect on the rate of reaction, so something else must be limiting. In this case it is probably the concentration of the *enzyme*. There are many other biological processes which show a similar pattern. Examples are the effects of light intensity and carbon dioxide concentration on the rate of *photosynthesis* and the rate of uptake of glucose by cells.

hint Curves similar in shape to the one shown in the graph aboveare very common. Make sure that you can explain why such curves go up to start with and why they then level out.

Lincoln index: an alternative name for the *capture-recapture* method of estimating the size of an animal population.

linkage: refers to two or more genes which are found on the same chromosome. This is important in genetics as it means that linked *alleles* are normally inherited together. During the first division of *meoisis*, the chromosomes come together in their pairs, with each chromosome split longitudinally into two daughter *chromatids*. If two alleles are linked, they will both be situated on a single chromatid, so they will be inherited together when these chromatids are eventually pulled apart. They will both go to the same daughter cell. The only way in which linked genes can be prevented from being inherited together is when *crossing over* occurs. Here the chromatids break and rejoin so that alleles from one chromatid become attached to alleles on another chromatid.

link reaction: a name given to the part of the respiratory pathway linking *glycolysis* with *Krebs cycle*. In the link reaction, pyruvate is converted into acetylcoenzyme A. This conversion involves the loss of a molecule of carbon dioxide. It is also an oxidation reaction, with hydrogen being lost and used to reduce a molecule of the coenzyme NAD. The link reaction is summarized in the following diagram.

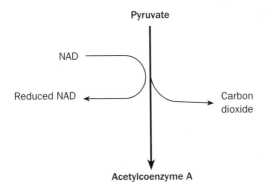

lipase: an enzyme which breaks lipids down into glycerol and fatty acids. Lipases are used in industrial processes to digest fats as, for example, in some biological washing powders. They are also used in the food industry, where they are involved in chemical reactions which swap fatty acid chains between different lipids. Cocoa butter is a very expensive, naturally occurring lipid, which is used in making chocolate. By using lipase and exchanging fatty acid chains, a cheaper substitute can be made from more readily available vegetable oils.

lipid: a large and varied group of organic substances which are insoluble in water but are soluble in organic solvents such as ethanol. A number of different molecules are classified as lipids. They include *triglycerides*, such as fats and oils, *phospholipids*, which play an important role in cell membranes, and *steroids*, which include hormones such as *progesterone* and *estrogen*.

lipoprotein: an association between protein and lipid molecules. Lipoproteins are found in the plasma of the blood and in lymph where they play an important part in the transport of *lipids*. Different sorts of lipoproteins are recognized. Low-density lipoprotein (LDL) contains relatively low proportions of protein; high-density lipoprotein (HDL) contains rather more protein, while very high density lipoproteins (VHDL) contain the highest proportion of all. These different lipoproteins have different functions in the body, LDLs, for example, play an important role in the transport of cholesterol.

liver: a very large, reddish-brown organ found in the body of a mammal. Like all other organs in the body, the liver is supplied with oxygenated blood. This reaches it through the hepatic artery. The liver, however, is also supplied with blood by another vessel. This is the *hepatic portal vein* which brings blood from the intestine. The hepatic vein returns blood to the heart. This organ has many different functions but it is possible to group these together:

- the liver is concerned with *homeostasis*. It helps to regulate the blood glucose level. If the concentration of glucose in the *blood glucose pool* rises, the excess is converted into *glycogen* in the liver. The hormones *insulin* and *glucagon* play a very important part in this process

- the liver synthesizes molecules such as *plasma proteins* and produces *bile*

- the liver stores molecules such as fat-soluble *vitamins*

- the liver is involved in *excretion*. It is in this organ that *deamination* takes place. In deamination the amino groups are removed from surplus amino acid molecules. The ammonia formed by this process is then converted into *urea*. In addition, the hemoglobin from broken down red blood cells is processed in the liver and the unwanted part is excreted in the bile.

lock and key hypothesis: a model used to explain the way in which enzymes work. An enzyme molecule has an active site formed by a group of amino acid molecules. This active site has a particular shape into which the substrate molecule fits to form an enzyme-substrate complex. By combining to form an enzyme-substrate complex, the *activation energy* is lowered and the reaction will take place under normal cell conditions. This complex then breaks down to form the products and release the enzyme molecule.

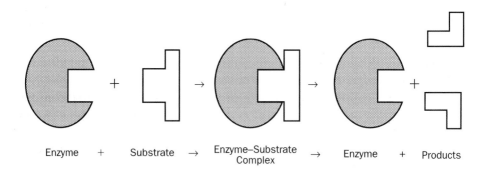

Enzyme + Substrate → Enzyme–Substrate Complex → Enzyme + Products

How an enzyme-controlled reaction takes place

It is important to realize that this is only a model. The actual situation is rather more complicated. However, a thorough understanding of the lock and key hypothesis enables us to explain a number of features which affect the rate of reaction of enzymes, such as substrate concentration and *competitive inhibition*. To find out more, see *induced fit hypothesis*.

hint Note that the active site of the enzyme and the substrate have shapes which are best described as complementary. It is incorrect to say that the shape of the active site is "the same as" the shape of the substrate.

locus: the position of a *gene* on a *DNA* molecule.

log phase: the part of a *population growth curve* where there is a rapid increase in numbers. Suppose a bacterium reproduces by dividing in two. In the conditions present in a particular culture, it does this once every hour. After an hour, there will be 2 bacteria present; after two hours there will be 4, then 8, 16, 32, 64, 128 and so on. Plotted on a graph this will give a curve which gets steeper and steeper. During this log phase, nutrients will be in plentiful supply and there will be no appreciable buildup of toxic waste products.

longitudinal study: a method of collecting information about growth by measuring the same individual over a period of time. A *cross-sectional study* involves measuring different individuals, each on a single occasion. Longitudinal studies have an important advantage over cross-sectional studies, because they show individual differences in growth patterns. Study of variation between individuals can produce evidence about the influence of genetic and environmental factors on growth.

However, there are some disadvantages as well:

- longitudinal studies can take a rather long time to carry out as it is necessary to wait for the individuals in the study to age
- it is necessary to keep careful track of individuals. This can be very difficult.

loop of Henle: part of a *nephron* which forms a loop into the medulla or inner part of the kidney. The different parts of the loop have different permeabilities to salt and to water. As a result of this, a gradient of salt concentration is set up in the medulla. The deeper into the medulla, the higher the salt concentration and, therefore, the lower the *water potential*. This means that the water potential of the fluid in the collecting duct is always higher than that in the medulla and water can be removed by *osmosis* all the way down the *collecting duct*. The length of the loop of Henle is associated with the amount of water that can be reabsorbed in the kidney. Small mammals which live in deserts have long loops of Henle and therefore produce small amounts of very concentrated urine.

lumen: the cavity inside a hollow structure. The lumen of the gut, for example, is the space in the gut through which food passes.

lung: one of a pair of gas exchange organs found in the chest of mammals and certain other chordates. When examined with a microscope, lungs are seen to be made up of three distinct types of structure:

- alveoli (see *alveolus*). These are the tiny thin-walled air sacs in the lungs where gas exchange takes place
- airways. Air is taken into and expelled from the lungs through the trachea and bronchi (see *bronchus*). The bronchi branch repeatedly finally leading into a system of smaller tubes known as *bronchioles*
- blood vessels. The lungs are well supplied with blood, and arteries, veins and capillaries can all be seen in suitably prepared microscope sections.

The total volume of air in the lungs of an adult male is about five liters, but the actual volume exchanged each time a person breathes depends on a number of factors (see *lung capacities*).

> **hint** When describing the adaptations of lungs for efficient gas exchange, use *Fick's law* to provide you with a framework for your answer. What are the adaptations that ensure a large surface area, a large difference in concentration and a thin exchange surface?

lung capacities: the volumes of air that are either contained in the lungs or are taken in or breathed out under particular circumstances. The diagram on page 164 shows these values in an average man.

163

For an average adult male, the total volume of air contained in the lungs is approximately five liters or 5000 cm^3. Its precise value will obviously vary with a number of factors, including the age of the individual, his physical fitness and state of health. When breathing quietly at rest, the volume of air taken in and given out at each breath is much less than this. It is only about 500 cm^3. This is the tidal volume. If, after having breathed in normally, the person continues to breathe in as deeply as possible, another 1500 cm^3 of air may be taken into the lungs. This is the inspiratory reserve volume. Similarly, if after breathing out normally, the person concerned continues to breathe out as much as possible, it is possible to expel another 1500 cm^3. The additional volume of air breathed out is known as the expiratory reserve volume. Finally the total of the amount of air that can be breathed out following the deepest intake of air possible is the vital capacity. This is the sum of the tidal volume, the inspiratory reserve volume and the expiratory reserve volume. It is not as much as the total volume of the lungs. There is always a small volume that cannot be expelled. This is known as the residual volume.

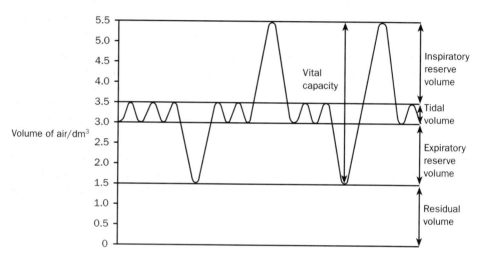

Human lung capacities

luteinizing hormone: see *LH*

lymph: *tissue fluid* after it has drained into the lymphatic system. Lymph is very similar to *plasma* in chemical composition but it contains less protein. *Lipids* are absorbed from the small intestine into the lymphatic system, and lymph from this part of the body may have a milky appearance due to the large amount of fat that it contains.

lymph node: one of a number of small swellings found in the *lymphatic system*. It contains channels in which there are white blood cells or *leukocytes*. Some of these engulf foreign material while others produce antibodies. Lymph nodes act as filters and help to prevent the spread of infection.

lymphatic system: the system of vessels which returns excess *tissue fluid* to the blood system. During the course of a day, more fluid leaves the blood *capillaries* as tissue fluid than drains back into them. If this tissue fluid accumulates it results in swelling, a condition known as edema. In a healthy person, the excess tissue fluid drains into a system of tiny, blind-ending tubes, the lymphatic capillaries. This fluid, now called lymph, is collected by a series of larger lymph vessels before finally returning to the blood in veins in the neck. Lymph vessels are surrounded by the skeletal muscles that allow movement. When these muscles contract they squeeze the lymph along the vessels. Valves in the larger lymph vessels, like those found in veins, ensure that the flow is in one direction. At intervals through the lymphatic system, there are swellings called *lymph nodes*.

lymphocyte: a white blood cell which has a relatively large nucleus and small amounts of cytoplasm. There are several different types of lymphocyte but they can only be identified by the way they respond to *antigens*. Broadly, they can be divided into *B-cells* and *T-cells*.

lysosome: an *organelle* containing digestive enzymes. These enzymes are separated from the rest of the cell contents by the membrane which surrounds the lysosome. This is essential otherwise these enzymes would digest the proteins and lipids normally found in the cell and destroy them. Lysosomes are usually associated with animal cells where they have a number of different functions:

* *phagocytes* are white blood cells which are able to surround and ingest bacteria. Once inside the phagocyte, these bacteria are destroyed by enzymes in the lysosomes. Food taken in by microorganisms like amoeba is also digested by lysosomal enzymes

* during the life of an animal most organs and structures increase in size. Sometimes, however, they may be broken down during the course of development. *Bone* is a tissue which has to be continuously remodeled to cope with the changing stresses which occur as an animal grows and increases in size. This can involve the breakdown of bone in some areas as well as its formation in others. Breakdown is carried out by lysosomes in special bone cells.

* *metamorphosis* is a process which occurs in the life cycle of some animals and involves a major change in the form and shape of the animal concerned. When a frog changes from a tadpole into an adult, its tail disappears as it is gradually reabsorbed. This is another process which is carried out by lysosomes.

165

malaria: an infectious disease caused by a protozoan microorganism that lives inside the red blood cells of its human host. The microorganism concerned is one of several species of *Plasmodium* and it is spread from person to person by anopheline mosquitoes. When a female mosquito bites an infected person, blood containing parasites is taken into her stomach. The parasites reproduce inside the mosquito and enter her salivary glands. When she bites another person, parasites are injected into the blood stream along with the saliva. Once inside a new human host, the parasites migrate to the liver where they multiply. After an incubation period which may vary from approximately two weeks to several months, they return to the bloodstream where they enter the red blood cells. A cycle of multiplication, destruction of blood cells and infection of further blood cells gives rise to the characteristic pattern of fever associated with malaria.

hint The organism that causes malaria belongs to the Protoctista. It is not a bacterium or a virus.

male: the organism, or part of the organism, which is responsible for producing male sex cells or *gametes*. From whatever organism they come, male gametes have three characteristic features:

* they are small in size when compared to female gametes

* they are produced in much larger numbers than the female gametes

* male gametes either move by themselves or are moved in some way to the female gametes.

malignant: a *tumor* which destroys the tissues around it and can spread to other areas in the body. Cells become detached from the primary or original tumor and are transported to other sites in the body by the blood or lymphatic systems. A malignant tumor in the breast, for example, can give rise to secondary tumors in the lymph glands or in the bones. If untreated, malignant tumors will result in the death of the patient.

maltase: an *enzyme* that breaks down *maltose*. It is a hydrolase and splits each maltose molecule by adding a molecule of water to produce two molecules of *glucose*. In a mammal, the maltase enzymes found on the plasma membranes of the epithelial cells in the small intestine break down the maltose produced by the digestion of *starch*. Maltase enzymes are also found in germinating seeds and are secreted by some fungi.

maltose: a *disaccharide*, that is a sugar which is made up of two sugar units. These units are both *glucose* and they are joined together by condensation. Maltose is a reducing sugar and therefore produces a positive result with *Benedict's test*. Maltose formed by the hydrolysis of *starch* during the *germination* of barley grain is fermented by yeast in the production of alcoholic drinks such as beer and whisky.

166

hint Maltose, sucrose and lactose are carbohydrates. Maltase, sucrase and lactase are the enzymes which hydrolyze these disaccharides. Make sure that you write clearly and distinguish between the carbohydrates and the enzymes.

mammal: a member of the class *Mammalia*. Care should be taken to distinguish between animals and mammals. The term mammal should only be used when referring to organisms such as humans, sheep, bats and whales that belong to this class. The term animal refers to any member of the animal kingdom, *Animalia*.

Mammalia: the class containing mammals (see *classification*). Mammals along with fish, amphibians, reptiles and birds belong to the phylum Chordata. In addition to its more familiar members, the Mammalia include kangaroos and other marsupials and some primitive egg-laying animals. Members of this class share the following features:

* they feed their young on milk secreted by mammary glands

* their body temperature fluctuates very little, although this is also a feature of birds

* they possess hair and sweat glands, distinctive mammalian features that play an important role in temperature control

* an individual mammal possesses different types of teeth which have different functions; other chordates have teeth which are all very similar to each other.

mark–release–recapture: see *capture–recapture*

mass flow: the transport of substances in bulk from one part of an organism to another. Large organisms cannot rely only on *diffusion* as this process is much too slow to meet their needs. They require a way of moving substances rapidly from one place to another. Mass flow systems are linked with exchange surfaces whose main function is to establish and maintain differences in concentration and pressure. It is because of these differences that substances can be moved by a mass flow system from one place to another. Examples of mass flow systems are the *blood system* of a mammal and *phloem* and *xylem* in a plant.

maximum sustainable yield: the largest mass of fish that may be caught without causing the population to fall. If no fishing were taking place, the total mass of fish in a population would be more or less stable. The mass of young fish added to the population and the increase in mass of the smaller ones as they grew to a larger size would be equal to the mass of those dying or being eaten by predators.

In a fishery where underfishing were taking place, there would be a lot of large old fish. These would not be growing very much themselves but they would be taking a lot of food from the younger ones which could, in the right circumstances, grow rapidly. On the other hand, if a fishery were overfished, few individuals would reach maturity and the breeding stock would fall dramatically. There must be an optimum somewhere between these two extremes. This is the maximum sustainable yield.

Although the term usually relates to fisheries, it could apply to other situations where wild populations are cropped on a continuous basis.

mechanoreceptor: a sensory *receptor* which responds to mechanical stimuli, such as touch, pressure, sound and gravity. Many mechanoreceptors are cells which have hair-like structures called cilia. When a cilium is moved, a nerve impulse is produced. These receptors can be used to give information about many different kinds of stimulus. For example, mechanoreceptors in muscles detect stretching, while others in the ear are concerned with balance and hearing.

medulla (brain): part of the hind brain. It contains the centers which are responsible for controlling heart rate, blood pressure and breathing.

medulla (kidney): the inner part of the *kidney*. As well as a large number of blood vessels, the medulla contains the *loops of Henle* and the *collecting ducts* of the kidney tubules. Many of the small mammals that live in desert regions have kidneys that can produce extremely concentrated urine. An animal such as jerboa, for example, can produce urine which is between two and three times as concentrated as that of humans. In order to do this, the loops of Henle are much longer than in closely related animals that live in areas where water is more readily available. This is reflected in the appearance of the kidney which has a much thicker medulla.

meiosis: a type of *nuclear division* in which the number of *chromosomes* is halved, resulting in the *haploid* number in each of the daughter cells. It involves two separate divisions, usually referred to as meiosis I and meiosis II.

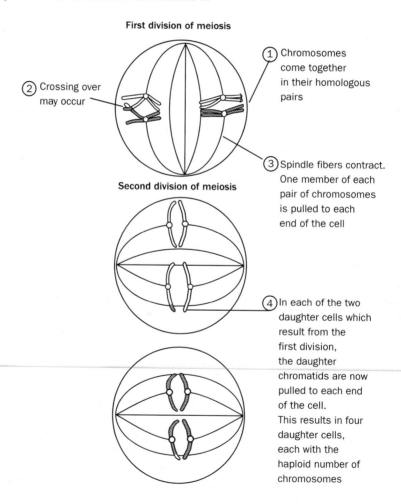

First division of meiosis

① Chromosomes come together in their homologous pairs

② Crossing over may occur

③ Spindle fibers contract. One member of each pair of chromosomes is pulled to each end of the cell

Second division of meiosis

④ In each of the two daughter cells which result from the first division, the daughter chromatids are now pulled to each end of the cell. This results in four daughter cells, each with the haploid number of chromosomes

168

In a normal body cell, there are a number of pairs of chromosomes. This number varies from organism to organism, but in humans there are 23 pairs. One member of each pair originally came from the female parent and one member from the male parent. In meiosis I, the chromosomes come together in their pairs, with each chromosome split longitudinally into two daughter chromatids. One member of each pair of chromosomes is now pulled to each end of the cell, so we arrive at a situation with one complete set of chromosomes at one pole of the cell and one complete set at the other pole. What is very important is that some of the chromosomes which have gone to a particular pole will have come from one parent and some from the other. This is called independent segregation and is one reason why the cells which are produced as a result of meiosis will differ from each other in the genetic information that they carry. Another process known as *crossing over* means that there are further genetic differences in these cells. At the end of meiosis I, there will be two daughter cells, each containing one complete set of chromosomes. Each chromosome will consist of two chromatids. In meiosis II, each of these daughter cells divides again. This time the *chromatids* separate and are pulled to the poles of the cell. The diagram shows how this process results in four daughter cells, each containing half the number of chromosomes that was present in the original cell.

Since *sexual reproduction* involves the fusion of the nuclei of two different cells, it is necessary to halve the number of chromosomes first, otherwise, there would be a doubling of their number with each successive generation. In most animals, *gametes* are formed by meiosis. Plants, however, have very complex life cycles involving an *alternation of generations* and meiosis takes place some time before gamete formation. The haploid cells which result from meiosis often go through several stages before the gametes are finally produced by *mitosis*.

> **hint** Meiosis does not take place in gametes. It leads to the formation of gametes.

memory cell: *lymphocytes* which remain in the blood system after the original infection has gone. When the *antigen* has been destroyed by the *plasma cells*, most of these cells die. The memory cells, however, remain in the blood system and respond rapidly if reinfection with the same antigen occurs. Multiplication takes place and more plasma cells are produced. In this way the response to a second infection is much more rapid.

menstrual cycle: the name given to the type of *estrous cycle* found in humans, apes and some monkeys. It differs slightly from the cycle found in other mammals, mainly as the lining of the uterus, the *endometrium*, is shed along with a certain amount of blood and lost from the body between cycles. In other mammals, this lining is usually reabsorbed back into the body if pregnancy does not take place.

meristem: a group of plant cells which are able to divide by *mitosis*. When an animal grows, each individual organ grows. The liver, for example, is present at birth. It contains fully functional liver cells which will divide and grow. In this way, the mature organ will be almost identical to that in the young animal; it will just contain more cells. Plant growth is very different. In the root tip, there is a group of unspecialized cells. These form the root meristem. The cells divide, and only then do they differentiate or become specialized for particular functions. Once they have differentiated, they no longer continue to divide. Another meristem near the tip of the shoot controls the growth of the stem. These two groups of cells are called primary meristems, as they are present when the plant begins its growth. They are concerned mainly with increase in height. Secondary meristems develop in woody plants and increase the diameter. An example of a secondary meristem is the *cambium*. The fact that there are some undifferentiated cells in a mature plant which still retain the ability to divide is important in allowing *micropropagation* of plants.

mesophyll: the types of cells found within a leaf. In leaves of plants which are not specialized for living in water or for particularly dry conditions, there are two sorts of cell present in the mesophyll. Palisade cells are found near the top of the leaf. They are tall and thin and packed tightly together. They contain many chloroplasts and have adaptations for efficient *photosynthesis*. Underneath the palisade cells is the spongy mesophyll. This consists of irregularly shaped cells in between which are large air spaces. These spaces play an important role in gas exchange between the leaf and the surrounding atmosphere. Plants found in full sunlight tend to have leaves with a thicker layer of palisade cells than those growing in the shade.

mesophyte: a plant which is adapted to growing in a place where there is an average supply of water, adequate for its needs. Mesophytes may be compared with *hydrophytes* which grow in very wet conditions and *xerophytes* which grow in very dry conditions.

mesosome: an infolding of the *plasma membrane* of a bacterium (see *bacteria*). Many functions have been suggested for mesosomes but there is still some doubt as to what they actually do. It is thought most likely that they are concerned with bacterial *respiration* but they are also considered to be involved in cell division.

messenger RNA: see *mRNA*

metabolic pathway: a series of biochemical reactions taking place in a living cell in which substances are either broken down to smaller molecules (*catabolic* reactions) or built up into more complex ones (*anabolic* reactions). The diagram shows a typical metabolic pathway in which substance A is converted to substance E via a number of intermediate steps. Each reaction in the pathway is controlled by a separate *enzyme*.

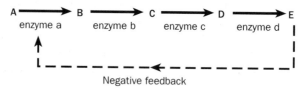

A metabolic pathway

Pathways can be regulated by *negative feedback*. When sufficient product, in this case substance E, has been produced, it acts as an inhibitor and slows down or stops the action of an enzyme at the start of the pathway. As a result, less product is formed. If there is not enough substance E, there will be no negative feedback and no inhibition of enzyme a. The enzyme will work at its normal rate so more substance E is produced.

metabolic rate: a measurement of the rate at which energy is released by the body. Since this depends to a considerable extent on body size, it is usual to give figures for metabolic rate in energy units per meter squared of body surface per hour. Metabolic rate is influenced by a number of factors. These include:

- physical activity. Minimum heat production occurs when the body is in a complete state of rest. It rises with an increase in physical activity. The average amount of energy released in an adult man lying down is about 4.5 kilojoules per minute. This can rise to as much as 70 kilojoules per minute with exceptionally demanding physical work

- reproductive state. The metabolic rate rises during pregnancy. This is partly due to an increase in the resting level of energy release, partly to the growth of the fetus and partly to an increase in the fat stores of the mother. Lactation involves an even higher level of energy expenditure

- *hormone* levels. Increased production of the hormone *thyroxine*, for example, causes an increase in metabolic rate

- time after eating. Heat production is increased immediately following a meal.

Since so many factors influence metabolic rate, it is usual to compare the heat given out under standard conditions. The *basal metabolic rate* (BMR) of a person is the metabolic rate at complete rest. It is a measurement of the amount of energy required for vital activities, such as the action of the heart and the muscles associated with breathing.

metabolism: all the chemical reactions which take place in living organisms. In a single cell, there are many hundreds of such reactions. Processes involving the breakdown of large molecules into smaller ones, such as many of those which occur in digestion and respiration, are known as catabolic reactions. Others, such as those involved in building proteins from amino acids and in *photosynthesis*, are concerned with combining smaller molecules to form larger ones. These are called anabolic processes. Catabolic reactions release energy, while anabolic ones require energy.

metamorphosis: the changes which take place in the transformation from the larval stage of an organism to the adult. In animals such as frogs, these changes occur progressively. The larval stage is a tadpole. Metamorphosis involves a series of gradual changes which are reflected externally by the growth of legs and the absorption of the tail. There are internal changes as well. These involve, for example, the transition from *gills* to *lungs* as gas exchange surfaces and modifications to the gut which are associated with the change from a vegetarian to a carnivorous diet. Many insects go through a more drastic change. The *larva* first molts into a pupa. Inside the pupa most of the existing tissues are broken down. This often involves *lysosomes*. The new tissues which make up the adult then develop. The changes which are involved in insect metamorphosis include the following:

- muscles. Larvae and adults often have different means of locomotion. Caterpillars crawl over the surface of their food plants, while adult butterflies rely mainly on flight as a means of locomotion. Larval muscles have to be broken down and new adult muscles formed

- the gut. Leaf-eating caterpillars develop into nectar-feeding butterflies. This involves changes both in the types of digestive enzyme secreted and to the structure of the mouthparts and the rest of the alimentary canal

- reproductive system. Most larval stages are sexually immature. The adults are the reproductive stages.

Metamorphosis is generally under the control of *hormones*.

metastasis: the process by which cancer cells can spread from the original *tumor* to other parts of the body. A clump of cells breaks free from the original, or primary, tumor and is carried in the blood or by the *lymphatic system* to another part of the body where it may form a secondary tumor.

microevolution: small-scale evolutionary changes. They usually involve changes in the frequency of *alleles* in a population, but do not, on their own, result in the formation of new species. One example is provided by the evolution of copper-tolerant grasses on contaminated soil. Some of the grass seeds which land in the area will have alleles which mean that they are better adapted to these conditions. They can survive on soils which are heavily contaminated with copper. They germinate, grow and breed, passing on these alleles to the next generation. This will produce a change in the proportion of alleles in the *population*. It is thought that a large number of small changes like this could eventually result in the evolution of a completely new species. The flowchart summarizes this process.

> Contaminated soil colonized by grass seeds.
> These seeds vary. A few of them will be
> copper-tolerant

↓

> The seeds that cannot tolerate high levels of
> copper die; those that can survive

↓

> The tolerant grass plants have less competition.
> They grow and reproduce, passing on the
> alleles associated with copper tolerance

↓

> Selection has operated. The population has
> changed from grasses which could not tolerate
> copper to grasses which could tolerate it

micrometer (µm): one thousandth of a millimeter. *Bacteria* are approximately 1 µm in diameter. In practice, they are about the smallest objects that can be seen with a light microscope. Animal and plant cells are larger. A human *red blood cell*, for example, is about 7.5 µm in diameter, while a palisade cell from the mesophyll of a leaf may be as much as 200 µm in length.

hint Many calculations involving size and magnification involve converting measurements to micrometers. Always measure in millimeters and remember that 1 mm = 1000 mm. Using centimeters leads to errors.

micronutrient: a substance required by an organism in amounts which are very small when compared to the other nutrients that it requires. Micronutrients include *vitamins* and mineral ions. The table shows some examples of mineral ions which are required as micronutrients by plants and animals.

Plant micronutrients	Main functions
Iron	• Essential for the synthesis of *chlorophyll*
	• Contained in cytochromes, molecules which form part of the *electron transport chain*. A cytochrome is a protein which is combined with another chemical group containing iron or copper
Molybdenum	• Needed for the functioning of the enzyme, nitrate reductase. This enzyme reduces nitrates during the synthesis of amino acids

Animal micronutrients	Function
Iodine	• Contained in *thyroxine*, the hormone secreted by the thyroid gland which controls the *basal metabolic rate*
Iron	• Contained in cytochromes, molecules which form part of the *electron transport chain*. A cytochrome is a protein which is combined with another chemical group containing iron or copper
	• Part of hemoglobin
	• Necessary for enzymes like catalase to work

microorganism: any organism which is too small to be seen without the use of a microscope. The three main groups, the *viruses*, the *bacteria* and the *fungi*, differ considerably from each other in their basic structure. Although they are responsible for many plant and animal diseases, they are also particularly useful in industrial processes. This is mainly because they have simple nutritional requirements, a very rapid growth rate and either produce or, with *genetic engineering*, can be altered to produce a whole range of useful substances.

micropropagation: a process by which large numbers of genetically identical plants can be produced from the growth region of a single parent. The growth region or *meristem* is carefully dissected from the apical bud and transferred to a tube containing nutrient agar to which certain *plant growth substances* have been added. The meristem grows rapidly into a mass of cells called a callus. After a short time, leafy shoots begin to grow from the callus tissue. These are separated and transferred to a different medium on which they grow roots and develop into complete plants. This process is summarized in the flowchart on the next page.

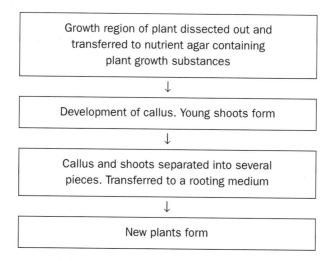

Growth region of plant dissected out and transferred to nutrient agar containing plant growth substances

↓

Development of callus. Young shoots form

↓

Callus and shoots separated into several pieces. Transferred to a rooting medium

↓

New plants form

The great advantage with this technique is that not only does it require relatively little space, heat and light, but that the plants that are produced are disease-free.

microvillus: a tiny, finger-like projection from the *plasma membrane* of an animal cell. Microvilli are found in large numbers on the epithelial cells that line the small intestine and on the cells forming the wall of the proximal tubule in a *nephron*. They help to increase the surface area over which *diffusion* and *active transport* take place. You should take care to distinguish a microvillus from a villus. Villi are much larger structures. Those in the small intestine, for example, are made up of many cells.

 Microvilli are found on cells in the small intestine and the first convoluted tubule in the kidney. There are no microvilli in the lungs.

migration: the movement of a population of organisms from one environment to another. The timescale over which this takes place can vary enormously. Small organisms in the plankton move up towards the surface waters during the daytime and move back into the depths at night. This is a daily movement. Swallows, on the other hand, undergo an annual migration. They leave the temperate regions of Europe in the autumn, spend the winter in Southern Africa and return to Europe the following spring. In addition, migration can involve many return journeys during the lifetime of the animal, a single return journey or even a one-way trip. The table on the next page shows some different patterns of migration.

mitochondrion: a cell organelle in which the biochemical reactions associated with *aerobic respiration* take place. Mitochondria are variable in size and shape, but often appear as small, elongated structures about a micrometer in length. Each mitochondrion is surrounded by an envelope consisting of two membranes. The inner one is folded to form structures called cristae. These cristae stick out into the matrix of the mitochondrion. The matrix contains the enzymes associated with *Krebs cycle*, while the *electron transport system* is located on the cristae. The presence of the electron transport system means that mitochondria are where most of the *ATP* is produced in a cell. Numbers of mitochondria vary considerably from cell to cell. Cells which contain particularly large numbers of these organelles usually require large amounts of ATP. This may be for one of the following reasons:

Organism makes more than one return journey

| Surface waters during the daytime | ⟷ | Deeper waters at night |

Small organisms in the plankton

| Europe during the summer months | ⟷ | Southern Africa during the winter months |

Swallow

| Tundra during the summer months | ⟷ | Conifer forests during the winter months |

Caribou

Organism makes a single return journey

| Rivers in Europe | ⟷ | The Atlantic Ocean |

Salmon

Organism makes a one-way journey

| Southern Europe | ⟶ | Northern Europe |

Several species of butterfly

- *active transport.* This process requires very large amounts of ATP. Active transport is important in the absorption of the products of digestion in the small intestine, reabsorption by the cells of the first convoluted tubule in a nephron and in loading sugar into the sieve tubes in the phloem. Cells associated with these processes all contain large numbers of mitochondria

- movement. The contraction of muscles, the separation of chromosomes in dividing cells and the swimming of sperms all require energy. Once again large numbers of mitochondria are associated with the cells concerned

- chemical reactions involved in producing new molecules also require energy. Examples of cells in which many such reactions take place include liver cells and rod cells in the retina of the eye.

mitosis: a type of *nuclear division* which results in each daughter cell having the same number of chromosomes as the parent cell. Although mitosis is a continuous process, it is convenient to divide it into four stages. These are:

- prophase. The chromosomes become shorter and thicker and show up clearly. Each one is seen to consist of a pair of chromatids which are joined together by a *centromere*. The nuclear envelope starts to break down and the spindle forms

- metaphase. The chromosomes are now arranged across the "equator" of the cell. They are attached at their centromeres to the spindle fibers

- anaphase. The pair of chromatids which make up each chromosome are pulled apart by the spindle fibers. One chromatid from each pair goes to each of the "poles" of the cell. Each chromatid has now become a chromosome

- telophase. The chromosomes form two groups, one at each pole. They gradually lose their form and can no longer be recognized as distinct structures. A new nuclear envelope forms round each group of chromosomes.

hint There are some very obvious differences between mitosis and meiosis but there is still a lot of confusion between these two processes. It might help if you tried to avoid revising these topics one after the other.

monoclonal antibodies: antibodies produced by the descendants of a single B-cell. Although the details of production may vary, monoclonal antibodies are normally made in the laboratory by the following technique.

A mouse or other small mammal is injected with a specific antigen. B-cells are then obtained from the animal

↓

The B-cells are fused with other cells so that they will effectively grow forever in cell culture

↓

The fused cells are diluted and placed individually in the wells on a culture plate

↓

Each individual cell is allowed to multiply so that a pure cell line or clone is established. This will only produce a single type of antibody, a monoclonal antibody

Monoclonal antibodies are extremely useful in biotechnology because they bind tightly to specific molecules. Because of this, they may be used to distinguish between similar molecules, for example, between human proteins and chimpanzee proteins or between molecules on the surface of cancer cells and molecules on the surface of normal cells.

monocotyledon: a member of the group of flowering plants characterized by having a single *cotyledon* in each of its seed. There are two main groups of flowering plants – the Monocotyledones or monocotyledons and the Dicotyledones or dicotyledons. They differ from each other in the number of important ways in addition to the number of cotyledons contained in their seeds. For further details, see *dicotyledon*.

monoculture: a type of farming in which a large area of land is devoted to the growth of a single crop. In contrast to this, more primitive methods of agriculture, such as subsistence farming, involve the growth of many different species of plant on a relatively small amount of land. Monoculture has a number of distinct advantages over agricultural techniques like subsistence farming. These include:

* monoculture is much less labor-intensive than subsistence farming. It relies on the use of machinery. Much greater profits may be made from monoculture

* because large areas are devoted to a single crop, it is much easier to use specialized machinery for crop management. Combine harvesters, for example, can only be used for harvesting wheat if it is grown in large fields

* different crops require different soil conditions. Such features as soil pH and the correct balance of nutrients can be managed so they are at an optimum for the crop concerned

On the other hand, monoculture has a number of distinct disadvantages. These include:

* large areas are devoted to the growth of single species of plants and this encourages the development of pests. At present, the only effective way of controlling these pests is with the repeated use of chemical *pesticides*

* the search for more and more productive crops results in little genetic variation. A lack of *genetic diversity* would create problems if circumstances were to change in the future

* devoting larger and larger areas to the growth of a single crop favors the removal of hedges and the development of bigger fields. This removes habitats for wildlife and may lead to soil erosion.

monocyte: a type of white blood cell. Monocytes are large cells which are made in the bone marrow. They engulf bacteria and other microorganisms by *phagocytosis*.

monohybrid cross: a genetic cross between individuals in which a single gene is being considered. For a worked example of how to set out a genetic cross, see *dihybrid cross*.

monomer: one of the similar small molecules that join together to form a polymer. There are three important biological monomers; *amino acids, monosaccharides* and *nucleotides*. For further details of the way in which these monomers are joined together by condensation, see *polymer*.

monosaccharide: a carbohydrate whose molecules consist of a single sugar unit. Monosaccharides are small molecules and dissolve readily in water. They are classified according to how many carbon atoms each molecule contains:

* trioses contain three carbon atoms per molecule. They are important intermediate compounds in biochemical pathways, such as those of *respiration* and *photosynthesis*

* pentoses are five-carbon sugars. Ribose and deoxyribose are important constituents of *RNA* and *DNA*. *Ribulose bisphosphate* is the molecule which combines with carbon dioxide in the *light-independent reaction* of *photosynthesis*

* hexoses are six-carbon sugars, such as glucose and fructose.

multiple alleles: a single gene which has more than two forms. An example of a gene which has multiple *alleles* is the one which controls the inheritance of the ABO *blood groups* in humans. There are four blood groups in this system, A, B, AB and O. Their presence is determined by a single gene with three alleles, I^A, I^B and I^o. The alleles I^A and I^B are *codomi-*

177

nant while the allele I^o is *recessive* to both I^A and I^B. Each gene occupies a particular place on a chromosome called its *locus*. Since chromosomes are found in pairs, it follows that there will be two places that can be occupied by alleles of this gene, one on one chromosome of the pair and the other at the corresponding place on the other chromosome. It is possible, then, to have two blood group alleles. They may be the same as each other or they may be different. The table below shows the various combinations that it is possible to have and the resulting blood groups.

Possible genotypes	Blood group
$I^A I^A$ or $I^A I^o$	A
$I^B I^B$ or $I^B I^o$	B
$I^A I^B$	AB
$I^o I^o$	O

muscle: an *effector* which contracts and brings about movement in animals. There are three types of muscle tissue which are found in mammals:

- *cardiac muscle* is found only in the heart

- striated or *skeletal muscle* is attached to bones in the skeleton and brings about limb movements. It also helps to maintain posture

- *involuntary muscle* is involved with the movement of organs which are not under conscious control. *Peristalsis* in the gut, the change in diameter of blood vessels and the alteration in the size of the pupil of the eye in response to changes in light intensity are controlled by involuntary muscle.

mutagen: an environmental factor which increases the rate of *mutation*. There are many different mutagens. They include certain types of radiation, many organic substances and some viruses. Mutations occur randomly and under natural conditions the rate at which mutation occurs is slow. Mutagens simply increase this rate. Exposure to mutagens in the environment is medically important because of the link between mutations and cancer (see *tumor*). Substances in tobacco smoke that act as mutagens also increase the likelihood of developing cancer.

mutation: a change in either the amount or the arrangement of the genetic material in a cell. Mutation can take place in any cell. If it occurs in a *gamete*, the resulting characteristic can be inherited. There are mutations, however, which occur in the normal body cells. These are called somatic mutations and they cannot be passed on to the next generation.

Mutations differ in their effects. Chromosome mutations involve large-scale changes and may involve parts of, or even whole, chromosomes. Point or *gene mutations*, on the other hand, are on a much smaller scale and are concerned with changes to relatively small numbers of bases in the genetic material. The rate at which mutation takes place can be increased by exposure to various mutagens, such as ultraviolet radiation, X-rays and a wide range of organic chemicals.

mutualism: a relationship between two different species of organism where both gain a nutritional advantage. In explaining examples of mutualism, it is important to identify nutritional rather than more general advantages, such as the provision of warmth and shelter. A good example is provided by the microorganisms which live in the *rumen* of cattle. The mammal provides a constant supply of food rich in *cellulose* for the microorganisms. In turn, the microorganisms provide the cattle with *fatty acids* obtained from the digestion of cellulose.

mRNA (messenger ribonucleic acid, messenger RNA): a nucleic acid which acts as a "messenger", taking a copy of the genetic code from the nucleus into the cytoplasm during *protein synthesis*. It is a type of RNA which consists of a single *polynucleotide* chain with a backbone built up of alternating ribose sugars and phosphate groups. One of four bases, adenine, cytosine, guanine or uracil, is attached to each ribose sugar. The mRNA molecule plays an important role in the process of *transcription*. Part of the *DNA* in the nucleus unwinds and acts as a template for the formation of mRNA. The mRNA then takes the code for the gene concerned from the nucleus into the cytoplasm, enabling amino acids to be assembled in the correct sequence for the required protein. A second type of RNA, transfer RNA or *tRNA*, is found in the cytoplasm. This molecule is important in assembling amino acids in the correct position on the mRNA during the process of *translation*.

mycelium: the mass of thread-like hyphae which make up the body mass of many *fungi*.

myelin: a fatty material which is produced by *Schwann cells* and forms a layer around the axons in many mammalian nerves. As it is an insulator, it only allows the electric currents which set up the next *action potential* to flow at the gaps where no myelin is present. These gaps are known as the nodes of Ranvier. Nerve impulses therefore travel along in a series of jumps from one node to the next. This is known as *saltatory conduction* and can lead to impulses moving much more rapidly than in nonmyelinated nerves.

myocardial infarction: the technical term for what is often referred to as a heart attack. The blood supply to the heart muscle may be interrupted in, for example, a person suffering from *atherosclerosis*. This results in the death of a large part of the muscle concerned, an event which is usually associated with a severe pain in the chest. The main danger is that a myocardial infarction may cause the heart to beat very rapidly without actually pumping any blood. This rapidly leads to death. With appropriate treatment, however, most patients can return to a normal, active life.

myofibril: one of the many smaller fibers which are arranged parallel to each other in a *skeletal muscle fiber*. When seen with a light microscope, each myofibril is made up of alternating light and dark bands. This is due to a regular arrangement of filaments made from the proteins *actin* and *myosin*. The diagram on the next page shows the structure of a myofibril as it would appear when seen with a light microscope together with a more detailed interpretation obtained by using an electron microscope.

myogenic: heart muscle is described as myogenic because it can contract automatically. It is not like most other muscle in the body which is called neurogenic because it will only contract when stimulated to do so by a nerve impulse. This automatic activity is particularly important in the *sinoatrial node* where it gives rise to the wave of electrical activity which controls the heartbeat.

myoglobin: an oxygen-storing pigment found in muscle. Chemically, myoglobin is very similar to *hemoglobin* as it is made up of a polypeptide chain which is associated with an iron-containing *heme* group. It differs though, in that it only contains one of these units, not four like hemoglobin. Although found in the muscles of a variety of animals, there are particularly high concentrations of myoglobin in diving mammals like seals and whales. The myoglobin acts as an oxygen store. It is saturated with oxygen at relatively low partial pressures and is able to hold on to this oxygen, only releasing it when the amount in the tissues falls to a very low level.

179

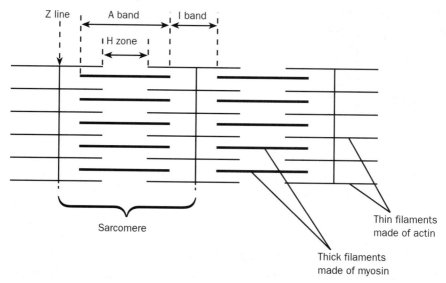

The structure of a myofibril

myometrium: the outer layer of the *uterus*. It is mainly made of *involuntary muscle*.

myosin: the protein of which the thick filaments in the *myofibrils* of *skeletal muscle* are made. Each myosin molecule consists of two parts and looks rather like a golf club with a long rod-shaped part and a globular head. The myosin heads go through a cycle of attaching to, changing their angle to the rest of the molecule and detaching from points on the *actin* filaments. This enables the two types of filament to slide between each other and the muscle to shorten. This is the basis of the *sliding-filament hypothesis* which explains how skeletal muscles contract.

NAD: a molecule which picks up hydrogens released during respiration. In simple terms, *aerobic respiration* involves a series of reactions in which glucose is oxidized step by step to carbon dioxide and water. Oxidation involves the loss of hydrogens. These are picked up by NAD which is reduced as a result. The reduced NAD then transfers these hydrogens to molecules in the *electron transfer chain*. As the hydrogens are passed from one molecule to the next along this chain, the energy which is released is used to form *ATP*. For each molecule of glucose which is oxidized, ten molecules of reduced NAD are produced.

NADP: a molecule which plays an important part in the transport of electrons. In the *light-dependent reaction* of photosynthesis, light is absorbed by chlorophyll molecules. This causes some of the electrons in the chlorophyll to become excited. They are picked up by NADP which is converted to reduced NADP in the process. This reduced NADP plays an important part in the conversion of carbon dioxide to carbohydrate in the *light-independent* reaction.

> **hint** The coenzyme involved in photosynthesis is NADP; that involved in respiration is NAD. Remember, P for photosynthesis.

natural selection: see *selection*

negative feedback: many substances or systems in living organisms have a set level. Negative feedback is the process in which any departure from this level sets in motion changes which lead to a return to the original value. The process plays a very important role in *homeostasis*. A good example of negative feedback is provided by the mechanisms associated with *temperature control* in a mammal. Humans, for example, maintain a body temperature within a degree or so of 37 °C. If the temperature rises too high, the resulting increase in blood temperature is detected by receptors in the *hypothalamus* of the brain. As a result, the heat loss center which is also in the hypothalamus sends nerve impulses to structures, such as arterioles and sweat glands in the skin, which bring about the necessary fall in temperature. Cold conditions, on the other hand, are detected by receptors in the skin. Impulses are sent to the hypothalamus. This time, the heat conservation center triggers mechanisms which conserve the body's heat, or even generate more heat by actions such as shivering. The outcome, once again, is a return to the normal level. This is summarized in the diagram on page 182.

There are many other examples of negative feedback operating in the body of a mammal. Blood glucose level, respiration rate and the level of some of the hormones which control reproduction are all controlled by negative feedback mechanisms. Care must be taken to distinguish negative feedback from *positive feedback*.

Negative feedback is one of the important synoptic themes that links different aspects of biology. The spider diagram shows some of these links. You can use the cross-referencing system in this book to add greater detail to the diagram.

181

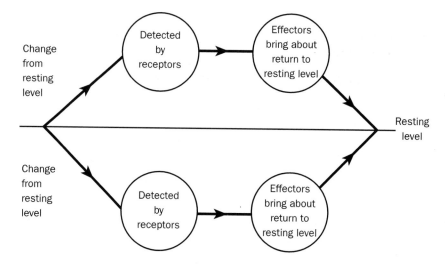

Negative feedback and body temperature

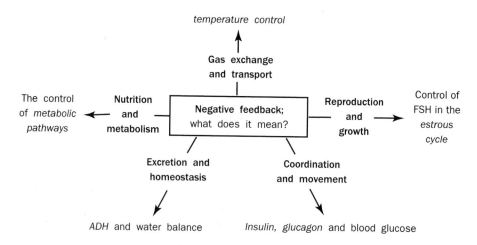

nephron: one of the many thousands of tubules present in the kidney which are responsible for the production of urine. Each nephron is divided into five main parts. In the *renal capsule* the process of *ultrafiltration* takes place. The composition of the liquid or filtrate produced is changed by the reabsorption of useful substances in its pathway along the first convoluted tubule. Finally, the contents are concentrated in the *loop of Henle* and the *collecting duct*. The second convoluted tubule plays an important part in controlling the *pH* of the blood.

The different parts of the nephron and their main functions are summarized in the diagram on page 183.

hint When you revise a complex subject like the formation of urine, try to get an overview of the whole process before you start looking at the detail. Read this entry before you look at what the individual parts of the nephron do.

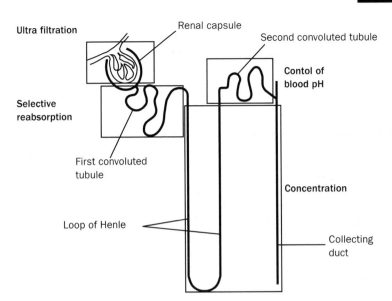

Ultra filtration

Renal capsule

Second convoluted tubule

Contol of
blood pH

Selective
reabsorption

First convoluted
tubule

Concentration

Loop of Henle

Collecting
duct

The structure and functions of nephron

nerve: a bundle of neurones or nerve cells enclosed in a sheath of connective tissue. Nerves may contain sensory *neurones*, which conduct impulses from sense organs, motor neurones, which conduct impulses to effectors, or they may be mixed, containing both sensory and motor neurones.

nerve impulse: the way in which information is transmitted along a nerve cell. The inside of the cell is negatively charged when compared with the outside, so that there is a potential difference across its *plasma membrane*. This is called the *resting potential*. When an impulse travels along the nerve, a wave of electrical activity called an *action potential* passes along. The charge is temporarily reversed and the inside of the nerve cell becomes positive. A nerve impulse is an *all-or-nothing* action. Either a full nerve impulse is generated or no nerve impulse at all. This is of importance as it means that the only way that information about the strength of a stimulus can be transmitted is by varying the frequency of the impulses.

hint Nerve cells convey impulses, not messages!

net assimilation rate: the rate at which the substances produced during photosynthesis accumulate in a plant. Net assimilation rate is measured in grams of dry matter produced per meter squared per day (g m^{-2} day^{-1}) and is a good indicator of the rate of growth. Dry matter is measured, as the amount of water in plant tissue can vary considerably. Net assimilation rate is sometimes referred to as net primary production.

net primary production: see *net assimilation rate*

net productivity: a measure of the rate at which new body substances accumulate in an organism. Organic molecules are produced in plants as a result of photosynthesis. In animals, food is digested and the products are absorbed. The term gross production is used to refer to all the organic matter gained, either from photosynthesis in plants or by

183

absorption in animals. Some of this, however, is used for respiration, so not all the material produced is available for growth. Net productivity is the amount of organic material which actually accumulates in the organism during the process of growth. It can be defined by the simple equation:

net productivity = gross productivity − respiration

neurone: a nerve cell. The main features of a neurone are:

* a cell body containing the nucleus

* numerous fine branches known as dendrites extending from this. They provide a large area for connections, or *synapses*, with neighboring nerve cells

* one or more longer processes called axons. These transmit *nerve impulses* away from the cell body. The axons of many neurones in mammals and other vertebrates are enclosed in a sheath of a fatty material called *myelin*. Nerve impulses travel faster in myelinated axons than in nonmyelinated ones.

Neurones which carry impulses from *receptors* towards the central nervous system are known as sensory neurones. Those which conduct impulses away from the central nervous system towards an *effector*, such as a gland or muscle, are motor neurones. Intermediate neurones provide connections between sensory and motor neurones.

neuromuscular junction: a *synapse* between a nerve cell and a muscle. Before a *skeletal muscle* contracts, it must be stimulated by a nerve. A *nerve impulse* travels down the *axon* to the point where it ends close to the muscle fiber. The events which follow are very similar to those which occur in a synapse between two nerve cells. *Acetylcholine* is liberated from small vesicles in the axon terminal. This diffuses across the gap between the axon and the membrane of the muscle fiber. In turn, this changes the membrane's permeability to sodium ions, causing a wave of electrical activity which leads to the contraction of the muscle.

neurotransmitter: one of a number of substances responsible for the transmission of information across a *synapse*. Two important neurotransmitters in mammals are *acetylcholine* and *noradrenaline* but a variety of others exist. The molecules of a neurotransmitter have a particular shape. This enables them to fit into receptor molecules on the postsynaptic membrane. Since other molecules, often used as drugs, may have similar shapes, they can interfere with synaptic transmission. If they mimic the action of the neurotransmitter, they are called *agonistic*; if they block its action, they are called *antagonistic*.

neutrophil: see *granulocyte*

niche: a description of the precise way in which an organism fits into its environment. In simple terms, any definition of an organism's niche must refer to where it lives and what it does there. Take as an example the two-spot ladybird, a small red and black beetle common on many plants where it feeds on aphids. Its niche would be described in terms of the *abiotic* aspects of its habitat, such as the temperature range it can tolerate and the height above ground at which it feeds. A full description would also refer to the biological aspects of its ecology, such as the particular species of plant on which it is found and the size of the prey that it normally eats.

A very important ecological principle is that no two different species have exactly the same ecological niche. The two-spot ladybird and the seven-spot ladybird both feed on aphids. The seven-spot ladybird, however, tends to be found on a much wider range of plants where it eats slightly larger aphids. The two species do not share the same niche.

> **hint** Use appropriate language when you write about ecology. Organisms have specific ecological niches in particular habitats. They don't live in "nests" or have "homes."

nitrification: the conversion of ammonium compounds to nitrites and nitrates, which takes place during the *nitrogen cycle*. The bacteria that do this are *chemoaurotrophs*. Nitrification is an *oxidation* reaction and releases energy. Nitrifying bacteria make use of this energy to produce their organic compounds from simple inorganic molecules.

nitrogen: a biologically important chemical element. Nitrogen forms part of the molecules which make up *amino acids* and *proteins*, *nucleic acids* and many other substances found in living organisms. The *nitrogen cycle* describes how nitrogen circulates or cycles in an ecosystem. The role of nitrogen in living organisms and in particular, its passage through the body of a mammal, forms one of the important synoptic themes that links together different aspects of biology. The spider diagram shows some of these links. You can use the cross-referencing system in this book to add greater detail to the diagram.

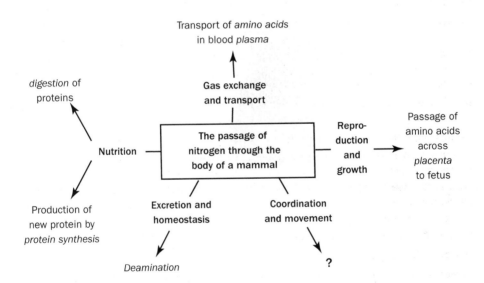

> **hint** Use the word "nitrogen" only when you mean the chemical element. Otherwise use terms such as "nitrates" or "nitrogen-containing compounds." This will help you to avoid confusion.

nitrogen cycle: the way in which the element nitrogen circulates or cycles in an ecosystem. The basic principles of the nitrogen cycle are exactly the same as for other nutrient cycles and the diagram on page 186 is based on this.

185

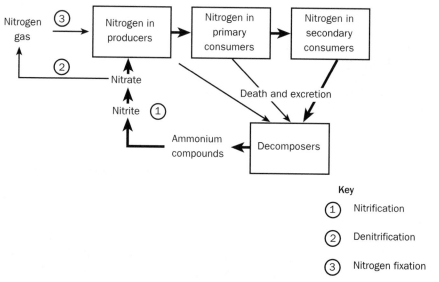

The nitrogen cycle

The particular points to note in this cycle are:

* nitrogen from animals is made available both on their death and through excretory products such as *urea*

* ammonia is produced when organic nitrogen-containing molecules such as protein are broken down by decomposers. This is oxidized to nitrites and nitrates in the process of *nitrification*

* some nitrates may be converted to atmospheric nitrogen. This is *denitrification*

* atmospheric nitrogen can be made available to plants by *nitrogen fixation*.

> **hint** Any examiner will tell you that the nitrogen cycle creates more difficulties than almost any other single topic in the biology examination. Make sure you understand the basic cycle described here.

nitrogen fixation: the incorporation of nitrogen gas present in the atmosphere into organic nitrogen-containing compounds. There are several ways in which this can occur and all of them require a considerable amount of energy. This is because the nitrogen molecules have to be separated to produce nitrogen atoms before fixation can occur.

* Lightning. During thunderstorms, electrical energy in lightning enables nitrogen and oxygen in the atmosphere to be combined to produce various oxides of nitrogen. These are washed into the soil by the rain and are taken up by plants as nitrates. In those parts of the world where violent electrical storms are common, this is an important method of nitrogen fixation.

* Nitrogen fixation by microorganisms living free in the soil. Some soil bacteria possess an enzyme called nitrogenase. With the aid of this enzyme, they are able to convert nitrogen in soil air into ammonium compounds using energy from ATP. They can then make the variety of organic substances which they require.

186

* Nitrogen fixation by microorganisms which live in the root nodules of leguminous plants. This is a similar process to the one described above. For more details, see *leguminous plants*.

* Very large amounts of ammonia are made by industrial processes each year. Nitrogen and hydrogen can be combined directly but this needs high temperatures and pressures and the presence of an inorganic catalyst. Much of this ammonia is used to produce inorganic fertilizers which are added to the soil. This can be taken up directly by plants.

node: the part of a plant stem from which a leaf comes.

node of Ranvier: a gap between the myelin secreted around a nerve cell by one Schwann cell and that secreted by the next. *Nerve impulses* travel along in a series of jumps from one node to the next. This is known as *saltatory conduction* and is responsible for impulses moving much more rapidly in myelinated than in nonmyelinated nerves.

nonactive site-directed inhibition: see *noncompetitive inhibition*

noncompetitive inhibition: occurs when the rate of reaction of an *enzyme* is slowed down or stopped by a molecule which does not combine with its *active site*. The inhibitor attaches somewhere else on the enzyme and in so doing changes the shape of the active site. As a result, the molecule with which normally reacts can no longer fit and form an enzyme-substrate complex. Compare noncompetitive inhibition with competitive inhibition.

nondisjunction: the failure of chromosomes to separate properly during *meiosis*. This may occur either in the first division or in the second and will result in gametes which contain an incorrect number of chromosomes. In humans, nondisjunction of chromosome 21 is the cause of *Down's syndrome*. During the first division of meiosis, the chromosomes belonging to pair number 21 fail to separate. This results in gametes which will contain an extra chromosome. When fertilization takes place, an individual will be produced with 47 chromosomes instead of the usual 46. There will be three copies of chromosome 21.

nonreducing sugar: a sugar which is not able to produce a positive result, an orange-red precipitate, when heated with Benedict's solution. In chemical terms, a *reducing sugar* contains either a free aldehyde group or a free ketone group. It is the presence of one or other of these chemical groups that enables it to reduce Benedict's solution. All *monosaccharides* and most *disaccharides* are reducing sugars. Sucrose is an exception. It is a nonreducing sugar. In order to test for the presence of a nonreducing sugar, it is necessary to hydrolyze it first so that it is broken down into reducing sugars. The test involves the following steps:

* confirm that there are no reducing sugars present by using Benedict's test. A negative result should be obtained

* hydrolyze the sample by heating with dilute hydrochloric acid

* neutralize with sodium hydrogencarbonate

* carry out Benedict's test again. A positive result will now be obtained.

nonshivering thermogenesis: the production of body heat by a method other than shivering. When a mammal moves into a colder environment, it can maintain its body temperature either by limiting heat loss or by the actual generation of heat. There are three basic ways in which it can produce heat. These are from metabolic reactions, from muscle

187

activity or from the process involved in the digestion and absorption of food. Under cold conditions, nonshivering thermogenesis results in extra heat being produced from metabolic reactions in the body. An important source of heat by this means is found in infants and hibernating mammals. This is from a special sort of fat known as brown fat. It acts rather like a radiator. It is respired, but the energy released is in the form of heat rather than as ATP. There is a lot of brown fat around the main arteries in the chest so the heat produced can be distributed rapidly round the body by the blood.

noradrenaline: a substance very similar in structure and function to *adrenaline*. It is a *neurotransmitter* produced by some of the synapses in the *sympathetic nervous system*. These synapses are known as *adrenergic* synapses.

normal distribution: a distribution which when plotted on a graph produces a bell-shaped curve.

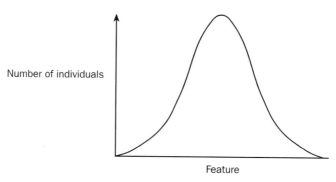

A normal distribution curve

When measurements are made of the variation shown by a particular feature in a population of mature animals or plants, and these figures are plotted on a suitable graph, they often show a normal distribution. A normal distribution curve has a number of mathematical properties. In simple terms, these include:

- the average value is the commonest value. In other words, the mean and the mode are the same

- the curve is symmetrical: 50% of all values will be greater than the mean and 50% will be less

- the way in which the results are spread out is the same for any normal distribution curve. In mathematical terms, approximately 95% of all measurements will be within two *standard deviations* of the mean.

nucleic acid: a molecule which is important in carrying genetic information or producing proteins from this. There are two different sorts of nucleic acid. *DNA* acts as the actual store of genetic information and carries the genes which determine the characteristics of the organism. The different sorts of *RNA* molecules, on the other hand, are involved in the production of proteins from the information coded on the DNA. Chemically, nucleic acids are *polynucleotides* made from *nucleotides* linked together by condensation reactions.

nucleotide: the basic unit or monomer from which *nucleic acids* are formed. The diagram on page 189 shows a nucleotide. It is made up of three components:

- a five-carbon or pentose sugar. This is either ribose, in the case of RNA, or deoxyribose in *DNA*

- a phosphate group

- a nucleotide base. There are five different bases. These are adenine, cytosine, guanine, thymine and uracil, which are usually abbreviated to A, C, G, T and U. The nucleotides which make up DNA contain one of the nucleotides, adenine, cytosine, guanine or thymine. In RNA, the thymine is replaced by uracil.

These three components are linked by means of condensation reactions to make a polynucleotide.

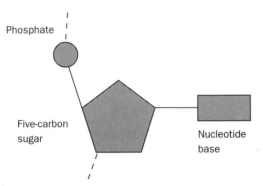

The structure of a nucleotide

nucleotide base: see *base*

nucleolus: a dark staining body found in the nucleus of a cell. It contains the *DNA* with the copies of the genes that code for the special sort of RNA found in *ribosomes*, ribosomal RNA. It is in the nucleolus that ribosomal RNA is produced and the early stages of ribosome formation take place.

nucleus: a large organelle which carries a cell's genetic material and is responsible for the control of its activities. Some cells do not have a nucleus. This is an important difference between the cells of *prokaryotes*, like bacteria, and those of *eukaryotes*, such as animals and plants. Among *eukaryotes*, there may be cells like *red blood cells* and some of the cells found in phloem which do not contain nuclei, but these will have always developed from nucleated cells.

A nucleus is surrounded by a nuclear envelope which consists of two membrane layers. The nuclear envelope has a large number of pores in it through which the nucleus is able to communicate with the cytoplasm. It is, for example, through the nuclear pores that molecules of mRNA pass during *protein synthesis*. Within the nucleus is the genetic material of the cell. This is *DNA*. The DNA is bound together with proteins to form structures known as *chromosomes*. The chromosomes in a cell can only be seen clearly when the cell divides. Between divisions, the genetic material in the nucleus is much more spread out so individual chromosomes are not visible. Within the nucleus is a dark-staining structure known as the nucleolus. This is involved in the production of RNA.

null hypothesis: a useful starting point in examining the results of a scientific investigation. It is based on the assumption that there is no significant difference between sets of observations. Some examples of null hypotheses are given on the next page:

- there is no difference in the mean surface area of leaves from nettle plants growing in the shade and from those growing in the open

- temperature does not affect the proportion of black ladybirds in a population

- the number of female finches visiting a bird feeder in winter is the same as the number of male finches. (In other words, there is not a significantly greater proportion of either males or females.)

Statistical tests can then be used to test these hypotheses. On the basis of the results of such a test, a decision can be made as to whether to accept the hypothesis, on the basis that any difference is likely to be due to chance, or to reject it. If the null hypothesis is rejected, it indicates that the results are more likely to be biologically significant and less likely to be explainable purely in terms of chance.

nutrient cycle: the way in which chemical elements pass through *ecosystems*. Whatever the element involved, the basic principles are the same. These are summarized in the diagram.

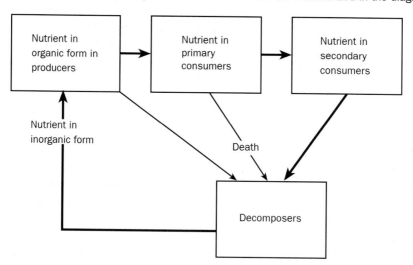

The element is taken up in an inorganic form and incorporated into organic molecules in plants. From the plant or *producer*, it is passed along the *food chain* to the *primary consumer* and *secondary consumer*. Death of any of these organisms results in their decomposition by microorganisms in the soil. The element is released in inorganic form once again and the cycle is complete. The *carbon cycle* and *nitrogen cycle* are specific examples of nutrient cycles.

nutrition: the way in which living organisms obtain the raw materials from which they build up their organic compounds. There are two main methods of nutrition. Autotrophic nutrition, is a method of nutrition in which an organism synthesizes the organic molecules that it requires from simple inorganic molecules such as carbon dioxide and water. To do this, an energy source is necessary. Autotrophic nutrition is characteristic of plants and other producers. Heterotrophic nutrition, on the other hand, involves organisms gaining their nutrients from complex organic molecules. These are broken down to simple, soluble molecules by digestive enzymes and then built up again to form the organic molecules which the organism requires.

oogenesis: the process in which the female gametes are formed in an animal. Towards the outside of each of the ovaries in a mammal is a layer of cells called the germinal epithelium. These are the cells which will develop into female *gametes*. The cells of the germinal epithelium divide by *mitosis* and produce a large number of immature follicles, each one consisting of a future female gamete surrounded by follicle cells. Under the influence of the hormones produced during the *estrous cycle*, one or more of these immature follicles develops to become a mature ovarian follicle. At the same time, the future female gamete that it contains will increase in size and divide by *meiosis*. The basic sequence of mitosis, growth and meiosis, are exactly the same as in the process of *spermatogenesis* in a male. The differences lie in the events which take place during meiosis:

- at the first division of meiosis, the nuclear material in the future gamete divides equally in two. The surrounding cytoplasm, however, almost all goes with one of the resulting sets of chromosomes. This gives a large cell which will become the female gamete and a small one, known as a polar body. The polar body will eventually disappear

- ovulation now takes place and the female gamete, which is still immature, travels down the oviduct

- the second division of meiosis now follows. Again, the nuclear material divides equally, but all the cytoplasm goes with the now mature, female gamete. The remaining cytoplasm goes to form another polar body.

For each female gamete to have the best possible chance of developing into a new individual, it needs to be as large as possible. This mechanism ensures that this is the case.

optic nerve: the nerve which conveys impulses from the retina to the brain.

oral contraceptive: pills consisting of one or more female sex hormones taken by women to prevent pregnancy. Most oral contraceptives are combined pills which contain:

- *Estrogen.* This prevents the development of a mature ovarian follicle by inhibiting the secretion of *FSH* from the pituitary gland. Without a mature follicle, ovulation will not take place and pregnancy cannot occur.

- *Progesterone.* This also blocks the action of hormones controlling the menstrual cycle but it has other contraceptive effects as well. It affects the lining of the uterus and the mucus around the cervix. If ovulation does occur, these changes make it much less likely that implantation will be successful.

Contraceptives containing progesterone only are called minipills. There is a slightly greater risk of pregnancy with them but the absence of estrogen means that there are fewer side effects.

191

oral rehydration therapy (ORT): a method of controlling diarrhea. While controlling this condition is not the same as treating the disease that causes it, it is important to do so, as untreated diarrhea will lead to severe dehydration. As a result of the secretions which are continually pouring into it, the gut contents of a healthy individual are liquid. Soluble molecules and ions are absorbed out of the gut into the blood. This gives rise to a *water potential* gradient. Water is drawn from the gut contents by *osmosis* and firm, dry feces are produced. In patients with diarrhea, there are problems with this process. Oral rehydration therapy involves giving the person concerned a balanced solution of salts and glucose which stimulates the mechanism leading to the reabsorption of water. Feces will then return to their normal consistency. Oral rehydration therapy was originally developed to combat cholera, but it has proved very useful in treating diarrhea resulting from other causes.

order: a level of *classification*.

organ: a structure, made up of different *tissues*, which has a specific function. The stomach is an example of an organ. *Epithelial tissue* lines it and secretes gastric juice. Surrounding this is a wall which has three layers of muscle tissue. Contraction of this muscle results in the stomach contents being continually mixed and churned. The functions of the various tissues are coordinated by nerve tissue, while nutrients are supplied by the blood which is an example of a *connective tissue*. The stomach is therefore an organ made up of four different tissues. Its function is the storage and digestion of food. A number of organs may work together to form a *system*.

hint It is better to define an organ as being made up of different tissues rather than as something which has a specific function. Cells, tissues, organs and systems all have specific functions.

organelle: a structure found within a cell. It has a specific function. Many organelles, such as mitochondria and chloroplasts, are surrounded by membranes. These are known as membrane-bounded organelles in contrast to those, such as ribosomes, which are not surrounded in this way. The table summarizes some of the properties of the main organelles associated with plant and animal cells.

Organelle	Main features	Function
Chloroplast	A large organelle, found only in cells in the green tissues of plants	The site of photosynthesis
Golgi apparatus	Found in the cytoplasm of both animal and plant cells, it is made up of a stack of flattened membranes	Has a variety of functions mostly concerned with the packaging and processing of molecules produced by the cell
Lysosome	A small sac surrounded by a membrane and containing digestive enzymes	Concerned with the breakdown of molecules and structures inside the cell
Mitochondrion	Found in all cells in animals and plants in which aerobic respiration takes place	The site of the reactions associated with aerobic respiration

Nucleus	Largest cell organelle, a single nucleus is found in most living cells. Some cells such as the red blood cells in a mammal do not have nuclei	Contains the cell's DNA. This is the material which forms the genes. These control all the activities of the cell
Ribosomes	Found in the cytoplasm of plant and animal cells and in other organelles such as chloroplasts and mitochondria	Site of protein synthesis

There are various techniques which have enabled biologists to investigate the structure and function of the different types of organelle which are found in a cell. Apart from chloroplasts and nuclei, most organelles are much too small to be seen with a light microscope. *Electron microscopes* have proved essential in allowing detailed structure to be observed. Investigation of the functions of a particular organelle are best carried out on preparations made by separating the organelle concerned from the rest of the cell. This is done by the process of *differential centrifugation*.

> **hint** Do make sure that you can identify an organelle. Neurones and lymphocytes are cells. Chlorophyll is a molecule. Cells and molecules are not organelles!

ornithine cycle: the biochemical pathway by which the *liver* converts ammonia to *urea*. Excess amino acids cannot be stored in the body. Instead they undergo the process of *deamination* where the amino group is separated off from the rest of the molecule and used to form ammonia. Ammonia is a very toxic molecule and, in most animals, has to be converted into a less poisonous substance before it can be excreted safely. In mammals, this molecule is urea. The ornithine cycle involves an acceptor molecule called ornithine taking up the ammonia resulting from deamination and carbon dioxide produced by respiration. In a cycle of reactions, urea is formed and ornithine is produced again.

osmoregulation: the ability of an organism to regulate the concentration of its body fluids. The overall concentration of ions in seawater is approximately the same as that in the body fluids of marine organisms. The simplest marine organisms have very little in the way of osmoregulatory mechanisms.

In most other habitats, the situation is very different. In freshwater, the total ion concentration is much lower. The *water potential* of the surrounding medium will therefore be higher and water will enter the organism by *osmosis*. If this is not controlled in some way, the cells concerned would rupture. This would clearly have disastrous consequences.

Terrestrial animals have different osmoregulatory problems. They live in an environment where water is constantly being lost by evaporation from *gas exchange surfaces* and in the removal of excretory products from the body. Water intake and loss need to be balanced and regulated. Different organisms have different mechanisms which allow them to do this. These range from the *contractile vacuoles* found in single-celled organisms like amoeba to the *kidneys* of vertebrates.

osmosis: is a special case of *diffusion* involving the movement of water molecules. The diagram shows two solutions separated by a *partially permeable membrane*.

193

The water molecules are able to pass freely through the membrane, while the solute molecules are too large and cannot pass through. The solution in side **A** has a greater concentration of solute molecules and a lower concentration of water molecules than the solution in side **B**. Since the membrane will only allow the water molecules to pass through, water molecules will diffuse from side **B** to side **A**. This is one way of defining osmosis: the net movement of water molecules through a partially permeable membrane from where they are in a higher concentration to where they are in a lower concentration. Since the concentration of water molecules is greater in side **B**, this solution will have a greater *water potential* than that in side **A**. A rather better way of defining osmosis is to say that it is the net movement of water through a partially permeable membrane from a solution of higher water potential to a solution of lower water potential.

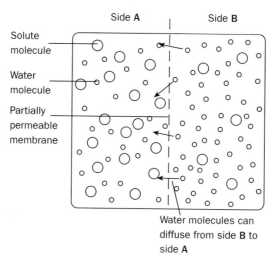

A simple model to show osmosis

 Osmosis involves the movement of water.

osteoarthritis: a disease affecting joints, particularly the knee and the hip. The cartilage is lost and this may result in damage to the underlying bone. The condition is painful and affected people experience difficulty in moving the joint concerned. Osteoarthritis commonly affects older people and may result from past injury or physical stress.

osteoporosis: a condition which accounts for a very high proportion of bone fractures among the elderly. As a person grows, *bone* tissue get denser as more bone substance is formed. After the age of about 40 though, there is a gradual decrease in the amount of bone substance present. Once this falls below a critical level, bones become brittle and there is an increased risk of fracture. This is part of the normal aging process. The situation is, however, more complicated in women. One of the female sex hormones, estrogen, plays an important part in maintaining bone density. Once a woman passes through menopause, she stops secreting estrogen. This can lead to a sharp fall in bone density and a greatly increased risk of arm, hip and leg fractures, and back pain resulting from fractures of the vertebrae in the spine.

ovarian follicle: a structure found in a mammalian ovary. At the start of an *estrous cycle* in a female mammal, the hormone *FSH* is secreted by the anterior lobe of the *pituitary gland*. This hormone stimulates one or more primary follicles found in the ovaries to develop into a mature ovarian follicle. Each mature follicle consists of a developing female gamete surrounded by a mass of follicle cells. The role of the follicle cells is to produce the hormone, *estrogen*, which has various effects on the reproductive system. In particular, it brings about the growth of the lining of the uterus. At ovulation, the developing gamete leaves the ovary and passes into the oviduct, while the remaining cells of the follicle develop into a structure known as the *corpus luteum*.

ovary (animal): the organ in a female animal which is responsible for the production of *gametes* or sex cells. Towards the outside of each of the ovaries in a mammal is a layer of cells called the germinal epithelium. These are the cells which will develop into female gametes. The cells of the germinal epithelium divide by *mitosis* and produce a large number of immature follicles, each one consisting of a future female gamete surrounded by follicle cells. Under the influence of the hormones produced during the *estrous cycle*, one or more of these immature follicles develops to become a mature *ovarian follicle*. When ovulation takes place, the developing gamete leaves the ovary and passes into the oviduct, while the remaining cells of the follicle develop into a structure known as the *corpus luteum*. The process by which female gametes are formed is known as *oogenesis*. In addition to gamete formation, the mammalian ovary has another important function. It secretes *estrogen* and *progesterone*. Between them, these hormones are responsible for the changes which take place in the female body during puberty and for controlling many of the processes associated with the estrous cycle and pregnancy.

 hint Make sure that you distinguish between an ovum, an ovarian follicle and an ovary. If you can't, look them up in this *Complete A–Z Biology* handbook!

ovary (plant): part of the flower that contains an ovule or ovules. The female part of the flower is known as the gynoecium. It is made up of one or more units called carpels. Each carpel has an ovary and a *stigma* which receives the pollen. The stigma is joined to the ovary by a style. After fertilization, the ovary develops into a *fruit*, while the ovules that it contains become the *seeds*.

oviduct: a tube which takes the female gametes from the ovary. In a mammal there are two oviducts, one from each *ovary*. These lead to the single *uterus* in humans or the paired uteri in many other species. Fertilization usually takes place near the top of the oviduct, and the fertilized egg is moved down into the uterus by the beating action of *cilia* on the epithelial cells which line the oviduct and contraction of muscle in its walls.

oxidation: a chemical reaction involving loss of electrons. Although oxidation is usually defined in these terms, it can also involve the addition of oxygen or the loss of hydrogen. Since the electrons that are lost must go somewhere, oxidation of one substance is always accompanied by *reduction* of another. In biological systems, oxidation and reduction are controlled by enzymes known as *oxidoreductases*. Oxidation of organic molecules is the basis of the process of *respiration*.

oxidative phosphorylation: the part of the respiratory pathway in which energy released in the *electron transfer system* is used in the production of ATP. In living organisms, there

195

are a number of ways in which ATP is produced. Two of the most important are oxidative phosphorylation and *photophosphorylation* which is an essential component of the *light-dependent reaction* of photosynthesis. There are a number of similarities between these two processes:

* they both involve the passage of electrons from one molecule to another through a system of carriers known as an electron transport system

* the energy released in this process is used to combine ADP with inorganic phosphate to produce ATP.

There are also some important differences between these two processes. These are summarized in the table below.

Difference	Oxidative phosphorylation	Photophosphorylation
Source of electrons	Chemical reactions during glycolysis and Krebs cycle	Excited electrons produced as a result of light energy being absorbed by chlorophyll
Electron acceptor	Usually NAD	NADP
Location of process	Inner membranes of mitochondria	Granal membranes in chloroplasts
Uses of ATP made in this process	Providing energy for many cell processes and reactions	Providing energy for the light-independent reactions of photosynthesis

oxidoreductase: a type of enzyme which is responsible for catalyszing a reaction involving *oxidation* and *reduction*, a redox reaction. Many important oxidoreductases are associated with the reactions that take place in *respiration*. Some, however, catalyze other biochemical reactions. The reason why fruits such as apples and bananas turn brown when cut open is because they contain certain substances that are oxidized on exposure to air. The enzyme involved in this process is an oxidoreductase.

oxygen debt: the amount of oxygen required to oxidize the lactate produced during *anaerobic respiration*. Muscles in the body of a mammal rely on the presence of oxygen in order to respire. During periods of vigorous exercise, however, there is insufficient oxygen available to meet their respiratory requirements and they respire anaerobically. Lactate is produced as a result and accumulates in the body tissues. When activity returns to normal, this lactate is removed. It is either converted to pyruvate or to carbohydrate in the liver. The oxygen needed for removing the lactate is the oxygen debt.

oxyhemoglobin dissociation curve: a graph showing the amount of oxygen carried by the *hemoglobin* in the blood plotted against the concentration of oxygen present. The graph on the next page shows the human oxyhemoglobin dissociation curve.

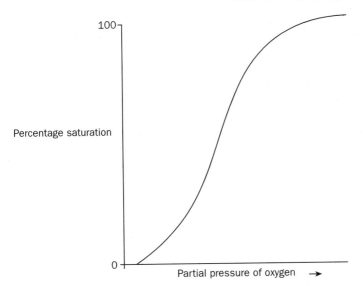

The important things to note are that:

- the hemoglobin is almost completely saturated with oxygen at the concentration normally found in the lungs
- the steep fall in the curve means that the blood will rapidly give up oxygen to respiring tissues.

There are many different types of hemoglobin which result from differences in the sequence of amino acids which make up their polypeptide chains. Each has slightly different oxygen-carrying properties which can be related to the environment in which the animal lives. The graph below shows the oxyhemoglobin dissociation curves of a number of different animals.

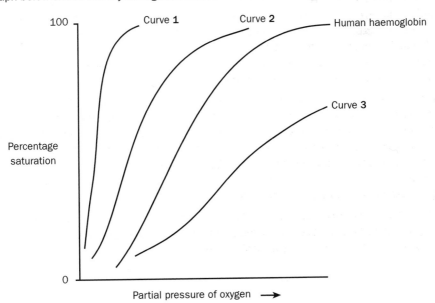

197

* Curve 1

  Pigments which have dissociation curves of this type usually act as oxygen stores. They are saturated with oxygen at relatively low partial pressures. They hold on to this oxygen, releasing it when the amount in the tissues falls to a very low level. The pigment myoglobin found in the muscles of seals and whales has a curve of this shape.

* Curve 2

  Found in animals which live in an environment which has relatively little oxygen present. Examples include the llama which lives high in the Andean mountains of South America, the prairie dog which spends much of its time in burrows deep underground and the mammalian fetus in the uterus of its mother.

* Curve 3

  Hemoglobin with a dissociation curve like this give up its oxygen readily. It is found in animals which have a very high respiration rate such as birds and small mammals like shrews.

oxytocin: a *hormone* released by the posterior lobe of the *pituitary* gland which causes contraction of the smooth muscle in the wall of the uterus during birth. It also brings about the release of milk from the mammary glands by causing muscle fibers in these glands to contract.

pacemaker: see *sinoatrial node*

Pacinian corpuscle: a type of pressure receptor. In humans, although they are found in other parts of the body, Pacinian corpuscles are most numerous in the skin of the fingers and toes. If a longitudinal section is cut through one of these receptors it looks rather like a similar section taken through an onion bulb. A nerve ending, consisting of a dendron from a sensory *neurone*, runs through the middle. This is surrounded by a capsule formed of layers of connective tissue separated from each other by a thick jelly-like substance. When pressure is applied to the capsule, these membranes are distorted. This affects the permeability of ion-channel proteins, allowing the sodium ions to flow into the sensory neurone which give rise to a nerve *impulse*.

palisade cell: tall, narrow, photosynthetic cell found in the mesophyll of a leaf. This shape enables them to be packed closely together and provides an excellent arrangement for absorption of light. Palisade cells contain many chloroplasts which are able to move within the cytoplasm. The chloroplasts can, therefore, absorb light very efficiently in low light intensities but will be protected from bleaching in bright light.

pancreas: a gland found in the abdomen of a mammal. It is an *endocrine* gland, as it secretes the hormones *insulin* and *glucagon*, which are important in the regulation of *blood glucose*. It is also an *exocrine gland*, secreting pancreatic juice along the pancreatic duct into the first part of the small intestine. Pancreatic juice contains many digestive enzymes, in particular *amylase*, *lipase* and *trypsin*, in a liquid made alkaline by the presence of hydrogencarbonate ions.

 The pancreas is a gland which secretes digestive juice into the small intestine. Food does not pass through it.

pancreozymin: a hormone which stimulates the *pancreas* to release digestive enzymes. A possible way in which it works is that the presence of amino acids in the first part of the small intestine stimulates cells in the wall to produce pancreozymin. This hormone travels in the blood to the pancreas. Here it causes digestive enzymes to be released from the cells of the pancreas into the pancreatic juice. It is one of several hormones which help to limit the release of digestive enzymes to the times when there is food actually present in the gut. It used to be thought that another hormone, called *cholecystokinin*, stimulated contraction of the gall bladder releasing *bile*. It is now known that only one hormone is involved. Because of this, this hormone is now usually called cholecystokinin-pancreozymin or CCK-PZ.

pandemic: an outbreak of disease that affects large numbers of people in many different countries. It is an *epidemic* on a global scale. Examples of pandemics include the Black Death which spread through Asia and Europe in the fourteenth century and the influenza

199

outbreak that followed the 1914–18 war. It is considered by many that AIDS should also be described as pandemic.

parasite: an organism that lives in or on a host organism. The parasite gains a nutritional advantage from this relationship, while the host suffers a disadvantage. Parasites are adapted in different ways to living inside or on their hosts but these adaptations frequently involve:

* a means of attachment to the host. This depends on the parasite concerned but may involve hooks and suckers as in tapeworms or the strongly curved claws of ticks and lice

* methods of resisting the defense mechanisms of the host. Parasites need to avoid the digestive enzymes of the host's gut or destruction by the immune system in its blood

* development of reproductive organs and enormous powers of reproduction. Only by producing very large numbers of young do parasites stand a chance of successfully completing their life cycles

* a complex life cycle. This is often an adaptation to enable infection of another individual to occur

* reduction of body systems other than those concerned with reproduction.

parasympathetic nervous system: is a division of the *autonomic nervous system* which generally controls the resting functions of the body and is more important when the body is at rest. Some of the important features of the parasympathetic system are that:

* synapses in this system secrete acetylcholine as a transmitter

* effects are usually inhibitory, for example, parasympathetic stimulation decreases the breathing rate and decreases the heart rate and stroke volume.

parenchyma cells are relatively unspecialized cells found in plants. Their main characteristics are that they have thin, permeable, *cellulose* cell walls and therefore have living contents. In addition, they are not as elongated as the cells in many other plant tissues. Parenchyma is often referred to as packing tissue, since it fills the spaces between other types of cells. This is rather misleading since it suggests that it does not have any real function. Parenchyma is, in fact, extremely important in providing support to young plants and can photosynthesize and store materials as well.

parthenocarpy: where fruit development in plants can occur without *fertilization* having taken place, as for example in cultivated bananas. Under normal circumstances, the seeds which develop as a consequence of fertilization produce *plant growth substances*, such as *auxins* and *gibberellins*. These substances control the growth of the fruit. By using these plant growth substances on unfertilized plants, seedless varieties may be produced. A commercial application of this idea is the use of gibberellins to produce seedless grapes.

partial pressure: a measure of the amount of gas present in a mixture of gases. It is a term frequently used when describing the transport of oxygen by *hemoglobin*.

partially permeable membrane: a membrane which only allows small molecules, such as those of water, to pass through. When a solution containing a particular solute such as sucrose is separated from distilled water by a partially permeable membrane, there will be a higher concentration of water molecules in the distilled water. Since only the water molecules can pass through the membrane, this will result in their *diffusion* from where they were in a higher concentration to where they are in a lower concentration. This special case of diffusion is known as *osmosis*.

passage cell: a cell in the *endodermis* of a root which allows substances to pass freely from the outer part of the root to the xylem. The endodermis consists of a ring of cells between the outer part of the root or the cortex and the vascular tissue in the center. A band of waterproof material called the *casparian strip*, runs round the walls of each of these endodermal cells. It prevents substances from moving through the cell walls along the *apoplastic* pathway. Passage cells do not have this band of material present and therefore allow water and other substances to pass much more readily through the endodermis.

passive immunity: when *antibodies* made by one organism enter the body of another organism. In mammals, this process occurs naturally as some antibodies are passed from the mother to the fetus through the *placenta*. In addition, the secretion produced by the mammary glands immediately after birth is not milk but a liquid called *colostrum* which is very rich in antibodies. Injections of antibodies are useful in circumstances where the body would not have enough time to produce antibodies of its own by active immunity. They are used to treat snakebite and certain diseases such as tetanus. Antibodies produced by one organism are foreign proteins when they are introduced into the body of another. They act as *antigens* and are usually destroyed rapidly. Protection by passive immunity is therefore only short-lived.

pathogen: a microorganism that is a *parasite* of an animal or plant and causes a *disease*. There are many different microorganisms found in and on the surface of the human body but not all of them are pathogens. The nineteenth-century biologist, Robert Koch, put forward a series of principles called *Koch's postulates* which can be used to confirm that a particular microorganism is a pathogen and causes a particular disease.

peak flow: a measure of the maximum rate at which air can be exhaled. *Asthma* is a condition in which the smooth muscle in the walls of the *bronchi* contracts. Together with the production of large amounts of *mucus*, this leads to a narrowing of the airways and the formation of a blockage which limits the rate at which air can enter and leave the alveoli. By using a peak flow meter it is possible to monitor this rate and control asthmatic attacks with the timely use of appropriate drugs.

pectin: a carbohydrate consisting of a mixture of polysaccharides. Pectins are found in and between plant *cell walls*, where they help to cement the *cellulose* fibers together. They are commercially very important in the extraction of fruit juices. When fruit ripens, the pectin molecules start to break down. Part of the molecule is insoluble and remains attached to the cell wall, while the part of it which is soluble extends into the cell, mixing with the juice. The result of this is that the juice is thick and difficult to squeeze out. The addition of pectinases to fruit such as currants considerably increases the amount of juice that can be extracted. The cloudiness of some extracted fruit juices is also due to pectin. In this case, treatment with pectinase *enzymes* helps to clarify it. Many *microorganisms* produce pectinases and the enzymes which are used commercially come from these sources.

pentose: a sugar that has five carbon atoms in each of its molecules. Pentoses are monosaccharides. Biologically important pentose sugars include deoxyribose and ribose, which form part of the structure of the nucleotides which make up *DNA* and *RNA*. Ribulose is a pentose which, when combined with two phosphate groups, becomes ribulose bisphosphate. This molecule acts as the carbon dioxide acceptor in the *light-independent reaction* of photosynthesis.

> **hint** If you are asked to identify the different components of a DNA nucleotide, make sure that you refer to the sugar as either a pentose or deoxyribose. Just calling it a sugar is not enough.

pepsin: a protein-digesting enzyme found in the stomach. Pepsin is secreted by cells in the lining of the *stomach* wall known as peptic cells. They produce the *enzyme* in an inactive form known as pepsinogen. This is activated on contact with hydrochloric acid in the stomach. Pepsin is an *endopeptidase*. It breaks the peptide bonds between certain amino acids in the middle of a polypeptide chain rather than peptide bonds at the ends. As a result, the products of digestion are a variety of smaller polypeptides. Pepsin works best in acid conditions and has an optimum pH of between 1.6 and 3.2. It can, therefore, only function in the stomach. Once the contents of the stomach enter the small intestine, they are neutralized by the alkaline secretion produced by the pancreas. Under these conditions, pepsin rapidly stops working.

pepsinogen: the inactive form in which the enzyme *pepsin* is secreted by cells in the lining of the stomach wall. Once in the stomach, some of the *peptide bonds* in the pepsinogen are hydrolyzed by hydrochloric acid. The pepsin produced as a result is now able to begin the digestion of protein. Once some pepsin has been produced, it can activate more pepsinogen, so the activation process can be thought of as being self-catalyzing.

peptide: a compound formed by the joining together with peptide bonds of a small number of amino acids. Peptides are difficult to distinguish from *polypeptides* and the two terms are used rather imprecisely. In general, a peptide is made up from a few *amino acids* and is therefore shorter in length than a polypeptide. Peptides and smaller polypeptides result from the digestion of proteins by *endopeptidases*, such as *pepsin* and *trypsin*. Some hormones are peptides. Among these are the hormones known as release factors which are produced by the *hypothalamus* in the brain. These are taken in the blood to the anterior lobe of the *pituitary gland*, where they stimulate the secretion of other hormones, such as growth hormone and follicle stimulating hormone, *FSH*.

peptide bond: the chemical bond formed when amino acids are linked together. The diagram shows two amino acids and the molecule produced when these are joined together by a peptide bond.

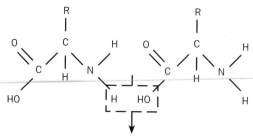

Removal of molecule of
water to form a peptide bond

This is a condensation reaction and involves the removal of a molecule of water. A peptide bond may be broken or hydrolyzed by the addition of a water molecule.

hint When an enzyme is heated, the bonds which maintain its *tertiary structure* break. This leads to it being denatured. Denaturation is not the result of breaking peptide bonds.

percentage cover: in an ecological study, this is the measure of the proportion of ground in the area being sampled that is covered by a particular species of plant. Percentage cover can be determined in one of two ways. A rough figure may be estimated from a simple inspection of the area enclosed by a *quadrat* frame. An alternative approach is to use *point quadrats* in which case the proportion of pins hitting the species concerned is recorded and expressed as a percentage.

Percentage cover has advantages over methods which rely on counting the number of individuals present. It probably gives a better idea of the ecological importance of a particular plant. A single large plant, for example, may have a much greater percentage cover than ten very small ones. It will also exert a much greater influence in the community. Another advantage is that it is often extremely difficult to identify individual specimens of plants such as grasses. By using percentage cover, there is no need to do so.

peristalsis: the wave of contraction of *involuntary muscle* by which food is pushed along the gut. After a meal, the increase in gut contents stretch the muscle in the gut wall. This initial stretching then brings about contraction of the circular muscles. The *lumen* of the gut narrows and squeezes the contents. As the muscle contraction moves in a wave along the gut, the squeezing pushes the food in front of it.

pesticide: a substance which is used to control pests. An ideal pesticide needs to have a number of properties. It should be:

* quick acting, rapidly killing the pest against which it has been used
* specific in its action and harmless to other organisms such as humans and other mammals, beneficial insects, soil bacteria and the crop itself
* effective at low dosages
* biodegradable so that, as soon as it has had the required effect, it is broken down and does not produce a long-lasting impact on the ecosystem.

Frequent use of pesticides may mean that pest populations become resistant, so there is a constant need to develop new, more effective chemicals. The most important agricultural pesticides are *herbicides* or weed killers, *insecticides* and fungicides.

hint There is a surprising lack of knowledge of the role of *fertilizers* and *pesticides* in agriculture. Fertilizers are substances which are applied to the soil to provide crops with the required nutrients. Pesticides include herbicides, fungicides and insecticides. They are used to kill pests. Treat the two as entirely separate and don't begin answers with statements such as "Pesticides and fertilizers...."

petal: one of an outer ring of structures in a flower which are often brightly colored and together make up the *corolla*.

pH: a scale which measures the concentration of hydrogen ions in a solution. Solutions with pH values below seven contain high concentrations of hydrogen ions and are acidic.

Solutions with values above seven have low concentrations of hydrogen ions and are alkaline. The pH of the surrounding medium can have a marked effect on protein molecules and hence on enzyme-controlled reactions. Soil and water pH are also important ecological factors which help to determine the distribution of living organisms.

hint Make sure you can relate pH, hydrogen ion concentration and acidity. The lower the pH of a solution, the greater the concentration of hydrogen ions and the more acid the solution.

phage: a particular type of virus which infects bacteria and fungi. The T2 bacteriophage which, as its name suggests, attacks bacteria, has been studied in detail and has contributed much to our present knowledge of the life cycle of viruses. Bacteriophages are important in *genetic engineering*, as they can be used as vectors to introduce genes into bacterial cells. With the growth of *biotechnology*, large numbers of bacteria are cultured in order to produce useful products. If a culture becomes infected with bacteriophages, it will be rapidly destroyed.

phagocytosis: the transport of large particles or even parts of cells through the membrane into the cytoplasm. The *plasma membrane* surrounds the particles concerned. A *vesicle* or *vacuole* is formed which is pinched off and moves into the cytoplasm. Although phagocytosis takes place in many cells, two particularly good examples are provided by amoeba and certain kinds of white blood cell. Amoeba changes its shape as it moves by extending its cytoplasm to form processes called pseudopodia. These cytoplasmic processes can engulf prey so that it becomes enclosed within a food vacuole in the cytoplasm. Enzymes are secreted into this vacuole and digest the prey. Various kinds of phagocytic white blood cell engulf pathogenic bacteria. Again, enzymes are released into the resulting vacuoles and the bacteria are digested.

phenotype: the characteristics of an organism which result both from the genes the organism possesses and the environment in which it has developed. The Himalayan rabbit provides a good example. These rabbits have white fur but black feet, ears and tail. Crossing two purebred Himalayan rabbits always results in animals with these characteristics. However, the black pigment will only develop on parts of the body that are at a low temperature. Himalayan rabbits clearly possess the *gene* for making black fur, but only in the right environment will the typical phenotype develop.

phenylketonuria: an inherited condition involving the lack of a specific enzyme concerned with *amino acid* metabolism. It is a *recessive* condition so children born with two copies of the faulty *allele* are unable to make the enzyme phenylalanine hydroxylase. This enzyme converts the amino acid, phenylalanine, into tyrosine. If the enzyme is missing, the concentration of phenylalanine in the blood increases leading to damage to the nervous system and severe mental retardation. Phenylketonuria is an example of a genetic disease which can be treated successfully. Newborn babies are screened for this condition. If they are found to be affected they can be put on a diet in which the amount of phenylalanine is carefully controlled, allowing the child to develop normally.

pheromone: a substance which is produced and released into the environment by one organism and affects the behavior of another organism of the same species. Pheromones have a wide range of different functions. For example, they are present on feces used to mark *territory* in badgers; they provide information about sex and reproductive state in dogs and can delay or hasten sexual maturity in mice. They should be distinguished from *hormones*.

phloem: is the plant tissue which is responsible for transporting the products of photosynthesis away from the leaves to other parts of the plant. It has a number of cell types, of which the most important are sieve tubes and companion cells. Sieve tubes are made up from individual cells arranged end to end. The walls between these cells are perforated and form the characteristic sieve plates. Sieve tubes do not contain nuclei and have only a little cytoplasm. Associated with the sieve tubes are companion cells, which are characterized by densely stained cytoplasm. The method by which phloem transports sugars through the plant is not well understood. Although most biologists consider that substances travel by a system of *mass flow*, there is still some debate over the mechanism involved.

> **hint** Do you know the positions of phloem and xylem in a stem?

phosphocreatine is a molecule which breaks down to give creatine. This liberates energy, which can be used to resynthesize *ATP* in muscles. At rest, there is very little ATP present and this is rapidly used up when muscles contract. The function of phosphocreatine is to replace and provide a supply of ATP for immediate use. Phosphocreatine can be replenished from ATP produced in respiration.

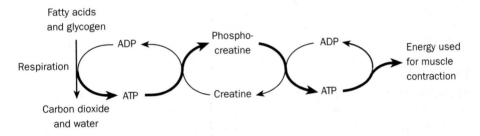

phospholipid: a type of *lipid* that contains a phosphate group. The diagram on the next page shows a common phospholipid.

The glycerol molecule has two fatty acids joined to it. Instead of a third fatty acid, there is a phosphate group. The phosphate "head" of the molecule is water soluble, unlike the fatty acid "tail." This property of the molecule is important in the structure of the *plasma membrane*.

photoautrotroph: an *autotroph* which uses light energy in order to convert simple inorganic molecules into organic ones. All green plants are photoautotrophs, as are a number of organisms belonging to the kingdoms *Prokaryotae* and *Protoctista*. Photoautotrophs are *producers* and form the basis of most *food chains* and food webs.

photolysis is the splitting of water molecules that occurs during the *light-dependent reaction* of photosynthesis. High-energy electrons are ejected from the *chlorophyll* molecules when light has been absorbed, leaving a positively charged chlorophyll ion behind. Water molecules break down to produce electrons which replace those that have been lost. The reaction also produces hydrogen ions and a molecule of oxygen. It can be summarized by the equation:

$$2H_2O \rightarrow 4H^+ + 4e^- + O_2$$

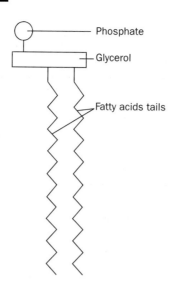

A simplified diagram of a phospholipid molecule

photophosphorylation: the part of the *light-dependent reaction* of photosynthesis in which energy released in the *electron transfer system* is used in the production of *ATP*. Light striking a *chlorophyll* molecule causes some of its electrons to become excited. These are lost from the chlorophyll molecule and pass through a series of electron acceptors. As a result, energy is progressively lost and some of this is used to make ATP. This is the process of photophosphorylation.

photoreceptor: a specialized cell which is sensitive to light. Plants and simple, one-celled organisms such as *Euglena* have areas in their cytoplasm which contain pigments which are sensitive to light. Animals, however, possess specialized light-sensitive cells, or photoreceptors, which are often grouped together to form photoreceptor organs. In their simplest form, such as in some worms, these organs are no more than flat sheets of cells which allow the animal to distinguish between light and dark. At the other extreme, the eyes of mammals are much more complex structures capable of detecting very low intensities of light, distinguishing between different colors, and forming sharply focused, clear images.

photosynthesis: the process by which plants and certain *photoautotrophs* are able to use light energy to convert carbon dioxide into carbohydrates. The diagram on the next page summarizes the main features of the biochemical pathway involved in photosynthesis. For further details of this biochemical pathway, see *light-dependent reaction* and *light-independent reaction*.

Although photosynthesis is often regarded as resulting in the formation of carbohydrates, it is important to appreciate that it is the starting point for the synthesis of all organic molecules in a plant. With the addition of mineral salts absorbed through the roots to the carbon part of the molecule produced in photosynthesis, molecules such as *amino acids, lipids* and *nucleic acids* are made.

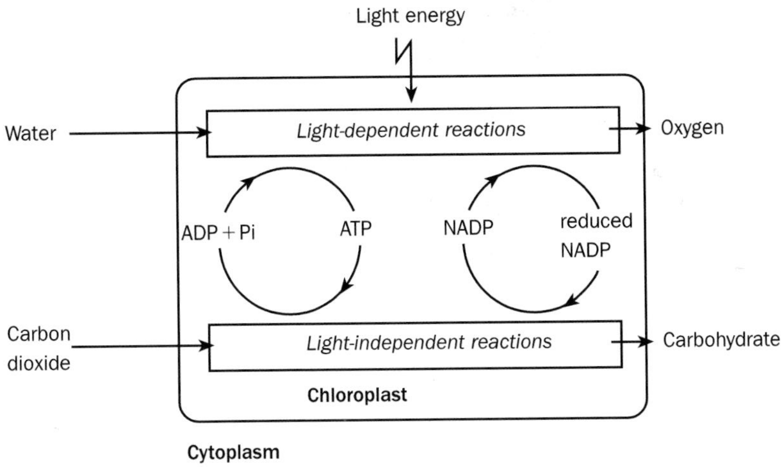

Light energy

Water → Light-dependent reactions → Oxygen

ADP + Pi   ATP   NADP   reduced NADP

Carbon dioxide → Light-independent reactions → Carbohydrate

Chloroplast

Cytoplasm

A summary of the biochemical pathways involved in photosynthesis

hint The equation:

$$6CO_2 + 6H_2O \rightarrow C_6H_{12}O_6 + 6O_2$$

is a very simple summary of a very complex process. It does not mean that carbon dioxide is converted into oxygen. Carbon dioxide is converted into carbohydrates and oxygen is produced as a waste product.

photosystem: a group of 300 or so chlorophyll molecules found on the membranes in a chloroplast. It functions as a sort of "light funnel", absorbing light and passing the energy on to one particular pigment molecule, the reaction center. The electrons which are involved in the *light-dependent reaction* and bring about the conversion of light energy to chemical energy come from these reaction centers.

phototropism: a growth movement of a whole plant organ in response to light. Stems are generally positively phototropic and grow towards the direction of the light. Roots are usually negatively phototropic, growing away from the light. For further details, see *tropism*.

phrenic nerve: a nerve which runs from the brain to the diaphragm. Regular bursts of nerve impulses pass from the respiratory center in the *medulla* down the phrenic nerve. They bring about the contraction and flattening of the diaphragm as a mammal breathes in.

phylum: a level of *classification*. Each kingdom is divided into a number of *phyla*.

physiology: the study of the way in which the organs and systems in a living organism function.

phytochrome: a pigment found in plants which is able to absorb light. It exists in two forms: one form is particularly sensitive to red light and is referred to as P_R; the other form, which is more sensitive to the far-red light that is just outside the normal visible spectrum is called P_{FR}. The diagram on the next page shows the relationship between these two pigments.

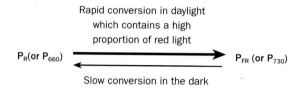

The relationship between the two forms of phytochrome

After a period in the daylight, most of the phytochrome will be in the form P_{FR} since the P_R will have had time to absorb red light and be converted into P_{FR}. However, the amount of P_{FR} will decline if the plant is left in the dark.

It is the P_{FR} which appears to be the active form and can promote various activities in the plant, such as *flowering*, or may stimulate *germination*.

pigment: a colored substance. A wide variety of different chemicals form pigments, and these pigments have many different functions in living organisms. The table shows some examples.

Pigment	Function
Anthocyanins	Pigments found in *cell sap* which are responsible for producing blue and red colors in flowers. Slight changes to the basic molecule will change the color of the pigment. The addition of a methyl (CH_3) group, for example, will produce a redder color
Chlorophylls	The group of pigments responsible for absorbing the light used in photosynthesis
Carotenoids	Responsible for the red, brown and yellow colors of plants. The yellow color of daffodils and the red of tomato fruits are due to the presence of these pigments
Rhodopsin	The light-sensitive pigment found in the rod cells in the eyes
Respiratory pigments	Molecules such as the red iron-containing *hemoglobin* and the blue-green copper-containing pigment, hemocyanin, found in the blood of many invertebrates, are involved in the transport of oxygen

pinocytosis: the way in which a cell takes fluids through the *plasma membrane* into the cytoplasm. The plasma membrane folds around and produces a small vesicle containing a minute drop of the solution or suspension of particles concerned. This vesicle is pinched off and moves into the cytoplasm.

pituitary gland: an *endocrine* gland attached to the underside of the *hypothalamus*. It really consists of two separate glands, the anterior lobe and the posterior lobe. The anterior lobe is connected to the hypothalamus by a *portal vein*. It secretes a number of different hormones and their secretion is under the control of the hypothalamus. This is summarized in the diagram on the next page.

The posterior lobe does not actually produce any hormones. It stores and releases *antidiuretic hormone*, which has been produced in special cells in the hypothalamus.

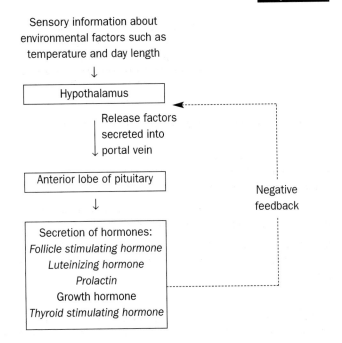

Sensory information about environmental factors such as temperature and day length

$\downarrow$

Hypothalamus

Release factors secreted into portal vein

Anterior lobe of pituitary

$\downarrow$

Secretion of hormones:
Follicle stimulating hormone
Luteinizing hormone
Prolactin
Growth hormone
Thyroid stimulating hormone

Negative feedback

placenta: an organ formed partly from membranes surrounding the developing mammalian fetus and partly from the lining of the *uterus*, the endometrium. It is the organ of exchange over which oxygen and nutrients diffuse into or enter the fetal blood system by active transport. Waste products such as *urea* and carbon dioxide are removed from the blood of the fetus to that of the mother via the placenta. Like all exchange organs, it has a number of adaptations (see *Fick's law*) which ensure efficient exchange. These can be summarized as:

large surface area	• the exchange surface of the placenta is folded into *villi and microvilli*
maximum difference in concentration on either side of the exchange surface	• fetal hemoglobin has a much higher affinity for oxygen than adult hemoglobin • the blood of the mother more or less flows in the opposite direction to that of the fetus, forming a *counter-current* system
thin membrane	• there is a reduction in the number of cell layers between the blood of the fetus and that of the mother.

In addition, the placenta has an important endocrine function and secretes a number of hormones, including *progesterone* and *estrogen*.

> **hint** The placenta functions as an exchange surface allowing the passage of substances between the blood of the mother and the blood of the fetus. It also secretes hormones. It does not hold the fetus in the uterus nor does it serve to protect the fetus from accidental bumps.

plankton: the name given to the microscopic organisms which float in the surface layers of open water. Phytoplankton consists of photosynthetic organisms which, when conditions are right, can multiply and increase in numbers extremely rapidly. They are mainly small unicellular algae. Zooplankton, on the other hand, are the animals of the plankton. They feed both on the phytoplankton and on other zooplanktonic organisms. A great variety of organisms make up the zooplankton, including small arthropods and the larval stages of many larger animals. Marine plankton is extremely important from an ecological point of view. It forms the food of many marine animals, including some of the largest fish and many whales. It also plays an extremely important part in the *carbon cycle* and is probably one of the most important ways in which the carbon in carbon dioxide can be converted into organic carbon compounds.

Plantae: the kingdom containing plants (see *classification*). It is important to remember that this kingdom contains not only the familiar flowering plants but also others such as mosses and ferns. Plants share the following features:

* cells have *cell walls*, of which the main component is *cellulose*

* they are photosynthetic and their chlorophyll is found in special organelles known as *chloroplasts*

* they have a life cycle which shows *alternation of generations* involving a sexually reproducing generation and an asexual generation.

plant growth substance: a chemical produced by a plant which is involved in the control of various aspects of its growth and development. Plant growth substances used to be called plant hormones but they differ from the *hormones* produced by animals in a number of ways. In particular, they are generally produced close to the site of action rather than being released from one organ and transported to another. There are five main groups of plant growth substance – *auxins*, *cytokinins*, *gibberellins*, *abscisic acid* and *ethene* – and they have a wide range of effects on plant tissues. Different concentrations of the same substance may have different effects on different tissues. They may even have different effects on the same tissue. Additionally, the effect of one substance may be modified by the presence of another. Despite the fact that relatively little is still understood about the way in which they act, they have widespread commercial applications.

plasma is the liquid part of the blood. It consists mainly of water but about 10% is made up of dissolved substances. Many of these dissolved substances, such as glucose, are maintained at roughly constant levels by the body's homeostatic mechanisms; others vary according to circumstances.

plasma cells are produced when *B-cells* in the blood are exposed to an infection by a *pathogen*. These lymphocytes divide rapidly by mitosis to form plasma cells. The plasma cells produce and secrete *antibodies*.

plasma membrane: the *cell membrane* found on the outside of a cell.

plasma proteins: the proteins that are normally present in the liquid part of the blood, the *plasma*. These proteins have a number of functions. They are *buffers* and help to maintain the pH of the blood at a constant level. They also contribute to the *solute potential* of the plasma and play an important part in the formation of tissue fluid and its return to the blood. There are three main types of plasma protein. The albumins are the commonest. Among their functions is the transport of other molecules in the plasma. Globulins are the proteins which include *antibodies*. *Fibrinogen* is extremely important in *blood clotting*. The *liver* plays an important part in the metabolism of plasma proteins, most of which are made in the liver. They circulate in the plasma and are eventually broken down again in the liver.

plasmid: a small circular piece of *DNA* found in bacteria and certain other organisms. In bacteria, most of the DNA in the cell is joined to make a loop. Despite the fact that this DNA is not associated with proteins, it is referred to as the bacterial *chromosome*. The bacterial chromosome is the largest piece of DNA in the cell but most bacteria also carry plasmids. Plasmids often contain genes which code for resistance to *antibiotics*. They can also be passed from one bacterium to another by a variety of processes. These two features give them considerable importance. Overexposure to antibiotics may result in otherwise harmless bacteria, such as those found in the gut, being selected for antibiotic resistance. This resistance can be passed via plasmid transfer to disease-causing bacteria, which would then be more difficult to control. In addition, the fact that plasmids are readily accepted by bacteria makes them useful as *vectors* in *genetic engineering*.

plasmodesmata: thin strands of protoplasm which penetrate the cell walls of plants and connect the cytoplasm of one cell to that of another. They are part of the *symplastic pathway*, by which substances are able to go from one cell to the next without having to pass through cell walls and *plasma membranes*. Most mineral ions move through the plant in this way.

plasmolysis: the shrinkage of cytoplasm away from the cell wall of a plant cell as a result of water being lost from the cell. If a plant cell is immersed in a concentrated sugar solution, the concentration of water molecules inside the cell will be higher than that outside. Water will move by *osmosis* from the higher *water potential* inside the cell to the lower water potential outside. The cell contents will shrink away from the cell wall and it will be plasmolyzed. If, on the other hand, the water potential outside the cell is higher than that inside, water will move into the cell and it will be turgid. At the point where the water potential outside is exactly the same as that inside, the cytoplasm will be just coming away from the cell wall. At this point, the cell is said to be in a state of incipient plasmolysis. Although this would appear to be a good way of measuring the water potential of a cell, there is a practical problem. It is not possible to see the point at which the cytoplasm is just losing contact with the cell wall. In practice the assumption is made that when 50% of cells in the tissue are plasmolyzed, the water potential of the solution inside the cell will be the same as the water potential of the surrounding solution.

The relationship between water potential, pressure potential and solute potential is given by the equation:

water potential = solute potential + pressure potential

At incipient plasmolysis, the cell wall will not be pressing on the cytoplasm so the pressure potential will be zero. If this is the case:

water potential = solute potential + 0

or:

water potential = solute potential.

So finding the water potential by this method also enables the solute potential of the cell to be determined.

plastid: an organelle found in the cytoplasm which is characteristic of plant cells. There are a number of different types of plastid as well as the familiar *chloroplast*. Plastids contain an internal system of membranes and pigments or storage materials.

platelets (thrombocytes): small cell fragments found in the blood which play an important part in the clotting process. They are formed as a result of the breaking up of large cells

211

found in the bone marrow. When the body is wounded and a blood clot forms, soluble *fibrinogen* is converted to insoluble *fibrin*. This forms a mesh of protein fibers over the surface of the wound which traps red blood cells and forms a clot. A complex mechanism controls the process of *blood clotting*. Fibrinogen can only be converted into fibrin in the presence of the enzyme, *thrombin*. Thrombin is normally present in the blood in an inactive form known as prothrombin. When the blood comes into contact with air or damaged tissue, the platelets break down and produce thromboplastin. Along with a number of other substances, thromboplastin converts the inactive prothrombin to thrombin.

Atherosclerosis is a disease of the arteries caused by *atheroma* or fatty deposits in the walls. These can result in rough patches to which platelets may stick and produce thromboplastin. If this happens, there is a risk that a blood clot may form and block an artery. This is thrombosis and it can cause the death of tissues which are supplied by the particular artery. When atherosclerosis affects the *coronary arteries* it may lead to a heart attack or *myocardial infarction*.

Platyhelminthes: the animal phylum containing the flatworms and two important groups of *parasite*, the tapeworms and the flukes. Platyhelminthes share the following features:

* they are dorsoventrally flattened, unsegmented worms

* they do not have a *coelom*

* their gut, when present, is branched and has only a single opening.

plumule: part of an embryo plant found in a seed. When *germination* occurs, the plumule grows upwards towards the light. It is the shoot of the embryo plant.

podocyte: a cell from the inner layer of the *renal capsule*. Podocytes have a large number of processes sticking out from them. Fluid entering the capsule from the blood goes through the gaps between these processes. This makes *ultrafiltration* more efficient as they do not provide a barrier to the movement of fluid from the blood system into the kidney tubule.

point mutation: see *gene mutation*

point quadrat: a *quadrat* reduced in size to a pinpoint. A point frame which is a group of 10 pins set about 5 cm apart on a frame is normally used. The pins are lowered in turn and the number of hits or misses on the particular species of plant under consideration is recorded. From this the *percentage cover* can be estimated.

pollen: the stage in the flowering plant life cycle which contains the male *gametes*. Pollen mother cells in the anther divide by *meiosis* to produce four immature pollen grains. The nucleus in each of these immature pollen grains then divides by *mitosis* to produce a pollen grain which, when it is mature, will contain three *haploid* nuclei.

One of these nuclei is the tube nucleus which controls the *pollen tube* and the other two will be the male gametes. The outer coat of the pollen grain is tough and very resistant to decay. This and the fact that the pollen grains from different species are very different in appearance can give us information about past climates.

 hint In a plant, the pollen grain is not the male gamete. It is more correct to say that the pollen grain contains the male gamete.

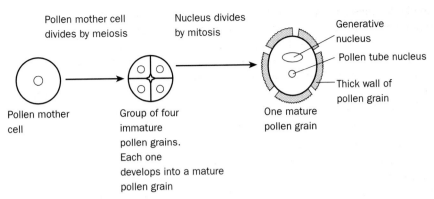

Pollen mother cell divides by meiosis

Nucleus divides by mitosis

Generative nucleus

Pollen tube nucleus

Thick wall of pollen grain

Pollen mother cell

Group of four immature pollen grains. Each one develops into a mature pollen grain

One mature pollen grain

The development of a pollen grain

pollen tube: a structure produced by a germinating pollen grain. It grows away from oxygen and towards moisture, down through the *style* to the ovules. It usually enters the ovule through a small pore called the micropyle. The male gametes travel down this pollen tube to the ovule.

pollination: involves the transfer of pollen from the *anther* to the *stigma*. It should be distinguished from *fertilization*, which involves the actual fusion of male and female gametes, and from seed dispersal which, as its name suggests, involves the dispersal of seeds. Although there are a variety of different methods by which pollen can be transferred, insect and wind pollination are the most frequent in temperate regions. The table shows some differences between the pollen of insect- and wind-pollinated plants.

Pollen from insect-pollinated plants	Pollen from wind-pollinated plants
Pollen grains larger in size, usually over 25 µm in diameter	Pollen grains smaller in size, usually between 10 and 25 µm in diameter
Produced in smaller amounts	Produced in larger amounts: a single hazel catkin, for example, may produce as many as 4 000 000 pollen grains
Pollen grains with sticky or rough surface	Pollen grains with smooth, dry surface

pollution: the contamination of the *environment* with harmful substances or in other ways such as with excessive heat, light or noise. The effects of pollution forms a theme that links together different aspects of biology. The spider diagram shows some of these links. You can use the cross-referencing system in this book to add greater detail to the diagram which is on the next page.

polygenic inheritance: where more than one gene *locus* is responsible for controlling a characteristic. Mendel's work showed that height in peas is controlled by a single gene with two alleles, with the allele for tall plants, T, being dominant to that for short plants, t. As a result, pea plants will be either tall or dwarf. They show *discontinuous variation*. Human height, however, shows *continuous variation*, with a range from the shortest to the tallest individual in a population. Height in humans is an example of polygenic inheritance, and the pattern of variation reflects the fact that there are many different combinations of alleles possible.

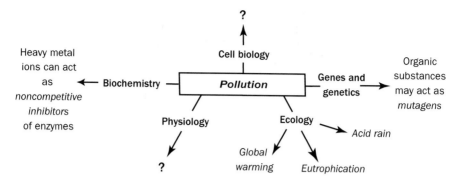

polypeptide: a *polymer* consisting of a chain of *amino acid* molecules joined by peptide bonds formed as a result of condensation reactions. One or more polypeptides form a *protein*.

polymer: a large molecule which is built up from a number of similar smaller molecules or monomers. There are three biologically important groups of polymers found in living organisms. Some of their characteristics are summarized in the table.

Polymers	Property		
	Monomers from which they are built up	Chemical elements they contain	Relative molecular mass
Nucleic acid	Nucleotides	Carbon	Between 10^4 and 10^{10}
		Hydrogen	
		Oxygen	
		Nitrogen	
		Phosphorus	
Polysaccharide	Monosaccharides	Carbon	Between 10^4 and 10^6
		Hydrogen	
		Oxygen	
Proteins	Amino acids	Carbon	Between 10^4 and 10^6
		Hydrogen	
		Oxygen	
		Nitrogen	
		Sulphur (in some amino acids)	

Biological polymers are all formed by condensation reactions which involve the linking of monomers by removing a molecule of water. They can be broken down to monomers from which they are built by hydrolysis, which involves the addition of a molecule of water (see the diagram on the next page).

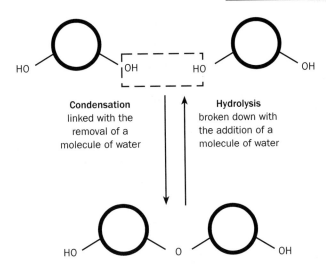

Condensation
linked with the
removal of a
molecule of water

Hydrolysis
broken down with
the addition of a
molecule of water

hint A polymer is built up from a large number of monomers. It may take the form of a straight chain or it may be branched. A triglyceride is not a polymer because it is not made up of a large number of monomers. It is incorrect to say that it is not a polymer because its molecules are branched.

polymerase chain reaction: a process used by biologists to make large amounts of identical *DNA* from very small samples. It is explained in the diagram below.

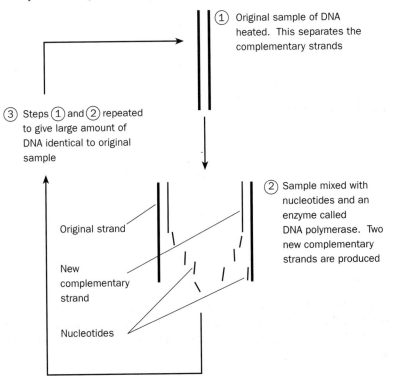

(1) Original sample of DNA heated. This separates the complementary strands

(3) Steps (1) and (2) repeated to give large amount of DNA identical to original sample

(2) Sample mixed with nucleotides and an enzyme called DNA polymerase. Two new complementary strands are produced

Original strand

New complementary strand

Nucleotides

polyploidy: the possession of three or more sets of *chromosomes* in each cell. Most organisms have two sets of chromosomes in a cell. At *meiosis*, haploid (n) gametes are usually formed which, on fertilization, will combine to give a new individual which again has two sets of chromosomes in each of its cells. Occasionally, a diploid (2n) gamete is formed. If this fuses with a normal haploid gamete, an individual would be formed which would have three sets of chromosomes. It would be triploid (3n). Because the *homologous chromosomes* of such an individual would not be able to pair during the first stage of meiosis, triploid individuals are usually sterile. However, if the species can reproduce asexually, a triploid population may become established. In fact, polyploid plants are quite common. Since they are often larger than diploid plants, with bigger cells, they can be important in agriculture. Some common examples are bananas (triploid and tetraploid), Bramley cooking apples (triploid) and wheat (hexaploid).

polynucleotide: a *polymer* made up from a number of *nucleotides* joined to each other by condensation reactions. *DNA* consists of two polynucleotide chains twisted around each other to form a spiral or helix. There is one complete turn of the spiral for every ten nucleotides. *RNA* is a shorter molecule and consists of a single polynucleotide chain.

polysaccharides: *carbohydrates* which are *polymers* of *monosaccharides*. They are made up of many sugar units joined by *condensation* reactions. Since they have large molecules, they are insoluble. Their main functions in living organisms are to act as storage molecules (*starch* and *glycogen*) or as structural materials (*cellulose*).

population: a group of individuals belonging to one species. In ecological terms, this usually refers to the area being studied, so it is possible to consider the population of frogs in a pond or of aphids on a sycamore tree. In evolutionary terms, however, it is usually necessary to add that members of a population are able to breed with each other.

population growth curve: a graph which shows the growth of a population of organisms over a period of time. The diagram shows a population growth curve for yeast cells growing in a sucrose solution.

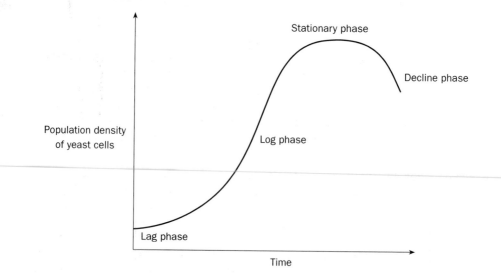

It is convenient to divide this curve into four separate stages or phases:

- *lag phase:* during this stage, the organisms do not increase in number; they are adapting to the medium in which they are growing

- *log phase:* a period of rapid population growth; nutrients are in plentiful supply and the amount of toxic waste products produced is very low

- *stationary phase:* the population remains more or less constant

- *decline phase:* a shortage of nutrients and a buildup of toxic waste products lead to the death of many organisms.

population pyramid: a way of showing the structure of a population graphically. A population pyramid shows the numbers of males and females in each age group in the population. Although most frequently used to describe human populations, it can also be used to represent the population structure of other animals. The diagrams show the shapes of two typical pyramids. Diagram A represents a fast-expanding population in which there is a large number of young individuals. Diagram B shows a stable population in which there are relatively fewer immature individuals and more adults.

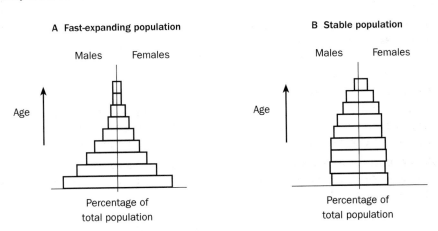

portal vein: a blood vessel that has a *capillary network* at both ends. The hepatic portal vein is a major blood vessel which transports the products of digestion from the intestine to the liver. The blood in most other parts of the blood system is remarkably constant in composition. For example, the blood glucose level is usually around 90 mg 100 cm^{-3} of blood. It is kept within very narrow limits by the action of hormones, particularly *insulin* and *glucagon*. The glucose level in the hepatic portal vein, however, may fluctuate much more and may rise significantly higher than this after a meal.

positive feedback: many substances or systems have a set level or norm within a living organism. Positive feedback results in any departure from this set level bringing about changes which produce further departures. A good example is provided by what happens in *hypothermia*. The human body temperature has a set level, generally around 37 °C. If the body temperature falls, then the *metabolic rate* of the body will fall. Less heat will be produced and the body temperature will fall further. Ultimately, this process may lead to the death of the individual concerned. For another example see *action potential*. Care must be taken to distinguish positive feedback from *negative feedback*.

potassium–argon dating: a method of dating fossil remains that has proved particularly useful for human fossils. It relies on the principle that radioactive potassium, or potassium-40, gradually breaks down into argon. Like all radioactive elements, potassium-40 breaks down at a steady rate; in this case it takes 1.3 billion years for half of the potassium-40 in a sample to break down. If the amount of argon can be measured and the rate of breakdown of the potassium-40 into argon is known, then the age of the sample can be calculated. There are some difficulties, however. Only volcanic rocks are suitable for this method of dating and they do not contain fossils. In addition, only very small amounts of argon are produced by decay of potassium-40, and it is easily contaminated by the argon that is normally present in the air.

potassium-movement hypothesis: a mechanism which has been suggested to account for the opening and closing of *stomata*. It is outlined in the diagram.

- Potassium ions pumped into guard cell
- Water potential of guard cell falls
- Water moves into guard cell by osmosis
- Thin outer walls of guard cell get longer
- Stoma opens

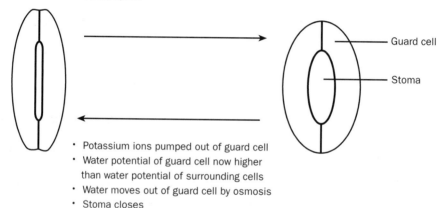

- Potassium ions pumped out of guard cell
- Water potential of guard cell now higher than water potential of surrounding cells
- Water moves out of guard cell by osmosis
- Stoma closes

predator: an organism which feeds on another organism, killing it before eating it. Most familiar examples of predators, such as lions and ladybugs, are animals and it is usual to think of prey as animals as well. However, predators and their prey can also be found in other kingdoms such as the *Protoctista*.

pressure potential, Ψ_p: the pressure produced by the cell wall pushing against the *plasma membrane*. In the cytoplasm, water molecules will be moving around at random. Some will collide with the membrane that surrounds them. They will exert a pressure on this membrane. This is the *water potential*, the symbol for which is the Greek letter psi, Ψ. The pressure produced by these randomly moving water molecules pushes against the membrane from the inside, while the pressure potential is pushing against the membrane from the outside. They are acting in the opposite direction to each other. Since all values of water potential are negative, values of pressure potential will be positive. The relationship between pressure potential, Ψ_p, water potential, Ψ, and *solute potential*, Ψ_s, can be summarized by the equation.

$$\Psi_P = \Psi - \Psi_S$$

primary consumer: an organism that feeds on plants. In the food chain:

nettle plant $\rightarrow$ large nettle aphid $\rightarrow$ two-spot ladybug

the nettle plant is the *producer*. It converts the energy in sunlight into energy in the organic molecules in the plant. The large nettle aphid feeds on the nettle and is the primary consumer.

primary immunological response: the response shown by the immune system when a person is first exposed to a particular *antigen*. Take as an example, infection with the virus that causes measles. The virus acts as an antigen. If an appropriate lymphocyte comes into contact with a measles virus, it is stimulated to divide rapidly and produces a large number of identical cells called a clone. This clone contains two sorts of cells, *plasma cells* and *memory cells*. The plasma cells produce and secrete *antibodies*. The rise in the level of specific antibodies after the first exposure to a particular antigen is the primary immuno-logical response. The memory cells remain in the blood system and respond rapidly if reinfection with the same antigen occurs. Multiplication takes place and more plasma cells are produced. In this way the response to a second infection, the secondary immunological response, is much more rapid.

primary metabolite: a substance which is produced by *microorganisms* when they are still actively growing. When microorganisms are grown in culture, they soon enter a phase of rapid growth. During this stage, enzyme-controlled reactions produce the substances required for growth. These substances are called primary metabolites. *Secondary metabo-lites* are substances which are usually produced later in the life cycle, as the microorganisms age.

primary structure (protein): the sequence of *amino acids* which makes up a *protein* molecule. A particular protein has a specific number of amino acids in each of its molecules. Typically this is somewhere between 100 and 650 in total. In addition to this, there are about 20 different amino acids which may occur in a protein molecule. They may be arranged in a variety of different ways. Variation in the number and arrangement of the amino acids in a molecule means that it is possible to have an enormous number of different proteins.

primate: a member of the order of mammals which includes lemurs, monkeys and apes. Humans also belong to this order. Most primates are tree-dwellers and many of the charac-teristics associated with the group are adaptations to living in trees. Primate characteristics include:

* forelimbs in which there is considerable freedom of movement. This is shown with the freely movable shoulder joint; an elbow joint which allows rotation of the forearm; and fingers and thumbs which are, to some extent, opposable. This should be compared with the degree of movement shown in the front limbs of mammals such as horses and dogs. The claws are flattened to form finger- and toenails

* the eyes are extremely well developed; primates can distinguish colors and have *binocular vision*

* the brain is relatively large in proportion to the size of the body

* young are usually born singly; the female has a single pair of mammary glands which are on the chest.

219

probability: a mathematical way of expressing the likelihood of a particular event occurring. Look at this simple genetical example. In humans, a single gene with two alleles affects the appearance of the earlobes. The allele for free earlobes, **E**, is dominant to that for fixed earlobes, **e**. If two heterozygous individuals produce children, the probability of having a child with fixed earlobes is one out of four, or 0.25.

When two or more factors are considered, it is important to remember that the probabilities of the individual events need to be multiplied. To use the same example, the probability of this couple having a child who is a girl with fixed earlobes can be calculated.

The probability of any child being a girl is one out of two, or 0.5.

The probability of the child having fixed earlobes is 0.25.

Therefore the probability of both events occurring is 0.5×0.25, or 0.125.

The results obtained in many experimental investigations may have a biological explanation. They may also be due to *chance*. Statistical tests can be used to determine the probability of results being due to chance.

hint Don't forget, combine probabilities by multiplying them.

producer: is an *autotroph*. All *food webs* and *food chains* ultimately depend on producers, organisms that can produce organic molecules from simple inorganic ones with the aid of an additional energy source. Most terrestrial food webs are based on *photoautotrophs*, organisms that depend on photosynthesis. There are some deep-sea ecosystems, however, that have developed around volcanic areas, where the producers are chemoautotrophic bacteria which use the energy released from chemical reactions to produce organic molecules.

progesterone: a female sex *hormone* produced during the second half of the reproductive cycle and during pregnancy. In the *menstrual cycle*, it is secreted by the *corpus luteum*. During the first weeks of pregnancy, the corpus luteum continues to secrete progesterone, but in the later stages this is taken over by the *placenta*. Progesterone has several effects on the reproductive system but in particular it maintains the lining of the uterus and inhibits contraction of the uterine muscle. It also exerts a negative feedback effect on the release of two more hormones, FSH and LH, from the anterior lobe of the pituitary gland. Progesterone can be used as an *oral contraceptive* either on its own or combined with estrogen.

Prokaryotae: the kingdom containing prokaryotic organisms (see *classification*). There are two main groups, true bacteria and blue-green bacteria, but they both differ from all the other kingdoms which have eukaryotic cells in a number of ways, some of which are summarized in the table at the top of the next page.

prolactin: a *hormone* secreted by the anterior lobe of the pituitary gland which stimulates the production of milk by the mammary glands.

protandry: occurs when the male part of a flower matures first. It is one of the ways in which *self-fertilization* can be prevented in flowers which possess both male and female organs. If the flowers of a rose-bay willow-herb are examined, for example, it will be seen that the *anthers* mature and shed their pollen before the lobes of the *stigma* open.

Prokaryotic cells found in the kingdom Prokaryotae	Eukaryotic cells found in all other kingdoms
Cells small with a mean diameter under 5 μm	Larger cells, often up to 50 μm in diameter
Circular strands of *DNA* not associated with proteins and found in the cytoplasm; no nucleus present	DNA linear and associated with proteins to form a true *chromosome*; chromosomes found within a nucleus
Few organelles present and none is surrounded by a membrane	Many membrane-surrounded organelles such as *mitochondria*
Flagella lack system of microtubules	Flagella (correctly known as undulipodia) have an internal arrangement of microtubules

protease: an *enzyme* which breaks down protein molecules by breaking the bonds which join the *amino acids* together. Proteases have a number of uses in *biotechnology*. They are used in biological washing powders and solutions such as those used for cleaning contact lenses, where they remove proteins; the food industry makes use of them in processes such as meat tenderizing and, since enzymes only alter the rate of a reaction, in certain circumstances they can join amino acids together. This property is used in making certain artificial sweeteners. Although some proteases are general in their action and can break the protein molecule at a variety of points along its length, most are very specific. They will only break *peptide bonds* between particular amino acids. Proteases can be divided into two categories, *endopeptidases* and *exopeptidases*, according to the point in the protein molecule at which they act.

protein: a *polymer* built from *amino acids* joined by peptide bonds. The resulting chain of amino acids (called a polypeptide) is then folded in different ways and to different extents, as shown in the diagram.

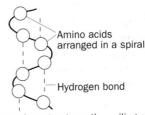

Amino acids arranged in a spiral

Hydrogen bond

Secondary structure–the coiling of an amino acid chain into a helix. This is only one example of secondary structure

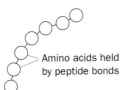

Amino acids held by peptide bonds

Primary structure–the sequence of amino acids

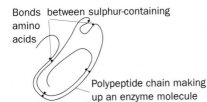

Bonds between sulphur-containing amino acids

Polypeptide chain making up an enzyme molecule

Tertiary structure–irregular folding of a polypeptide chain into a globular shape

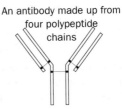

An antibody made up from four polypeptide chains

Quaternary structure–association of more than one polypeptide chain

protein synthesis

Many of the properties of proteins can be explained in terms of their shapes. Structural proteins are made from long parallel chains which form fibers or sheets. In *globular proteins* the tertiary structure is important in giving them their shape. This shape is important in determining how they function as, for example, *enzymes* or *antibodies*.

The importance of proteins is one of the important synoptic themes that links together different aspects of biology. The spider diagram shows some of these links. You can use the cross-referencing system in this book to add greater detail to the diagram.

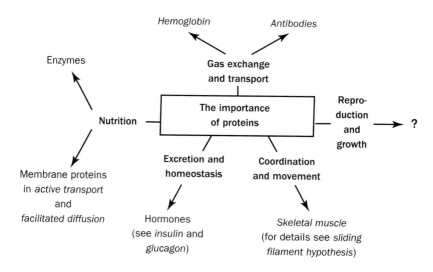

> **hint** All molecules have a three-dimensional shape, so it is not enough to define the tertiary structure of a protein as its "three-dimensional" shape. You should refer to the irregular folding of the whole polypeptide which gives the protein its globular shape.

protein synthesis is the process by which proteins are built up from amino acids using information carried by the genetic code on the cell's *DNA*. The diagram opposite summarizes the main features of the process. For further details, see *genetic code, transcription* and *translation*.

The genetic code representing a gene is first copied onto messenger RNA (mRNA) in the nucleus of the cell. This mRNA molecule moves out into the cytoplasm of the cell. Here the code on the mRNA molecule is used to control the production of a polypeptide chain by a ribosome.

Protoctista: the kingdom containing *eukaryotic* organisms which cannot be classified as members of the *fungi, Animalia* or *Plantae* (see *classification*). It contains a variety of different organisms ranging from those that are one-celled to algae such as seaweeds. It is not possible to give a list of features shared by all members of this kingdom; all that can really be done is to say that an organism belongs to the Protoctista if it doesn't belong to one of the other kingdoms!

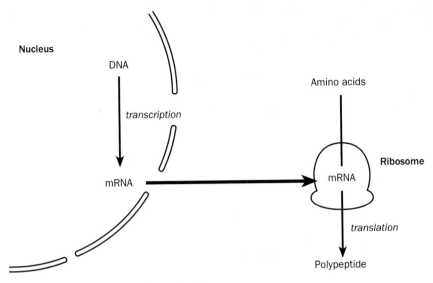

The main features of the protein synthesis

protogyny occurs when the female part of a flower matures first. It is one of the ways in which *self-fertilization* can be prevented in flowers which possess both male and female organs. In buttercup flowers, for example, the *stigmas* are receptive before the *anthers* mature and pollen is shed.

protozoa: a group of single-celled organisms which belong to the kingdom *Protoctista*. It includes a variety of free-living forms like amoeba as well as a number of important parasites such as *Plasmodium*, which causes malaria, and *Trypanosoma*, responsible for sleeping sickness.

proximal tubule: see *first convoluted tubule*

proximal convoluted tubule: see *first convoluted tubule*

puberty: the stage in the development of humans and other mammals when the sex organs mature and first produce gametes. Because of the difficulties in recognizing the first occurrence of these events, it is usually considered as involving the time over which the changes in the body which lead to sexual maturity take place. Puberty is controlled by hormones. There is a gradual increase in the amount of luteinizing hormone, *LH*, secreted by the anterior lobe of the *pituitary gland*. In females, this brings about an increase in the levels of *estrogen* secreted by the ovaries which, in turn, leads to the development of the *secondary sexual characteristics* such as the growth of breast tissue, the broadening of the pelvis and the growth of pubic hair. In males, the increase in LH leads to increased *testosterone* secretion by the testes. Again, this stimulates the development of secondary sexual characteristics such as the deepening of the voice and growth of facial and body hair.

Puberty takes place later in the life of humans than in that of any other mammal. This may well have been very important in human evolution since it prolongs the period of childhood dependency and allows greater opportunity for learning from the parents. In humans, there has been a recent trend towards earlier puberty. This is particularly apparent in more developed countries and can probably be explained in terms of nutrition.

pulmonary: an adjective meaning "to do with the *lungs*." The *pulmonary artery*, for example, is the blood vessel which takes blood to the lungs. The *pulmonary vein* returns it to the heart.

pulmonary artery: the blood vessel which takes the blood from the right *ventricle* of the heart to the *lungs*. There is a major difference between the pulmonary artery and other arteries in the body because it contains deoxygenated blood.

pulmonary vein: the blood vessel which takes blood from the *lungs* to the left *atrium* of the heart. After birth, it is the only vein in the body which contains oxygenated blood.

hint The pulmonary vein carries oxygenated blood from the lungs to the heart.

purine: one of the two chemical families of *nucleotide bases* found in a nucleic acid molecule. Adenine and guanine are the purines found in both *DNA* and *RNA* molecules. The important feature of a purine is that it contains two rings of atoms.

Purkinje tissue: see *Purkyne tissue*

Purkyne tissue: (formerly known as Purkinje tissue) is the specialized heart muscle that carries the wave of electrical activity that triggers contraction of the ventricles. The *bundle of His* runs in the wall between the right and left ventricles from the *atrioventricular node* to the base of the ventricle. This bundle splits into finer and finer branches, the Purkyne tissue, which conduct the electrical activity to the rest of the ventricle muscle.

pyramids of number, biomass and energy: various quantitative ways of describing the relationships between different *trophic* levels in an *ecosystem*. As the names suggest, they refer to the numbers of individuals, the *biomass* and the energy flow through the different trophic levels. If sufficient information is provided, it is possible to construct them to scale; in most cases, however, it is sufficient to indicate the relative proportions. The chart shows the shape of these pyramids for three simple food chains.

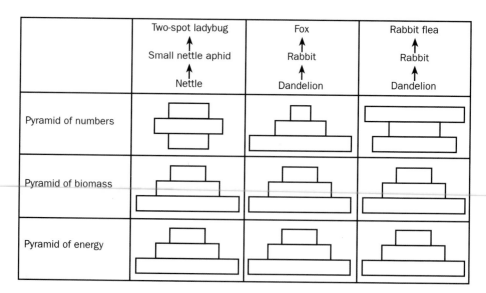

The shapes of pyramids for three simple food chains

It is worth noting the exceptions to the standard pyramid shape with pyramids of numbers. The shape is inverted where the producer is large in size compared to the primary consumers that feed on it, and again where small parasites are feeding on a larger host. Pyramids of energy are rather different from the other two. A pyramid of numbers and a pyramid of biomass refer to the organisms that are present at a particular time. A pyramid of energy, on the other hand, refers to the total amount of energy that flows through the different trophic levels over a period of time. It is for this reason that pyramids of energy are always pyramid shaped.

hint A pyramid of energy represents energy transfer through the different trophic levels of a food chain, so it is quite correct to say that more energy is transferred through the primary consumers than through the secondary consumers. It is wrong to say, however, that there is more energy in a primary consumer than there is in a secondary consumer. Using the examples in the diagram, it should be quite clear that there is more chemical potential energy in a large ladybug than in a smaller aphid.

pyrimidine: one of the two chemical families of *nucleotide bases* found in a nucleic acid molecule. Cytosine and thymine are the pyrimidines found in a molecule of *DNA*. Those found in *RNA* are cytosine and uracil. The important feature of a pyrimidine is that it contains a single ring of atoms.

Q_{10} (temperature coefficient): the increase which occurs in the rate of a process when the temperature is increased by 10 °C. At temperatures between 0 °C and approximately 40 °C, most *enzyme*-controlled reactions will have a temperature coefficient of approximately 2. In simple terms, what this means is that a ten degree rise in temperature will cause the rate of an enzyme-controlled reaction to double

A simple formula allows the temperature coefficient to be calculated from appropriate information.

Worked example: A solution containing the enzyme catalase and hydrogen peroxide gave off 63 cm³ of oxygen in one minute at 25 °C. At 15 °C, a second solution containing an identical mixture of catalase and hydrogen peroxide produced 29 cm³ of oxygen:

$$Q_{10} = \frac{\text{rate of reaction at (T + 10) °C}}{\text{rate of reaction at T °C}} = \frac{63}{29} = 2.1$$

Processes which do not rely on enzymes also take place in living organisms. These also increase in rate with an increase in temperature. However, the increase is nowhere near as much and the Q_{10} has a value of around 1.2 for processes such as *diffusion*.

quadrat: a sample area marked out in order to study the organisms that it contains. Quadrats are used in a variety of ecological investigations to obtain quantitative information about the distribution of animals and plants. They are frequently used to record the vegetation in a fairly uniform area. In using quadrats like this, the first step is to place quadrat frames at random or the results will not be representative of the area studied. Once the quadrat is in place, information can be collected concerning the species of organism present. Measures of *frequency* or *percentage cover* are commonly used.

Other studies, such as those involving the investigation of *succession* or the effects of grazing, make use of permanent quadrats which may stay in place for many years.

> **hint** You will not get a genuinely random sample by throwing a quadrat over your shoulder – even if you close your eyes and spin round three times first. You should:
> 1 place two tapes at right angles to each other along the sides of the area being studied.
> 2 use random number tables (or some method like taking numbers from a hat) to produce random pairs of numbers
> 3 use these numbers as the coordinates to locate your quadrat.

226

quaternary structure (protein): involves a *protein* being made up from more than one polypeptide chain. Two examples of proteins with quaternary structures are the *hemoglobin* and *antibody* molecules found in the blood of a mammal. Both of these molecules contain four polypeptide chains. Some *enzyme* molecules are even more complex. One of those involved in respiration, for example, contains 42 separate polypeptide chains.

radial symmetry: used to describe organisms which can be cut in more than one plane to produce a mirror image. This is because their parts are arranged in a circle around a line which goes down through the center. Most radially symmetrical animals are *sessile*, that is, they do not move from one place to another unless they are simply being carried by water currents. It is useful for such animals to be radially symmetrical as the stimuli to which they respond may come from any direction. Members of the phylum *Cnidaria* are sessile, radially symmetrical animals. In addition to animals, plants are also sessile and show radial symmetry to some extent.

radicle: part of an embryo plant found in a seed. When *germination* occurs, the radicle usually emerges first and grows downwards. It is the root of the embryo plant.

random sampling: the taking of experimental samples in a way which avoids bias. Random sampling is particularly important in ecology when placing *quadrats*. If samples are not taken at random there is always a danger that the results will be affected by deliberate choice. Throwing quadrats, whether with the eyes shut or not, does not result in a genuinely random distribution. You should use a method such as that shown in the diagram.

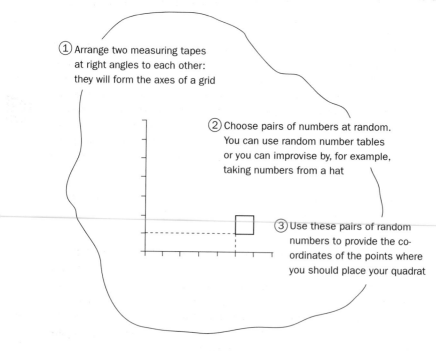

① Arrange two measuring tapes at right angles to each other: they will form the axes of a grid

② Choose pairs of numbers at random. You can use random number tables or you can improvise by, for example, taking numbers from a hat

③ Use these pairs of random numbers to provide the co-ordinates of the points where you should place your quadrat

receptor: a cell which is able to detect changes in its environment. There are many different types of receptor and they vary considerably in structure and function. Receptors are often grouped together to form complex sense organs. Receptors have a number of similarities:

* they are very specific. There are basically six different sorts of stimulus. These are produced by chemicals, electrical fields, light, magnetic fields, mechanical movement and temperature. A particular receptor will only respond to one kind of stimulus. A temperature receptor in the skin, for example, will not produce a response to pressure or light

* they are extremely sensitive: *rod cells*, for example, can detect a single photon of light

* they are often grouped together in sense organs which are able to magnify or amplify the stimulus. In the eye, for example, this is helped by the way in which a number of rod cells join to a single nerve cell in the optic nerve

* they convert sensory information into a form that can be transmitted to the central nervous system.

recessive: an *allele* is said to be recessive if it is only expressed in the *phenotype* of a diploid organism if the other allele of the pair is identical. In peas, the allele for yellow pods, g, is recessive to that for green pods, G. Since the g allele is recessive, only pea plants with the genotype gg will have yellow pods.

> **hint** You cannot tell whether an allele is recessive or dominant by simply counting the particular phenotypes. If you are given a family tree and five out of 12 individuals, for example, have green eyes, it does not prove that the allele for green eyes is recessive; green may simply be a rarer allele in this sample. You must look at the individual crosses for specific evidence.

recombinant: a cell which contains a different combination of *genes* or *DNA* from that found in its parent cell. Under natural conditions, recombinants often arise during *meiosis* as a result of *crossing over*. During the first stage of meiosis, the *chromosomes* come together in their homologous pairs, each chromosome consisting of a pair of *chromatids*. These chromatids may break and rejoin with other chromatids. This results in the genes on one part of a chromatid combining with the genes which were present on part of another chromatid. If one of the chromatids concerned originally came from the mother, while the other came from the father, there would be a new combination of genes produced and recombinants would be formed.

Recombination can also occur as a result of *genetic engineering*. A particular gene may be isolated from one organism, and a vector is then used to insert this gene into the DNA of another organism producing recombinant DNA.

red blood cell: a cell whose main function is the transport of respiratory gases in the blood. When seen with a light microscope, human red blood cells (or erythrocytes) appear as biconcave discs, 7 to 8 μm in diameter. Although each red blood cell contains large amounts of the oxygen-carrying pigment *hemoglobin*, it has no nucleus or mitochondria. Red blood cells are found in very large numbers. There are about five million in each cubic milimeter of blood. Their actual number varies from individual to individual and may be estimated with a specially marked slide known as a *hemocytometer*.

Red blood cells are produced in the bone marrow and their formation is controlled by a hormone secreted by the kidney. Since red blood cells do not contain either nuclei or ribosomes, they are unable to make new proteins. Partly because of this, they have a limited life span. At the end of a period of approximately four months they are broken down by the liver and certain other organs. Some of the material they contain, such as the iron found in the hemoglobin, is recycled, while the rest is excreted and forms products such as *bile pigments*. Red blood cells, together with the hemoglobin they contain, have three important functions:

* oxygen combines reversibly with hemoglobin to form oxyhemoglobin. In this way, each molecule of hemoglobin can transport up to four molecules of oxygen from the lungs to the respiring tissues

* the presence of red blood cells is also important in the transport of carbon dioxide. An enzyme, carbonic anhydrase, is found inside red blood cells. This catalyzes the reaction between carbon dioxide and water which produces carbonic acid. Some carbon dioxide is able to combine with hemoglobin to produce carbaminohemoglobin. Just over 20% of the carbon dioxide transported by the blood is carried in this form

* hemoglobin is an important *buffer* and prevents the accumulation of hydrogen ions in the blood. This keeps the pH of the blood more or less constant.

hint Oxygen diffuses into a red blood cell where it combines with hemoglobin. It is not carried in the depressions which give the cell its biconcave shape. Red blood cells do not change shape as they take up oxygen. The actual amount of oxygen transported by a red blood cell is very small and will make no difference to the size or shape of the cell.

reduction: a chemical reaction involving the gain of electrons. It may also involve the gain of hydrogen or the loss of oxygen. Reduction is an important process, occurring during both respiration and photosynthesis.

reducing sugar: a sugar which is able to produce a positive result, an orange-red precipitate, when heated with Benedict's solution. In chemical terms, a reducing sugar contains either a free aldehyde group or a free ketone group. It is the presence of one or other of these chemical groups that enables it to reduce Benedict's solution. All *monosaccharides* and most *disaccharides* are reducing sugars. Sucrose is an exception. It is a nonreducing sugar.

reflex: a very simple behavior pattern involving a rapid, automatic response to a *stimulus*. If the hand is accidentally placed on a hot object, it is quickly pulled away. This is a reflex action. Receptors detect the stimulus. They send impulses along sensory neurones to the spinal cord. An impulse from the spinal cord brings about muscle contraction and removal of the hand. Because the nervous pathway involved in a reflex involves few neurones, reflexes allow rapid responses to be made.

Although reflexes are automatic – the same stimulus always producing the same response – there are times when the conscious part of the nervous system can modify or even override a reflex. It is possible, for example, to grasp a hot object and maintain the grip even though it causes considerable pain.

Reflexes play an important part in:

- protecting the body and avoiding damage by removing the part concerned rapidly from the source of danger

- making rapid adjustments to external stimuli: the presence of food on the tongue, for example, stimulates the flow of saliva, while a bright light affects the pupil diameter of the eye. Reflexes associated with balance also come into this category

- enabling homeostatic adjustments to be made. Reflexes which involve the circulatory and breathing systems bring about the adjustments which are continually required to enable these systems to adapt to the changing needs of the body.

reflex arc: the nervous pathway associated with a *reflex*. The simplest type of reflex arc in a vertebrate involves just two *neurones*, a sensory neurone and a motor neurone. This pathway can be represented diagrammatically:

stimulus → receptor → sensory neurone → motor neurone → effector → response.

An example of a simple two-neurone or monosynaptic reflex is the knee-jerk reflex, which is important in helping to maintain posture.

Other reflexes involve the presence of additional neurones. The impulse is conveyed from the sensory neurone to the motor neurone by way of intermediate or relay neurones. An example of a reflex arc which has a pathway like this is the one involved in the rapid removal of the hand from a hot or sharp object.

reflex escape response: a *reflex* which involves the whole animal moving rapidly away from a source of danger. A reflex escape response may be shown, for example, by an earthworm which is lying on the soil surface, partly out of its burrow. If its front end is touched, its longitudinal muscles contract and it rapidly shoots back into the ground. Similar types of response are shown by snails retreating into their shells and by squid escaping from potential predators. All these animals have very large neurones which take *impulses* rapidly to all the muscles involved so that they can contract at the same time.

refractory period: the short recovery period which occurs immediately after the passage of a nerve impulse along the axon of a nerve cell. The sodium *ion-channel proteins* in the membrane are closed during the refractory period and the normal negative *resting potential* is reestablished inside the axon. During this stage, these ion-channel proteins cannot be made to open again, even if a much bigger stimulus is given. In other words, the membrane cannot be depolarized and a new *action potential* cannot be initiated. This has two important consequences:

- the refractory period separates nerve impulses from each other. Because it takes from 1 to 10 milliseconds, it limits the number of impulses that can be transmitted along an axon in a given time

- nerve impulses only pass in one direction along an axon. They go from an active region to a resting region. They cannot go the other way because the membrane immediately behind the impulse cannot be depolarized again.

relative growth rate: see *growth rate*

renal: an adjective meaning "to do with the *kidney*." The *renal artery*, for example, is the blood vessel supplying oxygenated blood to the kidney.

renal capsule: the first part of a *nephron* or kidney tubule. It is a funnel-like structure which surrounds a ball of capillaries called a *glomerulus*. It consists of an outer layer of cells which is separated by a space from an inner layer of specialized cells called *podocytes*. The glomerulus and renal capsule are concerned with the first stage in the process of urine formation. It is here that *ultrafiltration* takes place.

renin: a hormone secreted by the kidney which is involved in control of the concentration of sodium ions in the blood. If there is a fall in the salt concentration of the blood, its water potential will rise. As a result of *osmosis*, more water will now move into the body tissues, with the result that the blood volume and blood pressure fall. A fall in blood pressure is detected by a group of cells in the wall of the arteriole that takes blood to the *renal capsule*. These cells secrete renin. A series of stages then leads to the hormone *aldosterone* being secreted by the cortex of the adrenal gland. Aldosterone brings about an increase in the permeability of the kidney tubule to sodium so that more sodium is pumped back into the blood. The uptake of sodium ions is accompanied by the reabsorption of water.

reproduction: the production of offspring. It is a process in which genetic material is passed from the parent or parents to a new generation of individuals. There are two main types of reproduction. These are *asexual* reproduction, in which a single individual gives rise to genetically identical offspring, and *sexual* reproduction, which involves the fusion of genetic material, usually from different organisms.

reproductive isolation: two populations are described as being reproductively isolated when they are prevented from breeding with each other or the hybrid offspring produced fail to survive. It is convenient to divide the mechanisms involved into two kinds. Those that prevent organisms breeding and fertilization taking place are examples of prezygotic reproductive isolation. For example, the chiffchaff and the willow warbler are small insect-eating birds which spend the summer in a given area. They are so similar to each other in appearance that it is very difficult to tell them apart. However, their songs, which form an important part of courtship behavior, are completely different. Since successful mating will only result from the correct behavior pattern, chiffchaffs and willow warblers do not interbreed.

Postzygotic reproductive isolation occurs when the offspring which result from fertilization between individuals from the two parent populations are at a disadvantage. The two populations are reproductively isolated because the hybrid offspring do not survive as well. There are two common species of poppy in a given area, the common poppy and the long-headed poppy. Pollen from one species can be transferred to the other and fertile seeds result. However, these hybrid seeds do not have mechanisms which limit their germination to favorable conditions. Many germinate in unsuitable areas and die. The hybrids, therefore, do not survive, so the two species are reproductively isolated.

residual volume: the amount of air that cannot be expelled from the lungs however deeply a person breathes out. See *lung capacities*.

resistance: the ability of a pest, for example, to withstand the action of a particular pesticide. In a population of houseflies, there will be variation. A few of the flies may possess *alleles* which make them resistant to a particular *insecticide*. If, however, that insecticide is not in use, there will be no selection involved, either for or against the allele producing resistance. It is therefore likely that the allele will remain in small numbers in the population. If the insecticide is now introduced and used against the flies, those which are resistant will be at an advantage. They will be more likely to survive and will pass on their alleles to the next generation. In this way, a population of insecticide-resistant flies may arise. This has happened in many parts of Europe. Other important examples of resistance include:

* resistance associated with the control of malaria. Not only are many species of malarial mosquito resistant to the insecticides that were once used to control them but, in many parts of the world, the malarial parasite is resistant to drugs used to prevent and control the disease

* the development of resistance in rats to the poison warfarin, which was once used very successfully in their control

* the emergence of strains of the bacterium responsible for tuberculosis (TB) which are resistant to all antibiotics at present in use.

resolving power: the ability to distinguish between points that are close together. Consider a photograph which shows a crowd of people at a football match. It is impossible to make out the detail of individual faces. If more detail is required, it might be thought that the solution would be to take a small area of the photograph and blow it up. Unfortunately, when this is done, all that happens is that instead of having a small picture where it is not possible to see the detail, we simply have a large picture in which the detail cannot be seen. Magnification is only of real use if it is accompanied by greater resolution allowing more detail to be seen. The main problem with a light microscope is that the wavelength of visible light limits the visible detail. The resolving power of a good light microscope is about 0.2 μm so, if two objects are closer than this, they will only be seen as one. If greater detail is required, it is necessary to use an *electron microscope*. With this instrument, a beam of electrons gives much greater resolving power. With modern instruments, the resolution is about 1 nm.

hint *Magnification* refers to greater size; *resolution* refers to greater detail. Make sure that you can distinguish between these terms.

respiration: the biochemical pathway which takes place in cells and results in the release of energy from organic molecules. This energy is usually used to form *ATP*, although some is always lost as heat. The process of respiration is often summarized by the equation:

$$C_6H_{12}O_6 + 6O_2 \rightarrow 6H_2O + 6CO_2 + \text{Energy.}$$

It is important to appreciate that this equation is very much an oversimplification of what actually happens. Glucose is only one of a number of different substances which can act as a *respiratory substrate*. (It is possible to identify the likely respiratory substrate by calculating the *respiratory quotient*.) The process can take place either in the presence of oxygen, *aerobic respiration*, or in its absence, *anaerobic respiration*. If different substances are used, then the waste products will also vary. The only feature common to all respiratory pathways is the release of energy.

hint *Respiration* and *gas exchange* are terms which are used rather loosely. You will find it helpful to distinguish between them. Respiration is the chemical process which takes place inside cells and releases energy. Gas exchange, on the other hand, is the way in which the respiratory gases (oxygen and carbon dioxide) are exchanged between an organism and its environment. Lungs and gills should be described as gas-exchange organs rather than respiratory organs.

respiratory pigment: a substance found in animals which is involved in the transport of oxygen and, in some cases, carbon dioxide. *Hemoglobin* is an important respiratory pigment found in the blood of mammals and many other animals. Another respiratory pigment, *myoglobin*, is associated with muscle where it acts as an oxygen store.

respiratory quotient (RQ): the amount of carbon dioxide produced by an organism in a given time divided by the amount of oxygen consumed in the same time. Different *respiratory substrates* produce different values for the respiratory quotient, so its calculation gives an indication of the substrate being respired. The table below shows the RQ for some important respiratory substrates.

Respiratory substrate	RQ
Triglyceride (fat)	0.7
Amino acids and proteins	0.9
Carbohydrates	1
Anaerobic respiration of carbohydrate	>1

There are, however, several different interpretations of an RQ of 0.9. The organism might be respiring proteins, but a mixture of fats and carbohydrates could give the same value. Care must be taken in interpreting the figures obtained in a particular situation.

Worked example: Calculating the respiratory quotient from a chemical equation.

The equation represents the respiration of a typical fat:

$$2C_{51}H_{98}O_6 + 145O_2 \rightarrow 102CO_2 + 98H_2O$$

$$RQ = \frac{\text{amount of carbon dioxide produced in a given time}}{\text{amount of oxygen consumed in the same time}}$$

or, we could write this in another way:

$$RQ = \frac{\text{number of molecules of carbon dioxide produced}}{\text{number of molecules of oxygen consumed}}$$

So, from the equation, this would be

$$RQ = \frac{102}{145} = 0.7$$

hint Respiratory quotient is easy to calculate providing you remember which way up the equation goes. It gives an indication of the substance being respired (carbohydrate, lipid or protein). It does not provide any information about rate of respiration.

respiratory substrate: the organic substance which is being used for *respiration*. Note that oxygen is not a respiratory substrate.

resting potential: the difference in electrical charge across the *plasma membrane* of a nerve cell. There is a higher concentration of K^+ ions and a lower concentration of Na^+ ions inside a nerve cell than outside. The plasma membrane allows K^+ ions to diffuse out very readily. It is much more difficult, however, for Na^+ ions to diffuse in. This means that a large number of K^+ ions leave the cell but this is not matched by the number of Na^+ ions entering. As a consequence, there are fewer positive ions inside the cell than outside. In other words, in a resting nerve cell, there is a potential difference across the membrane of about -60 mV. This is the resting potential. During the passage of a *nerve impulse*, a small number of Na^+ ions enter the cell. A *sodium pump* in the membrane actively transports these Na^+ ions back out so that the resting potential is reestablished.

restriction endonuclease: one of a group of enzymes which are able to cut *DNA* molecules at specific points along their lengths. Restriction endonucleases are found naturally in many bacteria, where they are thought to help in protecting them from *viruses* by cutting up the DNA of the virus. Each restriction endonuclease recognizes a specific base sequence in a DNA molecule and will only cut it at this point.

These enzymes are extremely important in *genetic engineering*, where they can be used as molecular scissors to isolate, for example, an individual gene from a length of DNA. By using a particular restriction endonuclease, the cut can be made at exactly the right points.

GTT AAC
CAA TTG

GTTAAC
CAATTG

Restriction enzyme *Hpa* 1 recognizes this base sequence and cuts it as shown by the dotted line

Restriction enzyme *Eco* R 1 recognizes the same base sequence but cuts it in a different way

Two examples of restriction endonuclease

restriction enzyme: see *restriction endonuclease*

retina: the light-sensitive layer on the inside of an eye. The retina contains two sorts of light-sensitive cell, *rod cells* and *cone cells*. Rod cells are arranged more or less uniformly over the entire surface of the retina, although there are none present in the region known as the *fovea*. They are very sensitive and respond to light of extremely low intensities. Rods cannot, however, distinguish color. Cone cells are fewer in number and are concentrated at the fovea. They are not as sensitive to low light intensities as rod cells, but they are able to distinguish color. The retina also contains many sensory neurones with which rods and cones form *synapses*. These sensory neurones allow some sorting out and processing of information from the rods and cones before it is sent via the optic nerve to the brain.

retinal convergence: the way in which groups of *rod cells* form *synapses* with a single nerve cell in the retina. This arrangement helps to make the eye very sensitive to low light intensity. The stimulus affecting a single rod cell may not be strong enough to trigger a nerve impulse but the combined effects of a weak stimulus affecting a number of rod cells add together to produce an impulse in the nerve cell. This arrangement contrasts with that in *cone cells*, where each cone cell synapses with an individual nerve cell. Cone cells are less sensitive to low light intensities but can discriminate much better between separate stimuli.

235

retrovirus: a virus which has *RNA* as its nucleic acid. When it infects a cell from its host, it is able to make a *DNA* copy of this RNA using an enzyme called *reverse transcriptase*. This DNA copy then becomes incorporated into the genetic material of the host. *HIV* is an example of a retrovirus, while others are thought to be involved in the development of certain cancers.

reverse transcriptase: an enzyme which produces a *DNA* molecule from the corresponding mRNA. Reverse transcriptase enzymes are found naturally in certain viruses called retroviruses. These are viruses in which the genetic information is carried on an *RNA* molecule. When one of these viruses infects a host cell, it uses this enzyme to make a complementary DNA (cDNA) copy of its genetic information which is then incorporated into the host DNA. Reverse transcriptase is one of the enzymes used in genetic engineering, where it can be used to obtain a copy of a particular gene from the relevant mRNA. This has a number of advantages including:

* A cell which is making a particular protein will only contain two copies of the relevant piece of DNA, one on each *chromosome*. These are transcribed to produce large amounts of mRNA. It may be easier to locate the relevant mRNA than the DNA.

* The piece of DNA making up a gene is a lot longer than the piece of mRNA produced from it. This is because it contains some base sequences, called *introns*, which do not code for amino acids. When the mRNA is produced, it is edited so that these non-coding sequences are removed.

R_F **value:** a ratio used to identify substances separated from each other by *chromatography*. It is calculated by dividing the distance moved by the unknown substance by the distance moved by the solvent front. The actual distance moved by a particular substance will depend on how long it has been left. If the apparatus has only just been set up, then the substance is unlikely to have traveled any great distance. If it has been left for some time, then it will have moved farther. Because of this we cannot use the distance moved by the substance to identify it. The R_F value, however, can be used because the ratio involved will be the same, however long the chromatogram has been running.

Rhizobium: a genus of nitrogen-fixing bacteria found both free-living in the soil and in the root nodules of *leguminous plants*.

rhizoid: a thin, root-like structure found in primitive plants such as mosses and liverworts. The main function of rhizoids is to anchor the plant and, in this way, they are similar to roots. However, each rhizoid consists of a single cell or, in some cases, several cells. They do not contain either *xylem* or *phloem*.

rhodopsin: a light-sensitive pigment found in the *rod cells* of the eye. A molecule of rhodopsin consists of two parts: opsin, which is a protein, and retinine which is derived from vitamin A. When a photon of light strikes a molecule of rhodopsin, the pigment is broken down into its two parts in a process known as bleaching. The chemical events associated with bleaching lead to nerve impulses which convey sensory information along the optic nerve to the brain. Before it can be stimulated again, rhodopsin must be regenerated from opsin and retinine. This requires *ATP*, which is produced in the *mitochondria* present in rod cells. In bright light, nearly all the rhodopsin in the eye has been bleached. The eye is said to be light adapted. If a person moves from brightly lit conditions into dim light, very little can be seen as there is not enough rhodopsin present. Only after about half an hour in dim light conditions will enough rhodopsin have been generated for things to be seen. The eye is now said to be dark adapted.

ribonucleic acid: see *RNA*

ribose: the five-carbon sugar or pentose found in *RNA*.

ribosome: a very small organelle which plays an important part in making proteins. It is made of a mixture of protein and a special sort of RNA called ribosomal RNA. During *protein synthesis*, a molecule of messenger RNA (*mRNA*) carries the genetic code for a particular protein from the nucleus into the cytoplasm of the cell. Ribosomes now move along this mRNA molecule. This results in *amino acids* being assembled in the correct order to form a molecule of the protein concerned.

There are two basic types of ribosome. Those found in the cytoplasm of *eukaryotic* cells are slightly larger than those which are found in *prokaryotic* cells.

ribulose bisphosphate, RuBP: the molecule which combines with carbon dioxide in the *light-independent reaction* of photosynthesis. Ribulose bisphosphate is a five-carbon sugar which has two phosphate groups in its molecule. It combines with carbon dioxide and this results in the formation of two molecules of the three-carbon compound, glycerate 3-phosphate. Glycerate 3-phosphate is then reduced to triose phosphate. Some of the triose phosphate is used to produce carbohydrates such as glucose and starch while the rest produces more ribulose bisphosphate in a cycle of biochemical reactions known as the *Calvin cycle*.

RNA: a type of nucleic acid molecule. There are a number of different forms of RNA and two of these are very important in *protein synthesis*. During *transcription*, part of the *DNA* in the nucleus unwinds. It acts as a template for the formation of messenger RNA (*mRNA*). This takes the code for the gene concerned from the nucleus into the cytoplasm. A second type of RNA, transfer RNA (*tRNA*) is found in the cytoplasm. This molecule is important in assembling amino acids in the correct position on the mRNA during the process of translation. These two RNA molecules are both *polynucleotides* like DNA but they differ from each other and from DNA in a number of ways, which are summarized in the table.

Feature	mRNA	tRNA	DNA
Number of polynucleotide chains in the molecule	1	1	2
Shape of molecule	Single chain	Single chain twisted around itself	Double chain twisted to form a helix
Name of sugar present	Ribose	Ribose	Deoxyribose
Names of bases present	Adenine Cytosine Guanine Uracil	Adenine Cytosine Guanine Uracil	Adenine Cytosine Guanine Thymine
Bases linked by hydrogen bonds	No	Yes	Yes

In some *viruses*, for example HIV, the genetic material in the virus is RNA.

> **hint** tRNA is a type of RNA. It therefore contains the bases adenine, cytosine, guanine and uracil. It does not contain thymine.

RNA polymerase: an enzyme important in the process of *transcription*. In this process, one of the DNA chains acts as a template for the formation of a molecule of messenger RNA (*mRNA*). The bases of free mRNA *nucleotides* in the nucleus line up against the complementary bases on the DNA chain. They now join to form an mRNA molecule. RNA polymerase is the enzyme which helps to join these nucleotides together to produce the mRNA molecule.

rod cell: a type of light-sensitive cell found in the *retina* of the eye. Rod cells occur in enormous numbers: there are about 120 million in a human eye. Each rod cell consists of two parts, an outer segment and an inner segment. The outer segment has large numbers of flattened membranous vesicles which contain a light-sensitive pigment called *rhodopsin*. When a photon of light strikes a molecule of rhodopsin, the pigment is broken down chemically in a process known as bleaching. Bleaching leads to nerve impulses which convey sensory information along the optic nerve to the brain. The inner segment contains the nucleus of the rod cell as well as large numbers of mitochondria. These provide the ATP which is necessary for regenerating the rhodopsin after it has been bleached.

Rod cells are unable to distinguish color but they are extremely sensitive to low light intensities. The retina contains large numbers of bipolar neurones as well as receptors like the rod cells. A number of rod cells synapse with a single one of these neurones. This arrangement, known as *retinal convergence*, is largely responsible for the great sensitivity of rod cells to low light intensity.

root: the part of a plant which holds it in the soil and is involved in the uptake of water and mineral salts. The main external difference between roots and stems is that roots do not have either leaves or buds. There are differences in the internal structure as well. The *xylem* and *phloem* form a core which runs down the middle of the root rather than a ring towards the outside. This makes it better able to withstand the pulling forces that are exerted as it anchors the plant in the soil.

The other important function of the root is that it is concerned with the uptake of mineral ions and water. The concentration of most mineral ions is higher inside the cells of the root than in the surrounding soil. Movement of these ions into the root therefore involves *active transport* and results in the root cells having a lower *water potential* than the surrounding soil. Water flows down this water potential gradient into the root cells by *osmosis*.

It is not possible to identify a structure as a root just because it is found below ground. The potato tubers, for example, which are such an important part of our diet, are modified stems, even though they grow in the soil. This can be confirmed by the presence of tiny scale leaves, each of which is associated with a cluster of buds. There are some roots which arise above ground level, while some unusual rain-forest climbing plants even have upwardly growing roots which absorb the water and mineral salts which drip down from the canopy. Some roots are important storage organs and are commercially important. These include carrots, turnips and sugar beet.

root hair: an outgrowth from a cell in the *epidermis* of a *root*. Root hairs are usually found in young roots, in the region just behind the root tip. They provide a very large surface area through which the root can absorb water and mineral ions.

root nodule: a small swelling on the root of a leguminous plant which results from infection with nitrogen-fixing bacteria such as *Rhizobium*.

root pressure: a force generated by the roots of some plants which pushes water and mineral salts up the *xylem*. It can be demonstrated by cutting the top off an actively transpiring plant such as a tomato plant. Water oozes from the cut surface of the stem and may continue to do so for some time. The mechanism responsible is thought to involve *active transport* of mineral salts into the xylem of the root. There is good evidence for this since the process is halted by respiratory poisons or a lack of oxygen. Accumulation of mineral salts in the xylem leads to a lowering of the water potential inside the vessels. Water then flows in from the surrounding tissue by *osmosis*. Although the force generated by root pressure is not enough to account for the movement of water to the top of a plant, there is little doubt that this mechanism, along with *capillarity* and *cohesion-tension*, makes some contribution.

rumen: a chamber found at the front of the gut in some mammals in which live microorganisms that are able to digest cellulose. The relationship between these microorganisms and their hosts is an example of *mutualism*. The microorganisms gain as they receive a constant supply of nutrients, while the mammal benefits because it receives the products formed from the digestion of cellulose as well as amino acids from digestion of the microorganisms. The diagram on page 240 summarizes the digestive processes which take place in the rumen of a cow.

ruminant: a mammal which possesses a large chamber at the front of the gut called a *rumen*. Microorganisms live in the rumen and these are able to digest *cellulose*. Ruminants have a complicated way of feeding. Grass and other plant material is swallowed and goes into the rumen. Here, it is mixed with mucus, and microorganisms start the process of digestion of the cellulose which the food contains. This mixture can be brought back up the esophagus into the mouth where it is chewed as the "cud." The cud is then swallowed. It goes down into the stomach proper and on through the rest of the gut, completing the normal processes of digestion and absorption. Many important domestic animals, including sheep and cattle, are ruminants. As some of the energy in the food they take in is used by the bacteria, they are not such efficient energy converters as nonruminants like pigs. They can, however, live on foods which few other animals can utilize. In addition to cattle and sheep, wild mammals such as deer and antelope are ruminants.

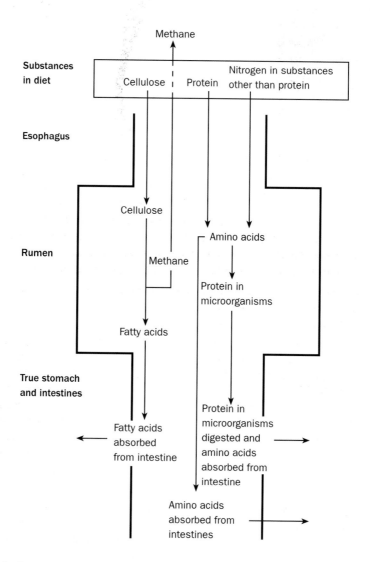

Digestion in the rumen

saliva: the digestive juice produced by glands which open into the mouth. In humans, saliva contains large amounts of *mucus* and an enzyme, *salivary amylase*, which starts the process of starch digestion. The secretion of saliva is controlled by a *reflex* involving the nervous system. The stimulus is provided by the presence of food in the mouth and is detected by the receptors which are the tastebuds. The salivary glands respond by increasing the production of saliva. The Russian physiologist, Pavlov demonstrated that a *conditioned reflex* was involved as well. The components of saliva differ from animal to animal. Anticoagulants, for example, are present in the saliva of blood-sucking animals like vampire bats and leeches.

salivary amylase: a starch-digesting enzyme produced in the *saliva*. Like all digestive enzymes, salivary amylase is a hydrolase. The main product of starch digestion by this enzyme is maltose.

***Salmonella*:** a genus of rod-shaped *bacteria* responsible for food poisoning, which is most likely to arise from food which has been kept in conditions which allow large numbers of bacteria to grow. Once contaminated food has been eaten, the bacteria enter the gut and multiply rapidly. Toxins are released as a result of the breakdown of bacterial cells and these cause gastroenteritis. The symptoms of *Salmonella* food poisoning affect the person concerned 12 to 72 hours after consumption of the infected food. The resulting sickness and diarrhea are often accompanied by fever. In some people, even though the symptoms may have disappeared, bacteria remain in the body and continue to contaminate the feces. Such people are known as carriers and can easily contaminate food. Other species of *Salmonella* cause typhoid and paratyphoid fevers.

saltatory conduction: the way in which nerve impulses travel along myelinated nerves. The *axons* in many mammalian nerves are surrounded by a fatty material called *myelin*. This does not, however, form a continuous layer. There is a small gap between each adjacent length of myelin called a node of Ranvier. Because of the insulating properties of the myelin, it is only at these nodes that a current can flow across the axon membrane and set up the next *action potential*. The nerve impulse therefore travels along in a series of jumps from one node to the next. This is known as saltatory conduction and can lead to impulses moving much more rapidly than in nonmyelinated nerves.

sampling: examining a small area or a small number of organisms rather than the whole area or entire population. In many ecological studies it is necessary to take samples as it would be much too time-consuming to examine the whole area. In determining the procedure to be used, the following points should be kept in mind:

* although limited by the time available, the sample should be as large as possible. Small samples are often unrepresentative of the area or population as a whole. For the same reason, as many samples as possible should be taken

241

* samples must be taken at random or the results will not be representative. They are likely to be biased and statistical tests cannot be used.

hint Look up *random sampling* to see how to take samples when you are using quadrats in ecology.

saprobiont: see *saprophyte*.

saprophyte: an organism which obtains its nutrients from dead or decaying organic matter. Many *bacteria* and *fungi* are saprophytes. They secrete digestive enzymes onto the surface of their food. These enzymes break down complex molecules such as those of starch and protein into smaller, soluble ones which are then absorbed. Saprophytes play an extremely important role in *nutrient cycles*. They break down organic matter and release simple inorganic molecules such as carbon dioxide and ammonia. Strictly speaking, the term saprophyte refers to plants. Because of this, some biologists prefer to use the term saprobiont when referring to organisms such as bacteria and fungi which have this method of nutrition.

sarcolemma: the membrane which surrounds a muscle fiber.

sarcomere: one of the basic structural units of which skeletal muscles are composed. A sarcomere is a section of a *myofibril* between *Z-lines*.

saturated fatty acid: a fatty acid with a long hydrocarbon tail in which there are no double bonds between the carbon atoms. Saturated fatty acids have higher melting points than fatty acids in which there are double bonds. *Triglycerides* which contain saturated fatty acids therefore tend to be solid at temperatures of around 20 °C. They are called fats and are found mainly in animals.

scanning electron microscope: a type of *electron microscope* which works by reflecting a beam of electrons off the surface of the object being examined. It produces a three-dimensional image which shows the surface in great detail.

sclerenchyma: a type of supporting tissue found in plants. Sclerenchyma cells have *cell walls* which are thickened with *lignin*. They do not have living contents. There is no cytoplasm present, for example. Two types of sclerenchyma are found. One type consists of long, thin cells tapered at both ends. These are known as fibers and, in the case of crops such as hemp and jute, may be of considerable economic importance. A second type has shorter, more rounded cells which are called sclereids. These form hard protective layers around nuts and other seeds.

Schwann cell: a cell which produces the fatty material called *myelin* that surrounds the *axons* of many mammalian nerve cells. Each Schwann cell wraps around and around the axon, covering it in a layer of myelin. Because a number of these cells occur next to each other along the axon, there are small gaps between the myelin produced by one cell and the myelin produced by the next. These gaps are known as the *nodes of Ranvier*. This arrangement is very important in allowing the fast passage of nerve impulses, which jump along from node to node, a process which is known as *saltatory conduction*.

sebaceous gland: a gland found in the skin of a mammal which opens into a hair follicle. It secretes an oily substance called *sebum* which plays an important role in making hair and skin waterproof, and also has an antibacterial effect.

sebum: an oily substance produced by sebaceous glands in the skin of a mammal.

second messenger: a molecule found inside a cell which responds to the presence of a hormone outside the cell by activating a particular enzyme. The first molecule to be identified with this function was *cyclic AMP*. It is produced when a hormone such as *adrenaline* binds with a receptor site on the *plasma membrane*. Cyclic AMP then activates the enzymes which control the biochemical pathways associated with the conversion of glycogen to glucose. A single hormone molecule leads to the production of many molecules of cyclic AMP inside the cell. In turn, each of these messenger molecules activates many enzyme molecules and each enzyme molecule affects large numbers of substrate molecules. This is referred to as the *cascade effect* and illustrates a second function of cyclic AMP, which is to amplify the effect of the hormone. Other second messengers have now been discovered and among the most important are Ca^{2+} ions.

secondary metabolite: a substance which is produced by *microorganisms* as they age rather than when they are still actively growing. When microorganisms are grown in a culture, they soon enter a phase of rapid growth. During this stage, enzyme-controlled reactions produce the substances required for growth. These substances are called *primary metabolites*. Secondary metabolites are substances which are usually produced later in the life cycle, usually as the microorganisms age. *Antibiotics* such as penicillin are examples of useful secondary metabolites which are produced by *biotechnological* processes.

secondary sexual characteristic: one of the distinguishing features between male and female animals which develops at the time of sexual maturity. In humans, secondary sexual characteristics include the growth of breast tissue, the broadening of the pelvis and distribution of subcutaneous fat in females, and the characteristic deepening of the voice and growth of facial and body hair in males. These characteristics develop as a result of the increase in the secretion of the hormones *estrogen* and *testosterone* around the time of puberty.

secondary structure (protein): the way in which the chain of *amino acids* which make up a polypeptide is coiled or folded. An α-helix is formed when the chain is coiled in a spiral shape. Each amino acid in the chain is linked to the others by *hydrogen bonds*. Many structural *proteins* are coiled in this way. An example is keratin, a fibrous protein found in hair.

Another common type of secondary structure is the β-pleated sheet. In this form, parts of the polypeptide chain run parallel to each other. These are held by hydrogen bonds and form flat sheets. The protein fibroin is the main component of silk. It has molecules which form β-pleated sheets. Because the amino acids in these sheets have been pulled into an extended form, silk cannot be stretched. The secondary structure of different proteins varies but is always determined by the *primary structure*.

secretin: a *hormone* produced when the acid contents of the stomach enter the small intestine. This stimulates cells in the upper part of the small intestine to produce secretin. Secretin increases the secretion of hydrogencarbonate ions by the pancreas, resulting in the pancreas producing large amounts of a watery, alkaline solution.

seed: the structure which develops from a fertilized ovule in a flowering plant. A seed usually contains an embryo and a store of food material. This food store may be in the *cotyledons*, which form part of the actual embryo, or it may be another tissue, the *endosperm*. The embryo and its food store are surrounded by a protective outer layer known as the *testa*, or seed coat. There are a number of advantages associated with the production of seeds. These include:

- protection of and a supply of nutrients for the developing embryo

- provision of a supply of nutrients which allows the seed to remain dormant. This may enable it to survive harsh conditions between growing seasons, such as winter weather or periods of very low rainfall

- an effective means of dispersal. Seeds, or the *fruits* that contain them, often have adaptations associated with dispersal. The seeds of soft fruits such as blackberries and tomatoes have outer layers which are resistant to the enzymes in the gut of the animal which eats them. They can therefore be transported long distances from the plant where they were eaten to the place where they are eventually deposited in the feces of the animal concerned.

hint Make sure that you can distinguish seeds from fruits. After fertilization, a seed develops from a fertilized ovule. A *fruit* develops from the tissues of the ovary which surrounds the ovules.

seed bank: a collection of seeds used as a way of banking or preserving plant genes. A suitable collection is made containing seeds from as many different varieties and areas as possible. These are kept in cool, dry conditions and will remain alive for many years. Seed banks are therefore very useful in the *genetic conservation* of plants as each seed will contain all the genes that will be present in the adult plant. Not only do seeds take up much less room than if the whole plant had to be grown but they also allow stocks of plants to be held in disease-free conditions. Unfortunately, the seeds of many commercially important tropical species such as rubber and cocoa cannot be stored under these conditions. The only possible way of conserving their genes is either to grow the actual plants or to use *tissue culture*.

selection: the process which results in the best-adapted organisms in a *population* surviving, reproducing and passing their genes on to their offspring. As a population increases in size, many factors in its environment become limiting. There may be *competition* for food in animals, or for light, water and mineral salts in plants. Some of the organisms in the population will have *alleles* which mean that they are better adapted to these conditions. These will survive and breed, passing these alleles on to the next generation. This will produce a change in the proportion of alleles in the *gene pool* and may result in evolution. The flowchart shows how antibiotic-resistant bacteria may evolve as a result of selection.

> Patient infected with bacteria. These bacteria vary. A few of them will be resistant to a particular antibiotic

$\downarrow$

> Patient treated with the antibiotic. The bacteria that are not resistant die; those that are resistant survive

$\downarrow$

> The resistant bacteria have less competition. They multiply and pass on the resistance gene

$\downarrow$

cont'd

> Selection has operated. The population has changed from bacteria which were nearly all killed by the antibiotic to bacteria which are mainly resistant to it

There are three basic ways in which selection can operate. These are *directional selection*, *stabilizing selection* and *disruptive selection*.

hint Do not confuse *selection* with ecological *succession*. In selection, one form changes into another; in succession, one species is replaced by another. Selection may take many millions of years and involves evolutionary change. Succession is a relatively short-term ecological process.

selective breeding: improving a particular variety of crop plant or domestic animal by breeding from individuals with desired characteristics. Plants and animals have been subject to selective breeding since they were first domesticated around ten thousand years ago. Early farmers, for example, tended to harvest and save the seed from cereal plants in which the grains had not fallen from the stalks. These would have provided the seed grain for the following year's crop. The plants that grew from this grain would have inherited the genes for stronger grain attachment. Selective breeding would have led to cultivated cereal plants which did not scatter their grains before the farmer had time to harvest them.

One of the first things that accompanied domestication in animals was a reduction in body size. Smaller cattle, sheep and goats resulted partly from selective breeding. Small parent animals would have been more likely to have produced small offspring. Selective breeding has been a continuous process carried out over many thousands of years. Modern examples include the selective breeding of maize plants to increase the oil and protein content of the grain and of dairy cattle to increase milk yield.

self-fertilization: *fertilization* where both male and female *gametes* come from the same organism. Self-fertilization is *sexual reproduction* because it involves the fusion of gametes, even if they do come from the same organism. It is quite common among flowering plants and has an advantage over *cross-fertilization* as it clearly increases the chances of male and female gametes meeting. Because both gametes come from the same individual, however, self-fertilization results in less genetic variation among the offspring.

hint Sexual reproduction involves the fusion of two sex cells or gametes. Self-fertilization is, therefore, sexual reproduction because it involves the fusion of two gametes from one individual. Self-fertilization is not a form of asexual reproduction.

semen: the sperm-containing fluid which is produced by male animals. Sperms are produced by the testes and stored in the epididymis, a mass of ducts just outside the *testis*. When a male mates or copulates with a female, some of these sperms are mixed with secretions from various glands in the male reproductive system to form a liquid called semen. As well as the sperms themselves, semen contains various substances which are needed to nourish and stimulate them. The volume of semen and the number of sperms it contains differ from animal to animal. In a bull, somewhere between 2 and 10 cm^3 containing up to two thousand million sperms per cubic centimetre are produced. Clearly there

245

are far more sperms than are necessary to fertilize a single egg cell. This makes it possible to collect bull sperm and dilute it for use in *artificial insemination*. In this way, the semen collected on a single occasion can be used to fertilize many cows.

semiconservative replication: the way in which a molecule of DNA can produce two exact copies of itself. The original molecule unwinds and each of the resulting chains acts as a template for the formation of a new chain. See *DNA replication* for more details of the process.

semilunar valve: a valve which prevents the backflow of blood. When the muscles of the ventricle wall relax during the *cardiac cycle*, the pressure in the left ventricle falls. As it drops below that in the aorta, the semilunar valves between the aorta and the ventricle shut and prevent blood in the artery from flowing back into the ventricle. A similar mechanism prevents blood from the pulmonary artery from flowing back into the right ventricle.

Blood in *veins* is at a low pressure and its return to the heart may be helped by semilunar valves. Veins returning from the limbs are situated between large muscles. When these muscles contract they squeeze the blood in the veins. Semilunar valves in the walls allow this blood to be squeezed towards the heart but do not allow it to go back in the other direction.

senile dementia: the symptoms and signs produced as a result of degeneration of the brain which occurs with age. People with dementia suffer from a loss of memory. As the condition becomes more severe, this is accompanied by a marked slowing down of general mental ability, difficulties with speech and, in its most severe forms, abnormal patterns of behavior. Although *Alzheimer's disease* is the commonest form of dementia, the condition may have a number of underlying causes. It affects, to some degree, almost a third of all people by the time they reach their late eighties.

sepal: one of the segments that forms the *calyx* of a flower.

seral stage: a recognizable stage in an ecological *succession*. The diagram summarizes some of the stages in succession on sand dunes described in the entry on succession.

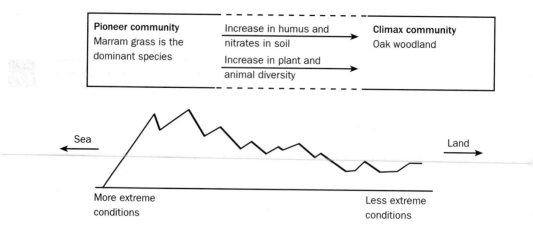

Each of the seral stages shown in the diagram above is made up of a distinctive *community* of organisms.

serum: the liquid left after blood is allowed to clot and the clot removed. Serum is very similar to *plasma* in its composition. The main difference is that it does not contain *fibrinogen* or certain other molecules involved in *blood clotting*.

sessile: sessile organisms are those that stay in one place. Plants depend on light, water and carbon dioxide in order to photosynthesize. They also require a number of inorganic ions. In suitable areas, these are all available in plentiful supply. There is no need to spend energy in moving to another place in order to exploit these resources more effectively.

Animals face slightly different problems. For most of them, food is distributed unevenly. Therefore it is an advantage for most animals to be motile. The few that are sessile, such as members of the phylum *Cnidaria*, tend to feed on small organisms which float in water currents. They are *filter feeders*. Sessile organisms have a number of adaptations associated with their habit:

- the stimuli to which they respond may come from any direction. Therefore they are usually *radially symmetrical*

- many sessile organisms have special adaptations involved in the transfer of gametes between individuals during *sexual reproduction*

- there are difficulties involved in colonizing new areas. Plants have *fruits* which can often be transported considerable distances from the parent. Many sessile animals have young stages which are free-swimming or can drift in water currents.

sex chromosome: one of the chromosomes whose presence determines the sex of an organism (see *sex determination*). In humans and other mammals, females have two *X chromosomes* while the male has one X chromosome and one *Y chromosome*. Although X and Y chromosomes differ from each other in appearance, they can pair with each other during *meiosis*. This is because part of each chromosome is identical, or *homologous*.

sex determination: in many organisms, sex is determined genetically. In mammals, for example, each body cell contains a pair of *sex chromosomes*. Females have two identical *X chromosomes* while males have one X chromosome and one *Y chromosome*. During *meiosis*, the two X chromosomes of the female can obviously pair with each other. In the male, the X chromosome is able to pair with the Y chromosome. This means that all the gametes produced by the female will be the same. They will all contain a single X chromosome. On the other hand, the gametes produced by the male will differ from each other. Half will contain an X chromosome and half will contain a Y chromosome. Since there is an equal chance that the sperm which fuses with the egg cell at fertilization will have an X chromosome or a Y chromosome, it would be expected that equal numbers of males and females will be produced. The same basic mechanism is found in birds but here the male is the homogametic sex. The male bird possesses identical sex chromosomes while the sex chromosomes of the female are different. The female bird is described as the heterogametic sex as she can produce eggs containing either an X chromosome or a Y chromosome.

In addition to the methods of sex determination described above, there are a number of other ways of determining sex. In bees, for example, females are *diploid* and have two sets of chromosomes in each body cell. The males or drones are *haploid*. They only have a single set of chromosomes in each cell. In certain reptiles such as crocodiles and tortoises, sex is not determined genetically at all. It is the temperature at which the eggs are incubated that determines whether the young will be male or female.

sex linkage: a characteristic is sex linked if the gene which controls it is found on one of the sex chromosomes. In most animals, the Y chromosome contains few if any genes. Sex-linked genes are, therefore, most likely to be found on the X chromosome.

An example of a sex-linked gene is that which controls the production of a pigment in the eye which enables discrimination between red and green. This gene has two alleles. The allele for normal vision, R, is dominant to that for red-green color-blindness, r. The genotype of a man with red-green color-blindness is normally expressed as $X^r Y$. This shows that, being a male, he has one X and one Y chromosome. The Y chromosome does not contain an allele of this gene. The allele for red-green color-blindness is r and this is situated on the X chromosome. Since only one allele for this gene can be present in a male, as there is only one X chromosome, he will be red-green color-blind.

The genotype $X^R X^r$, on the other hand, indicates a female since there are two X chromosomes. She is heterozygous and, since the R allele is dominant to the r allele, she will have normal vision. The example below shows the results of a cross between these two individuals.

Parental phenotypes: red-green color-blind male   female with normal vision

Parental genotypes: $X^r Y$ $X^R X^r$

Gametes: $(X^r)\ (Y)$ $(X^R)\ (X^r)$

Offspring genotypes:

		Male gametes	
		X^r	Y
Female	X^R	$X^R X^r$	$X^R Y$
gametes	X^r	$X^r X^r$	$X^r Y$

Offspring phenotypes: $X^R X^r$ female with normal vision
                                        $X^r X^r$ red-green color-blind female
                                        $X^R Y$ Male with normal vision
                                        $X^r Y$ Red-green color-blind male

sexual reproduction: reproduction which involves the fusion of *gametes*. One of the most important features of sexual reproduction is that the process gives rise to the production of genetically different offspring. The diagram opposite summarizes the steps in the process which contribute to this.

The production of genetically different offspring provides the material on which natural *selection* operates.

sickle-cell disease: an inherited condition in which affected individuals possess an abnormal type of *hemoglobin*. Sickle-cell hemoglobin has a very slight difference in the amino acid sequence in two of the polypeptide chains which make up each molecule. As a result, this abnormal hemoglobin crystallizes in low concentrations of oxygen and the red blood cells of an affected person collapse into a distinctive sickle shape. Sickle cells may clump together and block the flow of blood. They are also rapidly broken down so the affected person may suffer from severe *anemia*. The condition is determined by a single gene with codominant *alleles*. This gives rise to the three *phenotypes* shown in the table.

Genotype	Phenotype	Consequences
$Hb^A Hb^A$	Normal adult hemoglobin	Does not suffer from anemia but vulnerable to malaria.
$Hb^A Hb^S$	Both types of hemoglobin	Shows no symptoms of sickle-cell disease unless in conditions where there is severe oxygen shortage. Has some protection from malaria.
$Hb^S Hb^S$	Sickle-cell hemoglobin	Suffers severely from sickle-cell disease. Without medical help likely to die during childhood.

The table shows that a person with the genotype, $Hb^A Hb^S$, has a selective advantage in parts of the world where malaria is common. This is thought to have led to the high frequency of the Hb^S allele among people of African origin.

single circulation: a blood system in which the blood passes through the heart only once in its passage around the body. In fish, for example, blood flows from the single ventricle of the heart to the gills. After it has become oxygenated it goes to the rest of the body without going back to the heart again. A complete circuit therefore involves only passing once through the heart. A major disadvantage of a circulation like this is that blood is sent to the tissues at a very low pressure, much lower than is the case with a double circulatory system.

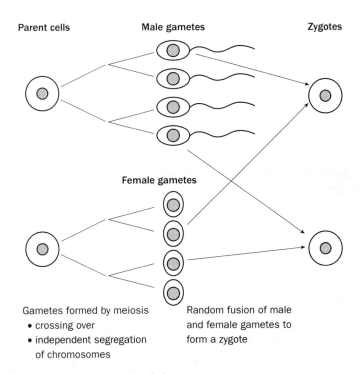

Parent cells   Male gametes   Zygotes

Female gametes

Gametes formed by meiosis
- crossing over
- independent segregation of chromosomes

Random fusion of male and female gametes to form a zygote

Sexual reproduction and variation

249

single-gene inheritance: a genetic cross between individuals in which only alleles of a single gene are considered. It is important in explaining any genetic cross to set out the explanation clearly and logically. The example shown under the entry on the *dihybrid cross* shows how to do this.

sink: part of an animal or plant where substances are removed from the transport system. These organisms are too large to rely on *diffusion* and have *mass-flow* systems in which substances are moved rapidly from one place to another. Mass-flow systems link exchange surfaces with each other. At some of these exchange surfaces, materials are taken up while, at others (sinks), they are unloaded. The term sink is usually used when referring to transport by the *phloem* in flowering plants. Here, roots and other underground storage organs, buds and developing fruits provide examples of sinks.

sinoatrial node: a small area of muscle in the wall of the right *atrium* of the heart which controls the events of the *cardiac cycle* or heartbeat. Heart muscle is not like muscle found in other parts of the body as it can contract and relax automatically. It does not need a nerve impulse to trigger this action. The sinoatrial node is the area of cardiac muscle which controls and coordinates contraction of the rest of the heart. A wave of electrical activity spreads from here over the walls of the atria to the *atrioventricular node*. This brings about contraction of the atria followed, after a short delay, by contraction of the ventricles. The rate at which the sinoatrial node sends impulses to the rest of the heart can be slowed down by impulses coming from the brain along the *vagus nerve*. It can be speeded up by impulses reaching the node along the *cardiac accelerator nerve*.

sinusoid: a blood-filled space in the *liver*. The liver is made up of structural units called lobules. Each lobule consists of a number of vertically arranged sheets of cells surrounding a central vein. Blood flows from branches of the *hepatic artery* and *hepatic portal vein* through the spaces, or sinusoids, between the liver cells to the central vein. As it flows past the liver cells, the products of digestion are absorbed.

skeletal muscle: muscle which, as a result of its contraction, usually brings about the movement of an animal. In mammals, skeletal muscles are arranged in *antagonistic* pairs attached to bones. They are under conscious control and, for this reason, are sometimes called voluntary muscles. A muscle contains many *skeletal muscle fibers*. Muscle fibers are of two different types, so-called *slow muscle fibers* which produce strong, sustained contractions, and *fast muscle fibers* which give rise to more rapid responses. Each individual muscle fiber is made up, in turn, of a large number of *myofibrils*. A knowledge of the structure of these myofibrils is important in understanding the *sliding-filament hypothesis*, the mechanism which explains the way in which a skeletal muscle contracts.

skeletal muscle fiber: one of the individual muscle fibers which make up a *skeletal muscle*. There are two types of muscle fiber, which differ from each other in their properties. These are fast or type 2 fibers and slow or type 1 fibers. The table summarizes some of the important differences between them.

Fast skeletal muscle fibers	Slow skeletal muscle fibers
Produce rapid, powerful contractions, which makes them very important during locomotion	Allow sustained muscle contractions. Play an important part in maintaining body posture

Muscle contains large amounts of stored *glycogen* which acts as the *respiratory substrate*. Since fast skeletal muscle fibers act before the oxygen supply to the time to increase, much of the ATP supplied comes from *anaerobic respiration. Oxygen debt* builds up rapidly and fibers fatigue quickly	Muscle does not contain much stored glycogen. Fat is often used as the respiratory substrate. ATP supplied by aerobic respiration, although these muscle has had fibers can respire anaerobically. Do not fatigue quickly

skeleton: a supporting structure. Skeletons can vary from the network of protein filaments which serve to provide support and shape to individual cells, to the very complex systems in mammals. Skeletons have three main functions:

- they provide support. Needs will vary according to the environment of the animal. Animals that live on land need much stronger supporting structures than those that live in the water

- they provide attachments for the muscles and act as a system of levers. This makes locomotion possible

- they have a protective function. In a mammal, for example, the skull protects the brain while the rib cage protects the organs in the thorax.

Three types of skeleton are found. An *exoskeleton*, such as that found in the *Arthropoda*, is on the outside of the body and encloses the soft internal structures; an *endoskeleton*, such as that found in mammals and other chordates, is inside the body. The third type of skeleton is the *hydrostatic skeleton*. This is based on fluid contained within the body and characterizes soft-bodied animals such as worms.

sliding-filament hypothesis: a hypothesis which explains the way in which *skeletal muscle* contracts. The thick *myosin* filaments and the thin *actin* filaments which make up the *myofibrils* slide between each other and, as a result, the muscle shortens. The mechanism which allows this to happen involves the formation of cross-bridges between the two types of filament. The head of myosin molecule attaches to a particular place or attachment site on an actin molecule. The head then rotates and pulls the two filaments into each other. It finally detaches and goes back to its original position. This cycle of attaching, changing position and detaching is repeated many times. Calcium ions are important in initiating the process. When the muscle is stimulated to contract, the release of calcium ions makes the attachment sites on the actin molecules available so that the cross-bridges can form. The process of muscle contraction requires a large amount of energy in the form of *ATP*.

slow muscle fiber: see *skeletal muscle fiber*

small intestine: part of the gut or alimentary canal between the stomach and the colon. It is a long coiled tube which is divided into two parts. The first part, into which *bile* and pancreatic juice are secreted, is known as the *duodenum*. The second, much longer, part is the *ileum*. This is the main site of absorption of the products of digestion.

smooth muscle: see *involuntary muscle*

solute potential, Ψ_s: the pressure produced by solute molecules in a solution. These molecules will be moving around at random and, as a result of this, some will collide with the membrane that surrounds them. They will exert a pressure on this membrane. This is the solute potential, the symbol for which is the Greek letter psi, Ψ_S. Like water potential, solute potential has a negative value. The relationship between solute potential, water potential and pressure potential can be summarized by the equation.

$$\Psi_S = \Psi - \Psi_P$$

source: part of a plant or animal where substances are taken into the transport system. These organisms are too large to rely on diffusion and have *mass-flow* systems in which substances are moved rapidly from one place to another. Mass-flow systems link exchange surfaces with each other. At some of these exchange surfaces, materials are taken up while, at others, they are unloaded. A source is the name given to a site where substances are produced and loaded into the transport system. The term is often used when referring to transport by the *phloem* in flowering plants. Here, a leaf is the most obvious example of a source.

speciation: the formation or evolution of two or more new species from existing species. This can happen in one of two different ways. Two different species may cross to form a hybrid which is different from both of its parents. This may be the case in plants. Modern varieties of wheat, for example, originated from the accidental crossing of two species of wild grass to form a hybrid, an event which took place many thousands of years ago.

The other method of speciation occurs when a single original species gives rise to two new ones. This occurs when populations of the parent species become separated from each other. If the separation is due to some form of geographical barrier, then this is known as *allopatric speciation*. On the other hand, if it occurs in two populations which occupy the same geographical area and some other factor prevents their interbreeding, it is *sympatric speciation*.

species: the starting point in the biological system of *classification*. A species is a group of similar organisms which are able to breed together and produce fertile offspring. At first sight it would appear easy to determine whether two organisms belong to different species. There are obvious differences between horses and donkeys. They can breed with each other but the offspring they produce, a mule, is sterile. Clearly, horses and donkeys belong to different species. However, there are examples where it is much more difficult to decide whether two organisms belong to the same or different species. Individual members of some species can look very different from each other; there may be considerable differences due to sex or geographical distribution. In addition, some species hardly, if ever, reproduce sexually, making it even more difficult to decide. The more that is known about *DNA* and *proteins*, the more it may become possible to involve these molecules in classifying organisms.

> **hint** Make sure that you use the term "species" correctly. Some students use the word in all sorts of other ways, all of them incorrect. Plant breeders, for example, cross different varieties together to produce a plant with, perhaps, a higher yield. They do not cross different species.

sperm: a male *gamete* from a mammal or other animal. In mammals, sperm cells are formed in the *testes* by the process of *spermatogenesis*. A mature mammalian sperm consists of a head containing a haploid nucleus. In front of the nucleus is the *acrosome*, a sac-like structure that contains enzymes. Just before fertilization the membrane surrounding the acrosome bursts and these enzymes are released. Their digestive action helps to separate the cells which still surround the egg, making it possible for fertilization to take place. A middle piece behind the head is packed with *mitochondria* which provide the ATP necessary for the sperm to swim with the aid of its flagellum.

sperm bank: a collection of frozen sperm cells used as a means of banking or preserving animal genes. Sperms are collected and, after suitable treatment, deep frozen in liquid nitrogen. They can survive in these conditions for long periods of time and may be used to fertilize eggs many years later. As with *seed banks*, sperm banks enable a wide variety of genetic material to be stored in a very convenient form and are valuable in the *genetic conservation* of domestic animals and of species which are in danger of extinction. Considerably greater success has been achieved storing sperms than storing eggs or embryos.

spermatogenesis: the process in which sperms are formed. The testes (see *testis*) in a mammal contain a large number of coiled tubules. It is in these seminiferous tubules that the sperms are formed. Around the outside of each one is a layer of cells called the germinal epithelium. The cells of the germinal epithelium divide by *mitosis* and produce a large number of cells, each of which is capable of increasing in size then dividing by *meiosis* to produce four sperms. The basic sequence of mitosis, growth and meiosis is the same as in the formation of gametes in a female during *oogenesis*. There are, however, some important differences in the way in which meiosis actually takes place. The process by which sperms are formed is also similar to that leading to the formation of female gametes in that it is controlled by hormones released from the anterior lobe of the pituitary gland.

Chance is loaded heavily against an individual male gamete meeting and fertilizing a female gamete. To ensure that the probability of this happening is as great as possible, the process of spermatogenesis produces very large numbers of sperms.

sphincter muscle: a ring of muscle which runs around the wall of a tubular organ. Its relaxation or contraction allows the organ to be opened or closed. There are a number of sphincter muscles along the gut of a mammal. The pyloric sphincter, for example, regulates the flow of the contents of the stomach into the small intestine, and the anal sphincter opens when feces are eliminated from the body. Other sphincter muscles lie between many *arterioles* and the *capillaries* which they supply. Their opening and closing regulates the flow of blood to the organ concerned.

spinal cord: part of the central nervous system of a vertebrate which is enclosed in and protected by the backbone or vertebral column. A cross-section through the spinal cord shows that it is made up of two distinct regions. There is a central area which is shaped rather like the letter H. This is called the grey matter and consists of a mass of nerve cells. Surrounding this is the white matter, which contains the axons which transmit impulses up and down the spinal cord. Between each of the vertebrae, there are paired spinal nerves, one on the right and one on the left. Each of these spinal nerves has a dorsal root which contains sensory neurones and a ventral root which contains the motor neurones. Apart from conducting impulses between the brain and the spinal nerves, the spinal cord plays an important part in coordinating many *reflexes*.

spindle: a system of *protein* fibers or microtubules that is involved in organizing *chromosomes* during cell division. In *mitosis*, the spindle starts to form as the chromosomes become visible during prophase. It is fully formed by the end of metaphase at which stage it fills the space that was originally occupied by the nucleus of the cell. The chromosomes are now attached by their centromeres to the spindle fibers and are arranged across the equator of the cell. Finally, during anaphase, the pair of chromatids that make up each chromosome are pulled apart by the spindle fibers to the poles of the cell.

spiracle: one of the openings in the side of an insect's body through which respiratory gases diffuse. The spiracles lead into a series of fine tubes known as *tracheae* which take air directly from the outside of an insect to its cells. The spiracles of most insects have a mechanism which allows them to open and close. They normally remain open for short periods of time. This allows efficient gas exchange but reduces the amount of water that can be lost. Both an increase in the concentration of carbon dioxide and a fall in the oxygen concentration can act as stimuli for the opening of the spiracles.

spirillum: a bacterium with spiral-shaped cells. Some of these may be of considerable size, up to 500 μm in length in extreme cases. Perhaps the best known of these bacteria is *Treponema pallidum*, the organism which causes syphilis.

spirometer: a device which measures the volumes of gas breathed in and out of the lungs. In its simplest form it relies on the subject breathing through a mouthpiece which is connected by a series of tubes and valves to a set of bellows. As air enters or leaves, the movement of the bellows is recorded by a pen on a moving chart. Records obtained from a spirometer enable the various *lung volumes* to be determined. If the carbon dioxide in the exhaled air is absorbed by a substance such as soda lime, the total volume of gas in the bellows gradually decreases as the oxygen is used in respiration. This allows the rate of oxygen consumption to be measured.

spore: a tiny structure, often only single celled, which is produced during reproduction and is dispersed from the parent to give rise to a new individual. Spores are produced by a variety of organisms and are found in all except the animal kingdom. They vary in the way that they are formed and in the details of their structure. For example, spores may result either from *sexual* or *asexual reproduction* and they may or may not be surrounded by a hard protective wall. Many spores are produced in enormous numbers and allow rapid increases in population or colonization of new areas. Those with hard outer coverings are often resistant to extreme physical conditions.

sporophyte: the diploid, spore-producing stage in the life cycle of a plant. Plant life cycles show an *alternation of generations* in which a spore-producing stage, or sporophyte, alternates with a gamete-producing stage, or *gametophyte*. The sporophyte produces the spores which, in turn, will grow and produce a gametophyte. In primitive plants such as mosses and liverworts, the gametophyte is the dominant stage. The sporophyte is the small capsule on the end of a stalk which develops on the moss gametophyte. It is completely dependent on the gametophyte. In flowering plants, however, the sporophyte is the dominant stage in the life cycle and the gametophyte is very much reduced. The adult plant, whether it is a cabbage or an oak tree, is a sporophyte.

stabilizing selection: selection which operates against the extremes in a population. Birth mass in humans provides an example.

The graph shows the pattern of variation in birth mass. Although there are many causes of variation in birth mass, some are genetic. Very small babies are less likely to survive. There is also a higher death rate among very large babies. Consequently selection is in favor of babies whose birth mass is around the mean, and genes responsible for very large and very small babies will tend to be eliminated. Stabilizing selection, therefore, prevents evolutionary change.

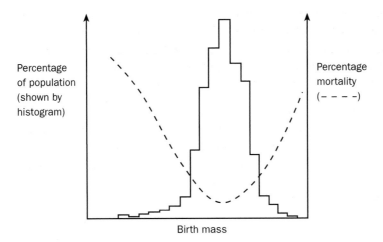

Stabilizing selection and human birth mass

stain: a substance used to show up part of a cell or tissue so that it can be seen clearly when examined with a microscope. When a light microscope is used, the stain is a colored substance. A basic stain has a colored cation. It stains the more acid parts of cells such as the nuclei which contain large amounts of nucleic acid. An example of a basic stain is hematoxalin which stains nuclei purple. Acid stains have colored anions and they are useful for staining cytoplasm. Eosin is an acid stain which colors cytoplasm pink.

Different sorts of stains are used with *electron microscopes*. These contain heavy metals such as uranium and lead. They bind to the chemical groups which are found in particular cell *organelles*. Since few electrons are able to pass through them, the structures they stain appear dark in color.

stamen: the male parts of a *flower*. Each stamen is made up of an anther in which the pollen develops and a stalk known as a filament. The anther has two or four lobes and, within each lobe, is a layer of cells which nourishes a mass of developing *pollen* grains. When ripe, the anthers split and release the pollen.

standard deviation: a mathematical term which gives a measure of how spread out a set of results are. In describing any set of results it is useful to know two things: where the middle is and how spread out the individual measurements are. The spread of results can be given simply as the difference between the largest reading and the smallest one. Unfortunately this can be misleading as it only gives an idea about these two particular values. The standard deviation, however, is much more useful as it gives us a measure in terms of the distances of all the results from the mean. Standard deviation can be calculated from a formula but, in most Biology examinations, it is enough to be able to press the appropriate buttons on a calculator.

starch: a large, insoluble carbohydrate which forms an important energy store in plants. Starch is a *polymer* and consists of a large number of α-*glucose* molecules joined together by *condensation* reactions. The exact chemical composition of starch varies from one species of plant to another as it consists of two main components which may be present in different proportions. Amylose forms long straight chains while amylopectin has branched chains. Starch has a number of features which make it an ideal storage compound:

255

- its molecules are tightly coiled, allowing a considerable amount of starch to be packed into a relatively small volume

- it is insoluble. It is therefore easier to store since it does not move out of cells readily, nor will it affect the *water potential* of a cell

- it is readily broken down by enzymes to glucose.

The test for starch is to add a drop of iodine solution. A blue–black color is produced.

starch–sugar hypothesis: an early idea suggested to explain the opening and closing of *stomata*. It was proposed that, in the daylight, starch present in the guard cells was converted to sugar. As the concentration of sugar increased, the *water potential* of the guard cells fell and water entered by *osmosis*. The stomata opened because of the resulting increase in turgor in these cells. However, although experimental work did show that there was a decrease in the amount of starch present, it proved difficult to obtain evidence which demonstrated the accumulation of sugar. This idea has now been largely discredited in favour of the *potassium-movement hypothesis*.

stationary phase: part of a *population growth curve* in which the rate of increase in the population is more or less balanced by the number of organisms dying. As a result the population remains more or less constant.

sterilization (microbiology): involves killing all the *microorganisms* present in a particular place. There are a number of physical agents which have a harmful effect on bacteria and other microorganisms. Chemical disinfectants and low temperatures will damage and kill many cells but will not necessarily achieve complete sterilization. Some methods commonly used to do this include:

- heat. Sterilization by heat is one of the most commonly used and reliable methods. Unlike the use of chemicals, it does not leave any toxic residues. Its only real disadvantage is that it could damage the material being sterilized. Exposure to moist heat is lethal at lower temperatures and in a shorter time than exposure to dry heat. The use of steam at lower than atmospheric pressures is particularly effective

- ultraviolet radiation. Certain wavelengths of ultraviolet radiation are absorbed by and cause damage to nucleic acids and proteins

- *ionizing radiation.* All types of radiation produced by atomic particles have a similar effect on living tissues. Electrons are knocked out of some molecules and gained by others. The ions which are produced as a result have extremely damaging effects. Radiation of this type is widely used in producing sterile syringes and Petri dishes.

sterilization (reproduction): the use of a surgical operation to produce sterility. In women, the modern way of doing this is to insert a small viewing instrument, or *endoscope*, through the wall of the abdomen. The lower parts of the two oviducts are then permanently closed with the aid of clips or small plastic rings. In males, sterilization involves vasectomy. The tubes which take the sperm from the testes are tied and cut. Although both of these procedures provide an effective means of contraception, they must be looked upon as permanent. They are very difficult to reverse.

steroid: one of a group of *lipids* whose molecules do not contain *fatty acids*. One of the most important steroids in the human body is *cholesterol*, which is made in the liver. Apart from being found in cell membranes, cholesterol can be converted into a number of other biologically important steroids such as the sex hormones, *estrogen* and *progesterone*.

stigma: the part of the *flower* which receives the pollen grains. In a wind-pollinated flower, the stigma is usually feathery and hangs loosely out of the flower. It is adapted to trap *pollen* which is present in the air around the plant. The stigmas of insect-pollinated flowers, however, are usually enclosed within the flower. They are much flatter and often have sticky surfaces, enabling them to trap the pollen carried on the body of a pollinating insect.

stomach: a large, muscular sac in the front part of the gut of a mammal which stores and digests food. Food is chewed in the mouth and mixed with *saliva*. It then travels down the *esophagus* into the stomach. The wall of this organ is unlike other parts of the gut in that it has three layers of muscle. Contraction of these muscles results in the stomach contents being continually mixed and churned. Cells in the wall of the stomach produce gastric juice. This is a mixture of mucus, hydrochloric acid and digestive enzymes, particularly important among which is the endopeptidase *pepsin*.

stomata: the small pores in the surface of a leaf through which gas exchange normally takes place. They may be found on either surface of the leaf but are more commonly located on the underside. However, in aquatic plants with floating leaves like the water lilies, stomata are only found on the upper surface. They are particularly important in allowing the carbon dioxide required in *photosynthesis* to diffuse from the atmosphere into the intercellular spaces of the leaf. Unfortunately, as gases diffuse into the leaf, so large amounts of water vapor will be lost via *transpiration*. There has to be a compromise between having the stomata permanently open so that the carbon dioxide can diffuse in and closed so that water loss is minimized. Stomata are surrounded by guard cells which are able to regulate the size of the stomatal opening. Various mechanisms have been suggested to explain how they do this. One of the earliest was the *starch–sugar hypothesis*, but the *potassium-movement hypothesis* is now thought to offer a more likely explanation. *Xerophytes* are plants which live in particularly dry places. The stomata on their leaves are often adapted by being sunken into pits or by being surrounded by hairs. This reduces water loss.

hint Stomata do not form the gas exchange surface of a leaf because they only allow gases to pass between the atmosphere and the air spaces inside the leaf. The gas exchange surface is the surface of the *mesophyll* cells.

stratigraphy: the study of the sequence of rock layers. Sedimentary rocks are formed as layers of material settle on top of each other. The older the rock, the lower it will be in a sequence; the more recently the rock was formed, the closer it will be to the surface. Knowledge of the stratigraphical sequence is particularly useful in comparing the age of fossils. The older the fossil, the more deeply it will be buried. This situation is complicated by processes such as weathering. The action of wind and water progressively erodes the land, exposing rocks of different age to the surface. Nevertheless, a good knowledge of the geology of an area still allows the various rocks in which fossils have been found to be placed in a time sequence. Unfortunately, stratigraphy can only provide information about the relative ages of particular fossils. To get an absolute date, other techniques such as *potassium–argon dating* must be used.

striated muscle: see *skeletal muscle*

stroke volume: the amount of blood pumped out each time the heart beats. The *cardiac output*, which is the total amount of blood pumped out by the heart per minute, is obtained by multiplying stroke volume by the heart rate, the number of beats per minute. The relationship

257

between stroke volume, cardiac output and heart rate is given by the equation:

$$\text{stroke volume} = \frac{\text{cardiac output}}{\text{heart rate}}$$

style: the part of the flower which connects the *stigma* to the *ovary*. Once a pollen grain has landed on the stigma, it produces a *pollen tube* which grows down through the tissues of the style.

suberin: a waxy, waterproof material found in plant cell walls. It is found in the outer cell layers of older roots and stems. The endodermis in a plant is a ring of cells between the outer part of the root and the vascular tissue in the center. A band of suberin, called the casparian strip, runs round the walls of each of these endodermal cells. It prevents water and dissolved substances from getting into the vascular tissue by going through the cell walls and intercellular spaces. As a result, these substances have to pass through the cytoplasm of the endodermal cells, which can therefore control their movement into the *xylem*.

substrate: in general terms, a substance which is acted on or used by something else. The term is used in a variety of contexts in biology:

* in biochemistry, it is the molecule on which an *enzyme* acts. Enzymes are extremely specific in their action. Only molecules with a particular shape will fit the *active site* of the enzyme concerned and form an enzyme–substrate complex

* a respiratory substrate is an organic substance which is used as a starting point in *respiration*. Note that oxygen is not a respiratory substrate

* in microbiology, a substrate is a medium on which *microorganisms* can grow. It may be either in a liquid or a solid form and will contain all the necessary nutrients

* *sessile animals* are those that stay in one place. They are attached to a particular substrate or surface.

succession: the way in which the different species of organisms which make up a *community* change over a period of time. Sand dunes are common in many coastal areas. The sand is very unstable, constantly being blown by the wind. Not only that, but there are very low levels of soil nutrients and such areas dry out rapidly. One of the first plants to colonize this harsh environment is marram grass. In time, this binds the sand particles together. Plants die and the resulting dead vegetation breaks down to increase the amount of humus present and the levels of important soil nutrients such as nitrates. Marram grass no longer thrives and it is replaced by other species which are better adapted to these changed conditions. There is a succession from pioneering species such as marram grass, through a variety of *seral stages* until, ultimately, a *climax community* is established. With the example chosen, this climax community might be oak woodland. A succession which is prevented in some way from reaching this natural climax is called a *deflected succession*.

> **hint** Succession is replacement of some species of organisms which make up a community by others over a period of time. It is an ecological concept and has nothing to do with the evolution of new species.

sucrose: a *disaccharide*, that is a sugar which is made up of two sugar units. These units are *glucose* and fructose and they are joined together by a *condensation reaction*. Sucrose is a *nonreducing sugar* and therefore will not produce a positive result with Benedict's test unless it has first been hydrolyzed. Sucrose is a particularly important molecule in plants. It is the form in which carbohydrates are usually transported in *phloem*.

summation: a process which occurs in *synapses* by which a number of weak stimuli can lead to the generation of a nerve impulse. A nerve impulse can only be triggered in the post-synaptic nerve cell when enough of the *neurotransmitter* has reached the receptor molecules on the postsynaptic membrane. This can happen in two different ways. If a large number of synapses are stimulated at the same time, the total amount of transmitter may be enough to trigger a nerve impulse even if the amount released at an individual synapse is not. This is called spatial summation. Temporal summation results from repeated stimulation of the same synapse.

supernatant: the liquid layer which is left on top when a suspension is centrifuged. *Cell fractionation* is a process in which cells are broken up and the different types of *organelle* separated from each other. The tissue is broken up and suspended in a buffer solution. When it is spun in a *centrifuge* at a relatively low speed, the larger, denser organelles such as nuclei and chloroplasts fall to the bottom to form a pellet. The smaller organelles will remain in the supernatant.

surface area–volume ratio: the relationship between the surface area of an organism and its volume. In biological terms, this is important because it means that the larger the size of an organism, the smaller its surface area in relation to its volume. Perhaps the simplest way to illustrate this principle is to consider the diagram below. If we take a number of cubes, each a little larger than the previous one, we can work out the surface area–volume ratio for each one. If this information is plotted as a graph, then the relationship between the surface area–volume ratio and the size of the cube can be clearly shown.

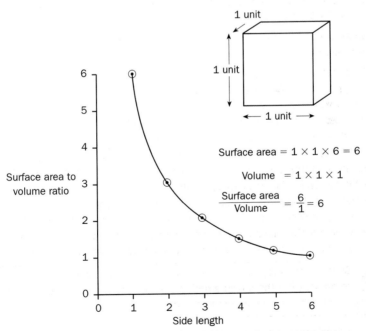

Surface area $= 1 \times 1 \times 6 = 6$

Volume $= 1 \times 1 \times 1$

$\dfrac{\text{Surface area}}{\text{Volume}} = \dfrac{6}{1} = 6$

Obviously, animals and plants are not cube shaped, but the principle is just the same whether the basic shape is a cube or that of a living organism.

This idea is of enormous physiological importance. Taking a single example, a small one-celled organism like amoeba has a very large surface area when compared to its volume.

Therefore, it is able to gain all the oxygen it requires by *diffusion* over its general body surface. A larger organism such as a fish, however, has a much smaller surface area to volume ratio. This explains why it needs a specially adapted gas-exchange surface.

> **hint** Remember, the larger the organism, the smaller its surface area to volume ratio. Make sure that you get this the right way round.

surfactant: a substance which lowers the surface tension of a liquid. If the inside of a polythene bag is wetted slightly, its surfaces will tend to stick together. This is due to surface tension, a force which pulls the water molecules together. The *alveoli* in the lungs are lined with a layer of liquid. If this liquid had a high surface tension, it would pull the sides of the alveoli together. They would stick to each other and reduce the surface area available for gas exchange. However, the liquid lining the alveoli contains surfactant. This lowers its surface tension and prevents the walls sticking together. There is a lower concentration of surfactant in the lungs of smokers than in those of nonsmokers.

Surfactant is very important in the lungs of a newborn baby. At birth, a baby takes several strong breaths which inflate the lungs. Without surfactant being present, the lungs would collapse again.

survival curve (survivorship curve): a graph which shows the percentage of individuals born, or starting to grow, at a particular time which survive at a given age. Three survival curves are shown below.

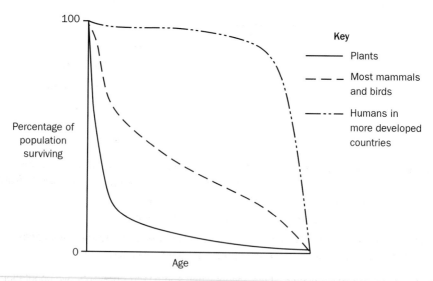

- Flowering plants usually produce very large numbers of seeds. Most of these fail to germinate or die very soon after they have started to grow. This accounts for the steep fall in the first part of the curve.

- Most mammals and birds, on the other hand, provide a considerable amount of parental care. Although there is still a high mortality early in life, relatively more young will survive. The curve does not fall as steeply. Curves of this pattern apply to human populations in many less developed countries.

- In more developed countries, a very high proportion of children survive to adulthood. The survival curve in this case shows a steep drop towards the end of life.

symbiosis: any nutritional relationship between two organisms of different species. The meaning of this term presents some difficulties since it may also be defined as a relationship between two organisms where both gain a nutritional advantage. It is better to keep the term symbiosis to refer to any nutritional relationship and to use *mutualism* when considering relationships where both of the organisms involved gain an advantage. The diagram below summarizes some different types of symbiotic relationship.

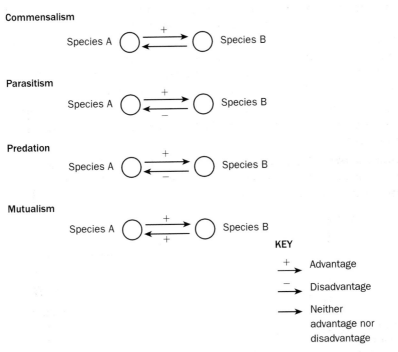

Commensalism

Species A ⟷ Species B

Parasitism

Species A ⟷ Species B

Predation

Species A ⟷ Species B

Mutualism

Species A ⟷ Species B

KEY

$+$ → Advantage

$-$ → Disadvantage

→ Neither advantage nor disadvantage

sympathetic nervous system: a division of the *autonomic nervous system* which generally acts in emergencies and controls the functions of the body during times of stress. Some of the important features of the sympathetic system are that:

- some of the *synapses* secrete noradrenaline as a transmitter
- effects are usually stimulatory, for example sympathetic stimulation increases the breathing rate and increases the heart rate and stroke volume.

sympatric speciation: *speciation* which occurs when some factor prevents two populations of a particular species living in the same area from interbreeding. The mechanism which prevents this is known as an isolating mechanism. In the case of sympatric speciation, it is something other than a geographical barrier. Some examples of isolating mechanisms are given in the table on page 262.

Isolating mechanism	Example
Differences in courtship behavior	The chiffchaff and the willow warbler are small insect-eating birds. They are so similar to each other in appearance that it is very difficult to tell them apart. However, their songs, which form an important part of courtship behavior, are completely different. Since successful mating will only result from the correct behavior pattern, chiffchaffs and willow warblers do not interbreed
Failure to transfer gametes	Bees can distinguish between flowers of the corn poppy and flowers of other species of poppy. Bees which are visiting corn poppies do not go to the flowers of other species. Pollen is not transferred between different species and interbreeding is prevented
The production of infertile hybrids	The carrion crow is found in the southern part of Britain; the hooded crow in the north. They interbreed and form hybrids but the hybrids are not as fertile as the parents. Selection therefore favors the parents

symplastic pathway: the pathway by which substances go through the cytoplasm of plant cells. The contents of neighboring cells are linked by thin cytoplasmic strands called plasmodesmata which pass through the cell walls. Substances are therefore able to pass from cell to cell along this pathway without having to keep on passing through cell membranes, which would require the expenditure of a great deal of energy. Water and mineral ions move through the roots into the *xylem*. They also move out of the xylem and through the leaf tissue. There are three separate pathways along which they can travel: the symplastic pathway described here; the *apoplastic* pathway through the cell walls and the intercellular spaces; and the *vacuolar* pathway which involves passing from vacuole to vacuole. The apoplastic pathway is probably the most important of the routes by which water moves through plant tissues, but the symplastic pathway is very important in the transport of mineral ions.

 Symplastic and apoplastic are two words which are frequently confused. Make sure that you know the difference between them.

synapse: a junction between nerve cells. Nerve cells do not actually join with each other. There is a very small gap between them, a gap which is only about 20 nm wide. This is called the synaptic cleft. On one side of this is the presynaptic membrane and on the other side, the postsynaptic membrane. A *nerve impulse* arriving at a synapse brings about the release of a *neurotransmitter*. This diffuses across the synaptic cleft and produces another nerve impulse on the postsynaptic side. The events which occur in a synapse where the neurotransmitter is *acetylcholine* are summarized in the flowchart on page 263.

Synapses play an important part in the working of the nervous system and they have a number of important properties:

* synapses only transmit information in one direction, from the presynaptic neurone to the postsynaptic neurone

Nerve impulse reaches synapse along presynaptic neurone

↓

Vesicles containing acetylcholine fuse with the presynaptic membrane. Acetylcholine is released into the synaptic cleft

↓

Acetylcholine diffuses across the synaptic cleft and attaches to receptor molecules on the postsynaptic membrane

↓

Ion-channels open allowing sodium ions to enter the postsynaptic neurone. This leads to a nerve impulse

↓

An enzyme called acetylcholinesterase brings about the rapid breakdown of acetylcholine molecules

* the process of *summation* means that a number of weak stimuli can lead to the generation of a nerve impulse

* the synapse in the flowchart is an excitatory synapse because the effect of the acetylcholine is to stimulate another action potential. *Inhibitory synapses* also exist. In these, the effect of the transmitter is to make it less likely that a new impulse will be produced.

synergistic: where the action of two things together is greater than that of their separate effects added together. Various enzymes are added to biological washing powders. Both lipases and proteases are quite effective at removing stains. Research has shown, however, that addition of both enzymes produces a greater effect than might be expected. Most stains are a mixture of biological compounds and often include both protein and lipid molecules. The proteins that hold the stain together and bind it to the fabric concerned are digested by the proteases. This allows lipases to penetrate more effectively and remove the lipid components of the stain. Drugs can also act synergistically.

synovial joint: a type of joint found between bones in the mammalian skeleton. The elbow, knee, shoulder and hip joints are all examples of synovial joints. The bones are held together by *ligaments*. These form a protective capsule round the joint. Lining this capsule is a thin synovial membrane. The cells which make up the synovial membrane secrete a fluid into the capsule of the joint. This fluid acts as a lubricant, helping to ensure smooth movement. The ends of the bones concerned are covered with a smooth layer of *cartilage* which further reduces the friction between them. *Osteoarthritis* is a disease affecting joints in which this cartilage is lost and the underlying bone is damaged as a result. The condition is painful and affected people experience difficulty in moving the joint concerned.

systemic herbicide: a substance used to control weeds that is transported through the tissues of the plant. Systemic herbicides are synthetic chemicals which are similar to naturally occurring *plant growth substances*. They are absorbed by leaves and roots and are transported to all parts of the plant where they cause death by interfering with the natural growth pattern. Systemic herbicides have two particular advantages over many other herbicides:

* they have little effect on monocotyledonous plants such as cereals and other grasses but they rapidly bring about the death of dicotyledonous weeds. They are very suitable, therefore, for controlling weeds on lawns or in cereal crops

* contact herbicides only kill those parts of the plant with which they come into contact. Systemic herbicides spread throughout the plant and can therefore ensure that the root systems and underground stems of perennial weeds are killed and will be unable to shoot again.

systemic insecticide: a substance used to control insects which works by first being absorbed into the tissues of the food plant. Systemic insecticides have two major advantages over other kinds of insecticide which only work when they come into contact with the insect concerned:

* since the insecticide spreads through the plant's tissues, the whole surface of the plant is effectively poisonous. An insect feeding on any part of the plant will be killed. This means that it is not necessary to spray the entire plant surface, reducing the amount of insecticide that it is necessary to use

* only insects which actually feed on the crop that has been sprayed will be killed. There is little risk to predators or insects which are simply resting on the plants.

systole: the stage in the *cardiac cycle* which involves contraction of the heart muscle. There are two steps involved. Atrial systole occurs when the muscle of the atrial walls contracts and forces blood into the ventricles. Ventricular systole is when the walls of the ventricles contract, pumping blood out through the arteries.

systolic blood pressure is the blood pressure in the main arteries measured when the *ventricles* are emptying (*systole*). At this stage in the cardiac cycle, blood pressure in the arteries will be at its highest.

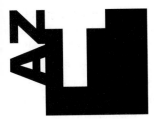

taxis: a form of behavior in which an organism moves either directly towards or away from a *stimulus*. A fly maggot, for example, will move away from the light. It will show negative phototaxis. On the other hand, adult flies will show positive phototaxis and move towards the light. Other stimuli will also produce this form of behavior. Blood-sucking leeches will respond positively towards a source of heat, while flatworms respond in the same way to water currents and chemicals. Behavior patterns like these clearly have survival value. The leach is a *parasite*. By moving towards a source of heat, it will be better able to find its warm-blooded host.

taxonomy: the scientific study of *classification*.

T-cell: a type of lymphocyte which does not secrete antibodies. There are many different types of T-cell, and their interactions with each other and with B-cells are very complex. Two of their important functions, however, are that they are responsible for killing cells which have been infected with pathogens and they are also involved in controlling the immuno-logical process.

temperature coefficient: see Q_{10}

temperature control: the system by which body temperature is controlled in animals. In order to function normally, animals must keep their body temperatures within certain limits. If an animal's temperature falls too low, biochemical reactions will be too slow for it to remain active. If it goes too high, there is the risk of enzymes and other proteins being denatured. To a certain extent, all animals exert some control over their internal tempera-tures. They vary in the way in which they do this. Mammals rely largely on physiological methods. Humans, for example, maintain a temperature within a degree or so of 37 °C. If the temperature rises too high, various internal mechanisms bring about loss of heat. If it falls too low, other mechanisms serve to increase heat production and reduce its loss from the body. Humans are, therefore, examples of *endothermic* animals. Crocodiles, on the other hand, rely largely on moving between land and water to remain in the environment with the most suitable temperature. They are referred to as *ectothermic* as they rely on the exter-nal environment to maintain a reasonably constant body temperature.

hint In humans and other mammals, as body temperature increases, arterioles dilate and more blood flows through the capillaries which lie close to the surface of the skin. The very thin walls of the capillaries do not contain any muscle so they cannot dilate or constrict. Large numbers of students writing about temperature control refer incorrectly to capillaries "dilating" or "moving up to the surface of the skin." Be careful!

tendon: a strip of *connective tissue* which attaches a muscle to a bone. Tendons contain a lot of closely packed fibers of a protein called *collagen*. The properties of collagen make it ideally suited to its functions in a tendon. It is flexible but it is very resistant to stretching. This is important as it means that when the muscle contracts it will move the bone, not just stretch the tendon.

territory: an area of land which an animal defends against other animals, usually those of the same species. A wide variety of animals, ranging from insects to mammals and birds, have territories but territorial behavior is very variable, even among individuals of the same species. For example, the spotted hyena lives in clans which fiercely defend feeding territories where their prey are numerous. This is the case in the Ngorongoro crater in northern Tanzania. On the nearby Serengeti plains, however, the antelopes on which the hyenas feed roam extensively, and here no territories are held.

Territories fulfill a variety of different purposes. In Britain, the pied wagtail defends a winter feeding territory along the side of a river or stream. The bird feeds by moving along the shore and eating small insects that have been washed up. When it has covered the length of the bank that forms its territory, it flies back to the beginning and starts again. The mistle thrush is another bird that defends a feeding territory – often, in this case, a holly bush.

Other animals hold mating territories. One example of an animal that does this is the three-spined stickleback, which defends the area around its nest from other males. Animals mark and defend their territories in a variety of different ways. Many mammals use *pheromones* produced from specialized scent glands or mark the boundaries of their territories with urine or feces. Birds, on the other hand, frequently use song or visual displays.

tertiary structure (protein): the irregular folding of a polypeptide chain. Globular *proteins* are proteins which are made of polypeptide chains which are folded so that the molecule is roughly spherical and has a compact overall shape. The tertiary structure of the polypeptide chain is determined by its *primary structure*. Its actual shape is held by chemical bonds between different amino acids. These bonds include:

* *hydrogen bonds*
* bonds between sulphur-containing amino acids
* ionic bonds.

The tertiary structure of a protein, then, gives the molecule its particular shape and this is very important in the function of the protein in the organism. Some examples of the way in which the shape of a protein molecule is related to its function are given below:

* the *active site* of an *enzyme* molecule has a specific shape. This enables it to bind to one particular substrate and, as a result, catalyze a specific reaction
* some proteins act as receptor molecules on *plasma membranes*. Again, their tertiary structure gives them a specific shape. It is because of this that particular hormones, for example, can only affect certain target cells
* *antibodies* are specific: they only respond to particular antigens. This specificity is determined by the precise shape of part of the antibody molecule.

Many of the bonds which maintain the tertiary structure of the protein molecule are not very strong and may be broken by heating. If this happens, the protein will be denatured and its three-dimensional shape changed. It will no longer be able to function.

> **hint** The tertiary structure is more than the three-dimensional shape of a protein. All molecules are three dimensional. You should define the tertiary structure of a protein as the way in which the whole polypeptide folds to give the molecule a compact shape.

test cross: a cross between an organism of unknown *genotype* and the relevant *recessive homozygote*. In peas, the allele for green pods, G, is *dominant* to that for yellow pods, g. A pea plant with green pods could have one of two possible genotypes, either GG or Gg. It is impossible to tell by simply looking at it. The only way of finding out the genotype of a particular plant would be to cross it with a yellow-podded pea. The genetic diagrams below show the expected results from this test cross.

If the unknown pea had the genotype GG:

parental phenotypes	green pods	yellow pods
parental genotypes	GG	gg
gametes	(G)	(g)
offspring genotypes	Gg	
offspring phenotypes	All with green pods	

If the unknown pea had the genotype Gg:

parental phenotypes	green pods	yellow pods
parental genotypes	Gg	gg
gametes	(G)(g)	(g)
offspring genotypes	Gg	gg
offspring phenotypes	green pods:yellow pods 1:1	

By comparing the results of the cross with the predicted ratios, it is possible to determine the genotype of the unknown green-podded plant.

testa: the protective outer layer of a *seed*. When an ovule from a flowering plant is fertilized, it develops into a seed. The female gamete inside the ovule is fertilized and becomes the embryo. The layers of tissue which surround the ovule are known as the integuments and these form the testa. Because the testa does not actually come from fertilized cells, it has exactly the same genetic makeup as all the other cells in the female parent. It does not contain any of the genes from the male parent.

Some seeds will not germinate even though conditions are otherwise suitable. They are said to be in a state of *dormancy*. The testa may be associated with dormancy. In some plants, for example, it may contain substances which inhibit growth, while in others it is impermeable to the passage of oxygen and water. Changes have to occur before germination can take place.

testis: the organ in a male animal which is responsible for the production of sperms. In mammals, the testis has two functions. It is packed with a mass of small tubes known as seminiferous tubules. Within the seminiferous tubules, the process of sperm formation, or *spermatogenesis*, occurs. When the sperm cells have been formed, they move to the epididymis. This is another mass of small tubes and is found just outside the testis. Sperms are stored in the epididymis. The process of sperm formation in many mammals works most effectively at a temperature a few degrees below body temperature. Because of this the testes are suspended in a sac, the scrotum, outside the body cavity. The second function of the mammalian testis is that it is the site of production of *hormones*. These are secreted by special cells which are found between the seminiferous tubules and known as interstitial or Leydig cells. The most important of the hormones produced in the testes is *testosterone*, the male sex hormone.

testosterone: an important male sex *hormone*. Testosterone is secreted by the cells which are found between the seminiferous tubules in the *testis*. These cells are known as interstitial or Leydig cells. Testosterone, like all hormones, is transported by the blood. It has a number of effects on the body. These include:

* initiation of the growth spurt at puberty by stimulating bone and muscle growth

* promoting the growth of male *secondary sexual characteristics* such as facial and body hair and bringing about the changes in the larynx which cause the voice to deepen

* maintaining the function of the testes throughout the reproductive life of the male.

Testosterone is an example of an anabolic steroid, anabolic (see *anabolism*) because it is involved in controlling reactions in the body which are concerned with the building up of tissues, and a *steroid* because of its chemical nature. Anabolic steroids such as testosterone have been used widely and illegally by sportsmen and women to encourage muscle development.

thermal pollution: pollution resulting from the release of heat from industrial processes. Electricity-generating stations are a major source of thermal pollution. Approximately half the energy they produce is in the form of heat, which is released into the environment. This is achieved by extracting cold water from a nearby river and passing it through cooling towers. The waste heat then warms this water, which is returned to the river. The returning water may have a temperature as much as 5 °C to 10 °C higher than when it was taken out. Thermal pollution has a number of effects on freshwater environments and on the organisms which live there. These include.

* the amount of dissolved oxygen present in the water will decrease. This may lead to a change in the structure of the animal and plant *community*. Those species which cannot tolerate the increased water temperature will disappear, while others that are able to do so may colonize the area

* respiration and growth rates of individual organisms will change and this may have an effect on feeding rates. It will also mean that organisms will be more likely to be affected by the presence of chemical pollutants in the river

* patterns of reproduction may change.

thermoregulation: see *temperature control*

thermostable enzyme: an *enzyme* that continues to function at relatively high temperatures without being denatured. Some microorganisms, called thermophiles, are able to grow in very hot conditions such as hot springs and around volcanic outlets on the sea floor. Thermophiles are of interest to biotechnologists because of the enzymes they contain. Many enzymes used in industrial processes are slow in their rate of reaction but the reaction rate cannot be increased by heating because the enzymes will denature. Enzymes from thermophiles are, however, thermostable and can be used at higher temperatures. Thermophilic microorganisms are difficult to grow in culture so the commercial production of thermostable enzymes usually involves genetic engineering, transferring the relevant gene to species of bacterium that can be grown in normal laboratory conditions. Immobilization, the binding of enzymes to inert particles or surfaces, not only allows enzymes to be recovered more easily, but also increases their thermostability.

thrombin: an enzyme responsible for converting fibrinogen to fibrin in the process of *blood clotting*.

thrombocyte: see *platelet*

thromboplastin: a substance involved in *blood clotting*. Thromboplastins are produced by blood *platelets* and damaged tissue. They are responsible for changing the inactive enzyme prothrombin to its active form, thrombin. Thrombin catalyses the reaction in which soluble fibrinogen is converted to the threads of fibrin which trap the red blood cells and form a blood clot.

thrombosis is a condition in which a blood clot forms inside a blood vessel. If this clot or thrombus breaks free, it can travel round the body and block a small artery. A heart attack (see *myocardial infarction*) occurs when one of the coronary arteries supplying the heart is blocked. A stroke (see *cerebrovascular accident*) may result from the clot interrupting the blood supply to the brain.

thylakoid: a system of flattened sacs formed by the membranes in a *chloroplast*. At various places, these thylakoids are stacked on top of each other rather like a pile of coins. These are the grana. The thylakoids contain the chlorophyll molecules which trap the light energy in *photosynthesis* and contain the various molecules needed for the *light-dependent reaction*.

thymine: one of the *nucleotide bases* found in nucleic acid molecules. Thymine is a pyrimidine, which means that it has a single ring of atoms in each of its molecules. When two *polynucleotide* chains come together in a *DNA* molecule, thymine always bonds with *adenine*. The atoms of these two bases are arranged in such a way that two hydrogen bonds are able to form between them. The base thymine does not occur in *RNA*. Uracil is found in its place.

thyroid gland: a gland in the neck of a mammal or other vertebrate which is responsible for the secretion of two hormones:

* *thyroxine* (thyroid hormone) helps to control the *basal metabolic rate* of the body
* calcitonin is one of the hormones which helps to regulate the blood calcium level. It stimulates the uptake of calcium by the bones.

Thyroxine contains iodine. If there is not enough iodine in the diet, the thyroid gland enlarges, producing a condition known as goiter.

thyroid stimulating hormone (TSH) is a hormone produced by the anterior lobe of the *pituitary gland*. It stimulates secretion of the hormone, thyroxine, by the *thyroid gland*. The concentration of thyroxine in the blood is maintained by a *negative feedback* system involving thyroid stimulating hormone. For details, see *thyroxine*.

thyrotrophic releasing hormone (TRH): a hormone involved in the control of thyroxine which is produced by the hypothalamus. It is transported in the blood to the anterior lobe of the *pituitary gland* where it stimulates secretion of another hormone, *thyroid stimulating hormone* (TSH). The concentration of thyroxine in the blood is controlled by a *negative feed-*

269

back system involving both thyrotrophic releasing hormone and thyroid stimulating hormone. For details, see *thyroxine*.

thyroxine (thyroid hormone): a hormone produced by the thyroid gland. It plays an important part in controlling the metabolic rate. It is also involved in the control of body temperature. If the body temperature falls, more thyroxine is produced. The energy released in the *electron transport system* is normally used to produce *ATP*. One of the effects of thyroxine, however, is to cause some of this energy to be released as heat rather than as ATP. The level of thyroxine in the blood is controlled by a *negative feedback* mechanism. This is summarized in the diagram.

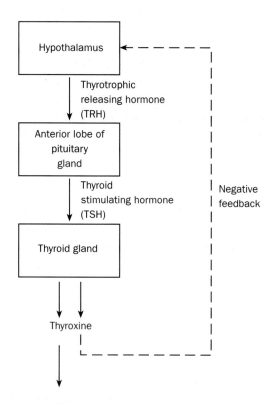

A high concentration of thyroxine in the blood will inhibit the production of *thyrotrophic releasing hormone (TRH)* from special cells in the *hypothalamus*. This means that there will be a fall in the amount of *thyroid stimulating hormone (TSH)* produced by the anterior lobe of the *pituitary gland*. As a result, less thyroxine will be secreted.

On the other hand, if the concentration of thyroxine falls, there will be no negative feedback effect. TRH will be secreted by the hypothalamus. This leads to secretion of TSH from the pituitary gland and there will be an increase in the concentration of thyroxine in the blood.

tidal volume: the volume of air taken in and given out at each breath when a person is breathing quietly at rest. For a normal healthy male, this is about 500 cm³. See *lung capacities*.

tissue: a group of cells which have a common origin and a similar structure, which enables them to perform a particular function. An example of a simple animal tissue that fits this definition is the layer of cells which lines the small intestine. This is an epithelial tissue. Other tissues, however, are often more complex and may consist of several types of cell. *Blood* is an example of a tissue like this. It consists of *red blood cells*, white blood cells and platelets. The red blood cells, however, all arise from cells in the bone marrow, so they have a common origin. They are very similar in structure and this structure is an adaptation to their function in the transport of respiratory gases. In mammals there are four basic tissue types. These are:

- epithelial tissue (see *epithelium*). This covers organs, either on the inside or on the outside. Glands are also formed from epithelial tissue

- *connective tissue.* This tissue has a variety of different functions. It can bind other tissues together or it may form *tendons, ligaments, bone* or *blood*. Many connective tissues have functions associated with support. In addition, connective tissue plays an important part in the defensive mechanisms in the body

- *muscle* tissue brings about movement and also helps to maintain posture. Muscle tissue is also involved with movement associated with organs which are not under conscious control, such as changes in diameter of blood vessels

- nervous tissue. This tissue is made up of specialized cells called *neurones* which provide links between *receptors* and *effectors*.

Plant cells are also organized into tissues. There are five main types:

- *parenchyma.* Relatively unspecialized cells which are important in providing support to young plants. They can photosynthesize and store materials as well

- *collenchyma.* A type of specialized supporting tissue whose cells have walls with extra cellulose thickening at the corners

- *sclerenchyma.* Supporting tissue in which the cell walls are thickened with lignin

- *xylem.* A plant tissue responsible for the transport of water and inorganic ions from the roots to the stem and leaves

- *phloem.* A plant tissue which is responsible for transporting the products of photosynthesis away from the leaves to other parts of the plant.

In both animals and plants, different tissues are grouped together to form *organs*.

tissue fluid: the fluid which surrounds the cells in the body. Tissue fluid is formed from *blood plasma*. At the arterial end of a *capillary*, the pressure of the blood forces water, small organic molecules and mineral ions out through the endothelial cells. This produces tissue fluid. Tissue fluid is very similar to blood plasma in composition but it does not contain blood cells or large protein molecules. At the venous end of the capillaries, water reenters the capillaries by osmosis. The blood at this point has a low *water potential*, due mainly to the presence of protein molecules which did not leave the blood at the arterial end. Water moves back from the higher water potential in the tissue fluid to the lower water potential in the blood plasma. As it moves, it takes with it waste products produced by the cells of the body. Some of the tissue fluid does not go back into the blood. Instead, it enters another system of vessels where it is known as *lymph*.

> **hint** Read this entry very carefully. A lot of students do not know what 'tissue fluid' is.

toxin: a poison produced by a living organism. Many disease-causing bacteria produce toxins and it is the effect of these substances on the body which produces the symptoms of the disease. *Endotoxins* are inside bacterial cells and are only released into the body of the host when the cell dies and the cell wall breaks down. *Exotoxins* are secreted by bacteria as they grow.

trachea (mammal): the windpipe, a tube which takes air down to the lungs. It is lined with ciliated epithelial cells. Particles in the air which is breathed in are trapped in *mucus* secreted by *goblet cells* in the airways of the lung. The *cilia* beat continuously and the mucus and the particles which have been trapped are carried upwards into the back of the throat. In the wall of the trachea, there are characteristic C-shaped rings of cartilage which prevent the tube from collapsing as air is breathed in.

trachea (insect): one of the many fine tubes which take air directly from the outside of an insect to its cells. Many insects have openings along the sides of the body. These are known as *spiracles* and they open into the tracheal system. The individual tracheae are lined with rings of *chitin*, a substance which is also found in the cuticle of the insect. These rings provide support, preventing collapse of the tubes. At the same time, they allow considerable flexibility. The tracheae form a series of branching tubules, getting smaller and smaller until they finally reach a diameter of about 2 μm. They then branch into a number of even smaller tubes known as tracheoles. The entire system allows oxygen to reach the respiring cells in the insect by the process of diffusion. It is well adapted for this function as:

* there is a large surface area over which diffusion can take place. This is provided by the very large numbers of small tubes

* the difference in concentration is maintained as high as possible. Not only is oxygen continually being removed by the respiring cells, but ventilation mechanisms replace the air in the tracheae

* the diffusion pathway is relatively short. Even in the largest insects, the distance from the outside air to the respiring cells is very short. Effectively, air is taken right to the cells.

tracheid: a long, thin, water-conducting cell found in *xylem*. The walls of tracheids are thickened and strengthened with a tough, waterproof material called *lignin*. Their end-walls, however, are perforated by a series of pores. These enable water to move up the plant from one tracheid to the next. Tracheids are unlike *vessels* as they are separate cells. Their end-walls do not break down and they do not fit together in the same way as vessels.

transamination: the way in which *amino acids* can be made by transferring an amino group from an amino acid to part of another molecule. In humans, about half the amino acids that we need must be taken in as part of the diet. These are known as *essential amino acids*. The others can be made from essential amino acids by the process of transamination which takes place in the *liver*. The diagram summarizes the process.

272

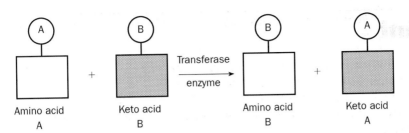

Transamination reactions, like all biochemical reactions in the body, require the presence of a specific enzyme. In this case, the enzyme is known as a *transferase* because it transfers an amino group from one molecule to another. Transamination is also an important process in the production of amino acids in plants.

transect: a line along which organisms are sampled in ecological studies. Transects are particularly useful in situations such as seashores and sand dunes where conditions and species vary across the area being studied.

transfer RNA: see *tRNA*

transferase: a type of *enzyme* which is responsible for catalyzing reactions which involve the transfer of chemical groups from one molecule to another. A good example is provided by the enzymes involved in the process of *transamination*.

transmission electron microscope: a type of *electron microscope* in which a beam of electrons is transmitted through the specimen that is being examined. The specimen has to be cut into very thin sections. These are stained with substances such as the salts of heavy metals which will scatter the electrons. The resulting pattern is detected on a fluorescent screen. Transmission electron microscopes have enabled *cell organelles*, which appear as no more than blurred dots when seen with a light microscope, to be examined in great detail.

transmitter substance: see *neurotransmitter*

transcription: the first stage of *protein synthesis* in which a messenger RNA (*mRNA*) molecule is produced from the *DNA* which makes up a particular gene. The main steps in the process are summarized in the flowchart.

> The hydrogen bonds holding the two DNA chains together break and the chains separate. This process involves enzymes and other proteins

↓

> One of the DNA chains acts as a template for the formation of an mRNA molecule. The bases of free mRNA nucleotides in the nucleus line up against the complementary bases on the DNA chain

↓

cont'd

273

> The nucleotides now join to form an mRNA molecule. This moves out of the nucleus, through the pores in the nuclear envelope, into the cytoplasm of the cell

translation: the second stage of protein synthesis in which the code on the mRNA molecule is used to control the production of a polypeptide chain by a ribosome. The main steps in the process are summarized in the diagram.

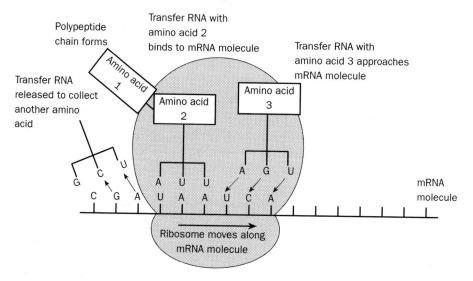

hint Don't confuse transcription and translation. It might help to remember that I

translocation: the transport of substances from one part of a plant to another. In flowering plants, there are two systems involved in this long-distance transport:

- water and mineral ions are transported from the roots to the leaves in the *xylem*. A number of mechanisms are thought to contribute to the movement of water inside these vessels. These include *capillarity, root pressure* and *cohesion-tension*

- the products of photosynthesis are taken from the leaves to the roots and the growing points in the *phloem*. Transport in phloem is in two directions: downwards to the roots and upwards to the growing buds and fruits. The method by which phloem transports sugars through the plant is not well understood. Although most biologists consider that substances travel by a system of *mass flow*, there is still some debate over the mechanism involved.

Although these two systems are quite separate from each other, there is evidence from using radioactive tracers that there is some movement between xylem and phloem, particularly in the younger parts of the plant.

transpiration: the loss of water which takes place from a plant. Although a small amount may be lost from the stem, most of this water escapes through the leaves and through the stomata (see *stoma*) in particular. In the case of stomatal loss, the air in the spaces within the *mesophyll* of the leaf is saturated, while the air outside the leaf has a lower relative humidity. In other words, the air inside the leaf has a higher concentration of water molecules and therefore a higher *water potential* than the air outside. Water molecules diffuse out of the leaf down this water potential gradient.

A variety of factors influence the rate of transpiration:

* factors which affect the supply of water to the leaves of the plant. If the soil dries out or roots are damaged by the feeding activity of an insect such as the cabbage root fly, less water will enter the plant. This will lead in turn to a reduction in transpiration rate

* factors which influence the opening and closing of the stomata. Since approximately 90% of water lost by a plant passes through the stomata, factors such as light intensity, which have a direct influence on the opening and closing of stomata, will also influence transpiration rate

* factors which affect the water potential gradient between the inside of the leaf and the outside. Here, such features as wind speed and relative humidity might be mentioned.

The potential loss of water by transpiration is extremely high in dry habitats. *Xerophytes* are plants which grow in such places. They usually show a range of adaptations, which include mechanisms which reduce their rates of transpiration.

tricuspid valve: the valve between the right atrium and the right ventricle in the heart of a mammal. See *atrioventricular valve*.

triglyceride: a lipid consisting of three fatty acid molecules which are linked by condensation to a molecule of glycerol. Triglycerides are the commonest naturally occurring lipids. Most of those found in animals are fats, that is they are solid at a temperature of about 20 °C. This is due to the fact that they have a high proportion of saturated fatty acids. Fats play a very important part in energy storage. A given mass of lipid will yield a greater amount of energy than the same mass of a carbohydrate like starch or glycogen. This is because they have a greater proportion of hydrogen atoms in their molecules. In addition to this, fats play an important role in providing insulation.

triploblastic: an animal body pattern in which there are three layers of cells. The outer layer of cells is called the ectoderm and gives rise to the animal's epidermis and nervous system; the inner layer, or endoderm, forms the lining of the gut and digestive glands; the remaining layer, between the ectoderm and the endoderm is the mesoderm. From this layer, all the other systems and organs in the animal's body develop. With the exception of the sponges and members of the phylum *Cnidaria*, all animal phyla which you are likely to encounter are triploblastic and have a body pattern which is based on three layers of cells. In some triploblastic animals, a split develops in the mesoderm and forms a body cavity known as a *coelom*. See diagram on the next page.

triploid: a cell or nucleus in which there are three sets of chromosomes. A number of commercially important plants are triploid. These include many varieties of apples and some bananas. Although they produce larger fruit, these varieties are sterile. This is because the *chromosomes* are unable to pair up during the first division of *meiosis*. The *endosperm*, a tissue which is found in seeds where it helps to nourish the developing embryo, is triploid.

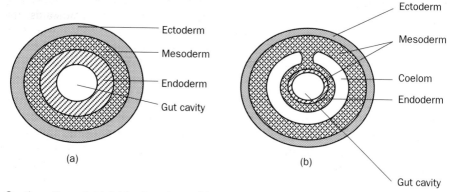

Sections through triploblastic animals (a) without and (b) with a coelom

tRNA: a type of nucleic acid which is important in assembling amino acids in the correct order during *protein synthesis*. A molecule of tRNA consists of a single *polynucleotide* chain. Each *nucleotide* comprises a five-carbon sugar called ribose, a phosphate group and a base. The base may be any one of four: adenine, cytosine, guanine or uracil. The polynucleotide chain which results from joining the individual nucleotides is twisted to form a three-dimensional structure which is often represented as being shaped rather like a clover leaf. A loop at the bottom of this molecule has a specific sequence of three bases on it. This is known as the anticodon. At the other end of the molecule an amino acid may be attached. A tRNA molecule with a particular anticodon will always attach to the same amino acid. The anticodon on the tRNA molecule enables it to line up in the right place on the *mRNA* during *translation*.

 tRNA is a type of RNA. It therefore contains the bases adenine, cytosine, guanine and uracil. It does not contain thymine.

trophic level: one of the steps in a *food chain*. In the following example of a food chain:

nettle plant → large nettle aphid → two-spot ladybug

there are three trophic levels. The nettle plant is the *producer*. It converts the energy in sunlight into energy in the organic molecules in the plant. The large nettle aphid is the primary *consumer* that feeds on the nettle, while the two-spot ladybug is the secondary consumer. When any of these organisms die, their remains provide energy for *decomposers*. Because energy is lost at each stage by processes such as respiration, it is rare for any food chain to have more than five trophic levels. Like many ecological terms, a trophic level is an oversimplification. Organisms often feed at more than one trophic level. Humans, for example, are both primary and secondary consumers. Insectivorous plants like the Venus flytrap may even be producers, primary and secondary consumers! In addition, some organisms feed at one trophic level at one stage in their lives and another as they grow. Young partridges are secondary consumers. They eat large numbers of insects. The adult birds, on the other hand, are primary consumers and feed mainly on plant material.

hint Use the terms "producer" "primary consumer" and "tertiary consumer" to refer to the trophic levels in an ecological pyramid. Do not use "first trophic level," "second trophic level" and "third trophic level."

tropism: growth made by a plant in response to an external stimulus. If a young seedling, for example, is fixed in a horizontal position, its root will curve and grow downwards. This is brought about by unequal growth: the cells on the upper surface grow more than those on the lower surface. Plants respond in a similar way to a number of stimuli. The response to gravity described in the example above is usually called geotropism, while that to light is phototropism.

Biologists are unsure about the mechanism which results in tropic responses. At one time, it was thought to be due to an uneven distribution of *plant growth substances* called *auxins*. However, many recent investigations contradict the results of the early experiments which led to this idea. Some biologists are now of the opinion that tropic responses can be explained mainly in terms of substances which inhibit growth. Others think that the stimulus has a direct effect on the cells concerned and that plant growth substances are not actually involved in controlling tropic responses.

true-breeding: an organism which has the same *genotype* as its parents. It is therefore *homozygous* for the *allele* in question.

trypsin: a digestive enzyme produced by the *pancreas*. Trypsin is an *endopeptidase* which breaks down proteins into smaller polypeptides. Like all digestive enzymes, it does this by *hydrolysis*.

Trypsin obtained from animal sources was one of the first enzymes to be used in *biotechnology*. The first biological washing powder was produced in Germany as long ago as 1913. It contained small amounts of trypsin. Traditionally, the removal of hair from animal's skins in the process of making leather also involved this enzyme. The skins were treated either with chicken or dog excrement. Both contained considerable amounts of trypsin which digested the proteins holding the hair in the skin.

trypsinogen: the inactive form in which the enzyme, *trypsin*, is secreted. Trypsin is a protein-digesting enzyme secreted by the pancreas. If it were released into the pancreas in its active form, it would obviously cause serious damage. It is therefore produced as an inactive molecule. Once it enters the small intestine, it can be converted into trypsin by enterokinase, an enzyme produced by cells in the wall of the intestine. Trypsin itself can also activate trypsinogen. This means that once some trypsin is formed, a chain reaction takes place and all the trypsinogen is rapidly converted to trypsin.

tumor: a swelling which may occur in any part of the body. It is made up of a mass of abnormal cells which keep multiplying in an abnormal way. There are two main types of tumor. Benign tumors are not cancerous. Although they may grow to a considerable size, they do not actually destroy the tissue in which they are situated or spread to other organs. This is in contrast to a malignant tumor, which both destroys the surrounding tissue and can spread through the blood or *lymphatic systems* to other sites in the body where it sets up secondary tumors.

tumor suppressor gene: A gene whose function is to help control multiplication of cells. *Mutations* to tumor suppressor genes may prevent them functioning properly. In some cases this leads to uncontrolled cell division and the development of a *tumor*.

turgid: a turgid cell is a cell which is full of water. The *water potential* of a plant cell is influenced by two factors, the concentration of dissolved substances inside the cell and the pressure of the cell wall pushing on the contents of the cell. The pressure produced by the

turgor

concentration of the cell contents is the *solute potential* or Ψ_S. The pressure produced by the cell wall is the *pressure potential*, Ψ_P. When a cell is fully turgid, solute potential and pressure potential are equal. If the solute potential of a turgid cell is -0.9 MPa, its pressure potential will be $+0.9$ MPa. The water potential of the cell can be calculated from the equation.

$$\Psi = \Psi_S + \Psi_P$$

In this case,

$$\Psi = -0.9 + 0.9 = 0$$

At full turgor, then, the water potential of a cell is zero. Since the water potential cannot be higher than zero, and water always moves from a higher water potential to a lower one, it follows that water cannot move into a turgid cell; it can only move out.

turgor: the pressure that the contents of a cell exert on the cell wall. In the parenchyma cells in a young plant stem, water moves into the cells by *osmosis* because there is a greater *water potential* outside than inside. The resulting pressure that the cytoplasm exerts on the wall is very important in providing support for the plant.

two-way chromatography is a way of separating substances present in a mixture. In its simplest form, substances are loaded onto a square of filter paper which is suspended with its bottom edge in a solvent and left for a period of time. Individual substances separate out in the usual way. The paper is now turned through 90° and a different solvent used. This separates out substances that are perhaps not separated by the first solvent.

ultrafiltration: filtration that is helped by the high pressure of the blood. It is a process which is involved in the formation of *tissue fluid*. The pressure of the blood in the *capillaries* helps to force fluid through the capillary walls into the surrounding tissues. Although this process occurs generally in capillaries, the term ultrafiltration is often linked with the first stage of urine formation in the kidney. The pressure of the blood in the capillaries that make up the *glomerulus* forces fluid through the basement membrane into the space in the *renal capsule*. This fluid has the same composition as blood *plasma* except that it does not contain protein. Protein molecules are too large to go through the basement membrane.

ultrasound: sound waves of very high frequency, so high that they cannot be heard by the human ear. Techniques based on ultrasound are frequently used in medicine. An ultrasound beam is aimed at the structure which is being examined and the reflected waves are used to produce an image which shows the various organs. One of its advantages is that it does not subject the patient to the harmful effects of X-rays. Partly because of this, it is often used to gain information about a developing *fetus* inside the uterus of its mother.

unsaturated fatty acid: a fatty acid with a long hydrocarbon tail in which there are double bonds between some of the carbon atoms. Unsaturated fatty acids have lower melting points than fatty acids in which there are no double bonds. *Triglycerides* which contain unsaturated fatty acids therefore tend to be a liquid at temperatures of around 20 °C. They are called oils and are found mainly in plants.

uracil: one of the *nucleotide bases* found in *RNA* molecules. Uracil is a pyrimidine, which means that it has a single ring of atoms in each of its molecules. In *protein synthesis*, when mRNA is formed or when the tRNA anticodon and the mRNA codon come together, uracil always bonds with *adenine*. The atoms of these two bases are arranged in such a way that two hydrogen bonds are able to form between them. The base uracil does not occur in *DNA*. Thymine is found in its place.

urea: the main chemical form in which nitrogen is excreted in mammals. Breakdown of excess amino acids during *deamination* produces ammonia. Unfortunately ammonia is very toxic and can only be safely excreted if large amounts of water are available to dilute it to safer levels. This is not possible in terrestrial animals such as mammals. They convert ammonia to the much less toxic urea. Urea has the molecular formula $CO(NH_2)_2$. While it is not necessary to learn this formula, it is worth pointing out that it is a very small molecule and is soluble in water. As such, it diffuses readily across cell membranes and may be found in small amounts in most body fluids. See the hint box at the top of page 280.

279

hint *Urea*, *urine* and *glomerular filtrate* are frequently confused. Urea is a substance formed in the liver. It is transported in the blood to the kidneys where it is excreted from the body in the urine. The glomerular filtrate is the fluid formed as a result of ultrafiltration in the renal or Bowman's capsule. The composition of the glomerular filtrate changes in its passage down the nephron until it ultimately becomes urine.

uric acid: the chemical form in which nitrogen is excreted in a number of terrestrial animals. When compared with *urea*, it is extremely insoluble and is excreted more as a sort of paste than a solution. This makes it an ideal excretory product in animals where maximum conservation of water is important. It is the main excretory product in many insects, reptiles and birds.

urine: the fluid produced by the *kidneys*. It contains water in which *urea* and other products of nitrogenous excretion are dissolved. Urine also contains significant amounts of sodium chloride as well as traces of many other compounds and ions. The actual composition of urine is variable and can be influenced by a number of features, including climatic conditions and diet. Analysis of urine can provide a lot of information about the health of the individual concerned. People suffering from *diabetes*, for example, may have significant amounts of glucose in the urine. Detection of the hormone human chorionic gonadotrophin, (*HCG*), in the urine is the basis of many tests used to confirm pregnancy.

urinogenital is an adjective meaning "to do with the excretion and reproduction." The urinogenital system includes the kidneys, bladder and reproductive organs.

uterus: the part of the mammalian female reproductive system in which the fetus develops. In many species of mammals the organ has two distinct parts, a right side into which the right *oviduct* opens and a left side into which the left oviduct opens. Humans and a few other species have a single uterus into which both oviducts enter. The uterus has an outer layer composed mainly of muscle and an inner layer, the endometrium, which contains many glands and distinctive spiral blood vessels. The endometrium undergoes a series of changes during the *estrous cycle*. During pregnancy, the uterus can expand enormously. In humans, the cavity can increase in size by almost five hundred times.

hint As a biologist, you should refer to the organ in which a mammalian fetus develops as the uterus not the "womb." It is particularly important in writing about reproduction to use the correct scientific terms.

vaccine: a preparation which, when given to a patient, stimulates some of the body's white blood cells to produce antibodies. Vaccines may incorporate whole microorganisms which have either been killed or *attenuated*, or they may be based on parts of microorganisms.

vacuole: an area in the cytoplasm of a cell which contains cell sap. It is surrounded by a membrane. Vacuoles are characteristic of many plant cells. In a young cell, there may be several small vacuoles but, as the cell gets older, these join to form a single large vacuole. This may take up as much as 90% of the space in the cell. The cell sap inside the vacuole contains water in which there are dissolved substances such as sugars and mineral ions. They are usually at a much higher concentration than they are in the surrounding cytoplasm. This means that there is a *water potential* gradient and water moves into the vacuole by *osmosis*. This is important in maintaining the cell in a turgid state and providing support to the plant. See also *food vacuole*.

vagus: a nerve which runs from the *medulla* in the brain to various internal organs. It is an important part of the *parasympathetic* nervous system. The branch that goes to the heart has an inhibitory effect and slows the heart rate. On the other hand, stimulation of the branch going to the stomach increases the rate of secretion of gastric juice.

variance: a measure of the amount of variation shown by a characteristic that demonstrates *continuous variation*. It shows how spread out the individual results are on either side of the mean. A character that has a high variance will show a wide range of variation. One with a low variance will have a much smaller range. Variance results from both the genes which the organism carries and from the environment in which it lives. The phenotypic variance, or the amount of variation shown by the different *phenotypes*, is usually a combination of both of these. It can be represented by a simple equation:

$$V_P = V_G + V_E$$

where V_P = phenotypic variance, V_G = the genetic component of variation, and V_E is the environmental component of variation.

variation: the differences that exist between living organisms. One of the fundamental characteristics of all living organisms is that they show a considerable range of variation. There is a saying "as alike as two peas in a pod," but even things that are supposedly this similar differ considerably when examined carefully and in detail. Within a particular species, there are two main causes of variation. It can result from the genes which the organism carries or it may come from the environment in which it lives. Variation may be described as being either *continuous* or *discontinuous*.

vascular system: a system whose main function is in transporting substances from one part of an organism to another. Examples of vascular systems are the *blood system* of a mammal and *phloem* and *xylem* in a plant.

281

vasoconstriction: the narrowing of blood vessels. The mechanism by which body temperature is controlled in a mammal involves vasoconstriction. The *arterioles* which supply blood to the network of *capillaries* in the skin have walls containing smooth muscle. In cold conditions, nerve impulses from the temperature regulatory center in the hypothalamus cause this muscle to contract. Blood is diverted from the surface capillaries and flows deeper in the skin. As a result, less heat is lost from the body by processes such as radiation.

vasodilation: the dilation of blood vessels. The mechanism by which body temperature is controlled in a mammal involves vasodilation. The *arterioles* which supply blood to the network of *capillaries* in the skin have walls containing smooth muscle. In hot conditions, nerve impulses from the temperature regulatory center in the hypothalamus cause this muscle to relax. Blood can then flow to the surface capillaries and heat can be lost from the body by processes such as radiation.

vasomotor center: a region in the medulla of the brain which is involved with the regulation of blood pressure. The flowchart summarizes the mechanism involved.

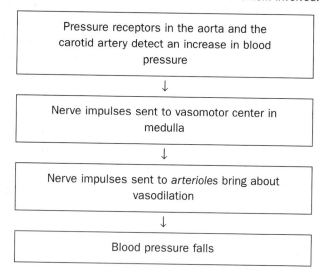

vector: a carrier. In biology, the term may be used in one of two different contexts, genetics and parasitology.

A genetic vector is used in *genetic engineering*. It is a molecule of DNA which acts as a carrier. The gene that has been isolated is spliced into the vector DNA to produce recombinant DNA which is then inserted into another organism. There are a variety of different genetic vectors and the one chosen will depend on the size of the isolated piece of DNA and the type of organism into which it is to be placed. Two of the most familiar vectors are *viruses* and *plasmids*.

When referring to parasites, the term still means a carrier. The malarial parasite is a microorganism which lives in the blood of its human host. In normal circumstances it would be very difficult, if not impossible, for an organism smaller than a single red blood cell to get from the blood of one individual into the blood of another. The transfer is made by a vector, in this case, an *Anopheles* mosquito.

vegetative propagation: a form of asexual reproduction found in plants. It often involves the growth of a bud or a stem which eventually becomes separated from the parent plant to form a new individual. The table gives information about some common examples of vegetative propagation.

Structure	Examples	Notes
Bulb	Onion, daffodil, hyacinth	A bulb is a shoot with a very short stem and fleshy leaves which act as a food store
Rhizome	Iris, many important weeds such as couch grass and stinging nettles	Rhizomes are underground stems which grow horizontally below the soil surface. Buds on the rhizome give rise to leafy shoots
Runner	Strawberry	A stem which grows horizontally above ground and gives rise to new plants

vein (animals): a blood vessel which takes blood from the *capillaries* back to the heart. In mammals, veins usually contain deoxygenated blood, but there is one important exception to this. The blood that returns to the heart along the pulmonary vein has come from the lungs. Blood in this vein is therefore rich in oxygen. Blood flowing into veins is at a low pressure and its return to the heart may be helped by muscles and valves. The larger veins returning from the limbs are situated between large muscles. When these muscles contract they squeeze the blood in the veins. Valves in the walls allow this blood to be squeezed towards the heart but do not allow it to go back in the other direction. This helps to maintain the flow of blood in the veins. Like all blood vessels, veins have a lining of epithelial cells. Their walls, however, contain much less muscle and elastic tissue than do those of *arteries*.

 hint A vein is a blood vessel which returns blood to the heart. Use the word only in this context. Use the term "blood vessel" if you want to refer to an artery or a vein.

vein (plants): a structure found in leaves. Veins help to provide support and contain the conducting tissue. This is the *xylem*, which supplies the leaf with water and mineral ions, and the *phloem*, which transports the products of photosynthesis to the rest of the plant.

ventilation: the mechanism by which the air or water in contact with a gas exchange surface is changed. Oxygen is taken up across all gas exchange surfaces by *diffusion*. A ventilation mechanism ensures that the difference in concentration across such a surface is kept as high as possible. If ventilation did not occur, the oxygen level in the external medium would gradually fall and diffusion would become less and less effective. Different organisms have different ventilation mechanisms:

- in large, active insects, contraction of muscles allows the abdomen to be flattened. When these muscles relax, the abdomen returns to its normal shape. This helps to change the air in the *tracheae* which supply oxygen to the body tissues

- in fish, water is moved over the gills by a pumping mechanism which produces a combination of increased pressure in front and reduced pressure behind

- mammals have a diaphragm which separates the thorax from the abdomen. Contraction of the muscles of the diaphragm and of those between the ribs causes the chest cavity to enlarge. This results in a lower pressure in the lungs than in the surrounding atmosphere so air is able to move in until the pressures are equal.

ventral root: part of a spinal nerve which joins with the *spinal cord*. The spinal nerves contain both sensory and motor *neurones*. Just before each nerve joins with the spinal cord, it splits in two, producing a ventral root and a *dorsal root*. The ventral root contains the motor neurones which carry nerve impulses from the spinal cord to effectors such as muscles and glands.

ventricle: one of the chambers of the heart. The walls of the ventricles are much thicker than those of the atria (see *atrium*). When the muscle in these walls contracts, blood is forced out into the arteries. Chordates such as fish have a *single circulation*. The heart of a fish has only one ventricle. When it contracts, blood is forced out to the gills where it picks up oxygen and then continues round the body. Mammals, on the other hand, have a *double circulation*. Their hearts have two ventricles. The right ventricle receives blood from the right atrium and pumps this out through the pulmonary artery to the lungs. Oxygenated blood returns to the left side of the heart from where the left ventricle pumps it to the organs of the body.

hint It might seem very simple, but make sure that you do not confuse the right and left sides of the heart. The right ventricle takes blood to the lungs; the left ventricle transports it to the remaining organs of the body.

ventricular diastole: the stage in the *cardiac cycle* where the muscles of the heart relax. Blood flows into the atria. Some goes into the ventricles.

ventricular systole: the stage in the *cardiac cycle* where the thick muscular walls of the ventricles contract and force blood out of the heart through the aorta and pulmonary arteries.

vesicle: a small sac in the cytoplasm of a cell, surrounded by a membrane. The *Golgi apparatus* is an organelle which is responsible for the processing and packaging of substances produced by a cell. These substances are enclosed in small vesicles which are continually being formed by being pinched off from the flattened sacs which make up the Golgi apparatus. Other vesicles are formed as a result of *phagocytosis* and *pinocytosis*, where particles or small droplets of liquid are taken through the membrane into the cytoplasm. The *plasma membrane* surrounds the particles concerned. A vesicle is formed which is pinched off and moves into the cytoplasm. Apart from the fact that vesicles are smaller than *vacuoles*, there is no difference between these structures.

vessel: one of the *xylem* tubes which transport water and inorganic ions from the roots to the stem and leaves. Vessels are made up from dead cells which have lost their cross-walls and fit together rather like a series of drainpipes. The walls of these cells are thickened and strengthened with a tough, waterproof material called *lignin*. As the vessel gets older, lignin is laid down on the original cell wall. At first, this is done in a series of rings or spirals which prevent the vessel from collapsing but still allow the flexibility needed in a young stem. Later this lignin forms an almost complete layer.

viability: the ability of *seeds* to germinate when exposed to suitable conditions. If seeds are stored under conditions where they are kept cool and dry, they may remain viable for very long periods of time. Some seeds excavated from a grave in South America, for example, were still able to germinate although they were found by carbon dating to be over 600 years old. Seeds from many garden plants, however, lose their viability very rapidly. This is particularly true if the packet of seeds has been opened and left under conditions where the temperature varies and they are exposed to air with a high moisture content. Many important *weeds* have seeds which remain viable for very long periods of time. This enables their seeds to remain buried in the soil for many years and still germinate when they are exposed to suitable growing conditions. Viability is also a term which can be used when referring to *microorganisms*.

vibrio: a bacterial cell which looks like a curved rod. The best known example is probably *Vibrio cholerae*, the bacterium which causes cholera.

villus: a small finger-like structure which sticks out from the inside of the *small intestine*. Villi are present in enormous numbers and greatly increase the surface area of the intestine. Each villus has a lining of epithelial cells. The *plasma membrane* of these epithelial cells is increased by the presence of *microvilli*. Together, the villi and microvilli produce an enormous surface area which allows for the very efficient absorption of the products of digestion. Each villus contains a network of capillaries into which amino acids and simple sugars diffuse, and a lacteal which absorbs the products of fat digestion.

Villi are also found at the exchange surface between the fetal and maternal parts of the *placenta*, where they have a similar function in increasing the surface area and allowing more effective exchange between the blood of the mother and that of the fetus.

> # hint
> Make sure that you distinguish between villi and microvilli.

virus: an extremely small particle which is only capable of multiplying once it is inside a living cell. A virus consists of a molecule of *nucleic acid*, either DNA or RNA, surrounded by a protein coat. Some of the larger viruses have an outer lipid layer as well. Outside the cells of their host, they are completely inert. They cannot feed, respire or multiply, so they are best thought of as nonliving particles. The life cycle of a virus involves the takeover of the host cell and the use of its biochemical processes to make more virus particles. All viruses are, therefore, to some extent harmful. They cause a wide range of human diseases including the common cold, influenza, AIDS and some cancers. They can also produce a number of economically important plant diseases. Viruses may be used as *vectors* in *genetic engineering*.

visceral muscle: see *involuntary muscle*

visual purple: see *rhodopsin*

vitamin: one of a group of chemically unrelated organic substances which are needed in very small amounts in the diet. Many vitamins are obtained from plant foods but some may be obtained from the tissues of other animals or are produced by microorganisms living in the gut. They can be divided into two groups. Fat-soluble vitamins can be stored in the liver. Water-soluble vitamins, however, cannot be stored and are removed from the body in the urine. It is essential, therefore, to maintain a constant supply of the water-soluble vitamins. Vitamins have a variety of different functions, although many of them act as *coenzymes*.

vital capacity: the total amount of air that can be breathed out following the deepest intake of air possible. It is not as much as the total volume of the lungs as there is always a small volume that cannot be expelled. See *lung capacities*.

voluntary muscle: see *skeletal muscle*

water potential, Ψ, is a measure of the ability of water molecules to move. The diagram shows water molecules enclosed by a membrane.

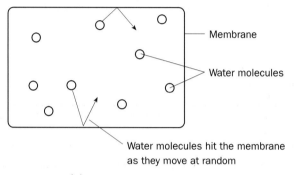

Understanding water potential

These molecules will be moving around at random and, as a result of this, some will collide with the membrane that surrounds them. They will exert a pressure on this membrane. This is the water potential, the symbol for which is the Greek letter psi, Ψ. Clearly, the higher the concentration of water molecules, the greater the pressure that will be exerted and the higher the water potential. Pure water has the greatest concentration of water molecules and therefore the highest water potential. The value of the water potential of pure water is, in fact, zero and so, as all other solutions will have a water potential less than this, they will be negative. The table explains some typical water potential values found in and around a typical plant.

	Typical value of water potential/MPa	Explanation
Water in the soil around plant roots	−0.03	Water in the soil has small amounts of dissolved substances in it. The concentration of water molecules will be less than in pure water but still very high
A xylem vessel	−0.3	There is a much higher concentration of dissolved substances, therefore a lower concentration of water molecules than in the soil
Humid air in the atmosphere around a leaf	−30	The concentration of water molecules in the air is very low, even in humid air

Water will move from a higher (less negative) water potential to a lower (more negative) water potential. This is what happens in *osmosis*. The water potential of a cell is influenced by two other factors, the concentration of dissolved substances inside the cell and the pressure of the *plasma membrane* or cell wall pushing on the contents of the cell. The pressure produced by the concentration of the cell contents is the *solute potential*, or Ψ_S. The pressure produced by the plasma membrane or wall is the *pressure potential*, Ψ_P. Water potential is the sum of the solute potential and the pressure potential. This can be summarized by a simple equation:

$$\Psi = \Psi_S + \Psi_P$$

Calculating water potential: The solute potential of a cell from a potato tuber is 20.8 MPa and its pressure potential is 0.5 MPa. Calculate the water potential of this cell.

Using the equation,

$$\Psi = \Psi_S + \Psi_P$$
$$\Psi = -0.8 + 0.5$$
$$= -0.3 \text{ MPa}$$

hint A solution with a lower water potential is a solution with a more negative water potential.

weed: a plant growing in a place where it is not required. Weeds are economically very important as they can severely reduce crop yields by competing for light, water and inorganic ions. The control of weeds is very important in agriculture and there are a variety of chemical *herbicides* which may be used for this purpose. Plants which have become important weeds are often those that normally occur early in the process of ecological *succession* and are therefore well adapted to growing in disturbed soil. They share a number of features:

* no special requirements for germination. Seedlings grow rapidly
* seed production begins after a very short growth period and continues for a long time
* produce large numbers of seeds in ideal conditions and at least some in poor conditions
* seeds can often live for a long time in the soil and have a variable period of *dormancy*.

white blood cell: see *leukocyte*

white matter: the outside layer of the *spinal cord* which encircles the inner *grey matter*. White matter consists of the axons of nerve cells which conduct impulses up and down the spinal cord. Many of these axons are surrounded by a sheath of *myelin*. It is the white color of this fatty material from which the name, white matter, is derived.

wilting: a condition in which the leaves and young stems of plants droop. It occurs when the amount of water lost through *transpiration* is greater than that absorbed by the roots. The factors which can result in a plant wilting include:

* environmental factors such as high temperature and low relative humidity which promote a rapid loss of water through transpiration
* factors which prevent uptake of sufficient water from the soil. These include a lack of soil moisture and damage to the root system by insect pests or by diseases caused by microorganisms.

287

X chromosome: one of the *sex chromosomes*. In humans and other mammals, each body cell in a female contains two identical X chromosomes. The body cells of males contain a single X chromosome and a *Y chromosome*. In most mammals, the X chromosome is larger than the Y chromosome. However, in other animals such as the fruitfly, the Y chromosome is the larger one. Each X chromosome is made up of two distinct regions, the so-called *homologous* part, which is identical to the same region on the Y chromosome, and the nonhomologous part, which is different and found only on the X chromosome. The nonhomologous region carries a number of genes. In humans, for example, there are genes associated with blood clotting and color vision. These genes are said to be *sex linked*.

xerophyte: a plant which is adapted to dry conditions. Although xerophytes are often associated with desert regions, plants showing some similar adaptations may be found in other places. In the Arctic, for example, there is a water shortage in winter as the ground remains frozen. General features shown by xerophytic plants include:

* structural adaptations of the leaves which reduce *transpiration*. Sunken stomata, reduced leaf surface area and thickened cuticles are often found

* adaptations which enable them to collect and store water. Root systems may cover a very wide area or penetrate deep into the soil. Stems and leaves may be fleshy, storing large amounts of water

* protective mechanisms such as spines. These prevent damage and loss of water-storing tissue by feeding animals

* modified life cycles. Some seeds germinate and the plants grow and reach maturity before all the water from a short wet season evaporates.

xylem: a plant tissue responsible for the transport of water and inorganic ions from the roots to the stem and leaves. The xylem cells which carry out this function are known as *vessels*. They are dead cells which have lost their cross-walls and fit together rather like a series of drainpipes. A number of mechanisms are thought to contribute to the movement of water inside these vessels. These are *capillarity, root pressure* and *cohesion-tension*. Xylem also plays an important part in supporting plant stems. The vessel walls are thickened and strengthened with a tough, waterproof material called *lignin*. As the vessel gets older, lignin is laid down on the original cell wall. At first, this is done in a series of rings or spirals which prevent the vessel from collapsing but still allow the flexibility needed in a young stem. Later this lignin forms an almost complete layer.

hint Do you know the positions of xylem and phloem in a stem?

288

Y chromosome: one of the *sex chromosomes*. In humans and other mammals, each body cell in a female contains two identical *X chromosomes*. The body cells of males contain a single X chromosome and a Y chromosome. In most mammals, the Y chromosome is smaller than the X chromosome. However, in other animals such as the fruitfly, the Y chromosome is the larger one. There are very few genes carried on the Y chromosome, but this does not mean that it has no function. In humans, the Y chromosome is important in the embryo while it is in the uterus. Its presence brings about the development of the testes in the male.

Z-line: a thin line in the center of the *I-band* on a *myofibril*. It holds the thin *actin* filaments in position.

zygote: the cell formed in the process of *sexual reproduction* by the fusion of two *gametes*. The zygote will have the *diploid* number of chromosomes. It contains all the genetic information that will be present in the mature organism into which it will eventually grow.